Distribution Generated Taste-and-Odor Phenomena

The mission of the Awwa Research Foundation is to advance the science of water to improve the quality of life. Funded primarily through annual subscription payments from over 1,000 utilities, consulting firms, and manufacturers in North America and abroad, AwwaRF sponsors research on all aspects of drinking water, including supply and resources, treatment, monitoring and analysis, distribution, management, and health effects.

From its headquarters in Denver, Colorado, the AwwaRF staff directs and supports the efforts of over 500 volunteers, who are the heart of the research program. These volunteers, serving on various boards and committees, use their expertise to select and monitor research studies to benefit the entire drinking water community.

Research findings are disseminated through a number of technology transfer activities, including research reports, conferences, videotape summaries, and periodicals.

Distribution Generated Taste-and-Odor Phenomena

Prepared by:
Djanette Khiari[1], Sylvia Barrett, and Russell Chinn
Metropolitan Water District of Southern California
La Verne, California 91750
[1]Presently employed by Awwa Research Foundation

Auguste Bruchet and **Philippe Piriou**
ONDEO Services, Technology and Research Centre
Le-Pecq sur Seine, France 78230

Lleonard Matia and **Francesc Ventura**
AGBAR
Barcelona, Spain

Irwin (Mel) Suffet
University of California
Los Angeles, California
and

Tom Gittelman and **Preston Leutweiler**
Philadelphia Suburban Water Company
Bryn Mawr, Pennsylvania

Jointly sponsored byby:
Awwa Research Foundation
6666 West Quincy Avenue
Denver, CO 80235-3098
and
UK Water Industry Research Foundation
London, UK

Published by the
Awwa Research Foundation and
American Water Works Association

Disclaimer

This study was jointly funded by the Awwa Research Foundation (AwwaRF) and UK Water Industry Research Limited (UKWIR). AwwaRF and UKWIR assume no responsibility for the content of the research study reported in this publication or the opinions or statements of fact expressed in the report. The mention of trade names for commercial products does not represent or imply the approval or endorsement of either AwwaRF or UKWIR. This report is presented solely for informational purposes.

Library of Congress Cataloging-in-Publication Data has been applied for.

Printed in the U.S.A.

ISBN 1-58321-227-4

Printed on recycled paper

CONTENTS

TABLES

FIGURES

FOREWORD

The Awwa Research Foundation is a nonprofit corporation that is dedicated to the implementation of a research effort to help utilities respond to regulatory requirements and traditional high-priority concerns of the industry. The research agenda is developed through a process of consultation with subscribers and drinking water professionals. Under the umbrella of a Strategic Research Plan, the Research Advisory Council prioritizes the suggested projects based on current and future needs, applicability, and past work; the recommendations are forwarded to the Board of Trustees for final selection. The foundation also sponsors research projects through the unsolicited proposal process; the Collaborative Research, Research Applications, and Tailored Collaboration programs; and various joint research efforts with organizations such as the U.S. Environmental Protection Agency, the U.S. Bureau of Reclamation, and the Association of California Water Agencies.

This publication is a result of one of these sponsored studies, and it is hoped that its findings will be applied in communities throughout the world. The following report serves not only as a means of communicating the results of the water industry's centralized research program but also as a tool to enlist the further support of the nonmember utilities and individuals.

Projects are managed closely from their inception to the final report by the foundation's staff and large cadre of volunteers who willingly contribute their time and expertise. The foundation serves a planning and management function and awards contracts to other institutions such as water utilities, universities, and engineering firms. The funding for this research effort comes primarily from the Subscription Program, through which water utilities subscribe to the research program and make an annual payment proportionate to the volume of water they deliver, and consultants and manufacturers subscribe based on their annual billings. The program offers a cost-effective and fair method of funding research in the public interest.

A broad spectrum of water supply issues is addressed by the foundation's research agenda: resources, treatment and operations, distribution and storage, water quality and analysis, toxicology, economics, and management. The ultimate purpose of the coordinated effort is to assist water suppliers to provide the highest possible quality of water economically and reliably. The true benefits are realized when the results are implemented at the utility level. The foundation's trustees are pleased to offer this publication as a contribution toward that end.

This study, conducted by a diverse group of international experts, created a vast new knowledge base about taste-and-odor issues in drinking water distribution systems. The study is focused on causes of taste and odor in distribution systems and potential solutions to specific problems. Overall, this study provides guidelines to help utilities in control or prevention of taste-and-odor problems in distribution systems and thus helps them to provide a high quality drinking water to their customers. It is hoped that this information will help decrease customers' complaints about the taste and odor of their water and consequently will result in better customer relations.

Edmund G. Archuleta, P.E.	James F. Manwaring, P.E.
Chair, Board of Trustees	Executive Director
Awwa Research Foundation	Awwa Research Foundation

ACKNOWLEDGMENTS

The authors of this report are indebted to the following water utilities for their cooperation and participation in this project:

- Calleguas Muncipal Water District (Thousand Oaks, California)
- City of Los Angeles, Department of Water and Power (California)
- City of Pasadena, Water and Power Department (California)
- Detroit Water and Sewerage (Michigan)
- Eastern Municipal Water District (Perris, California)
- Long Beach Water Department (California)
- Portland Water Bureau (Oregon)
- Southern California Water Company, Central District (Santa Fe Springs) and Orange County District (Anaheim, California)
- Other utilities that wished to remain anonymous

In addition, the authors would like to thank David Kimbrough, Castaic Lake Water Agency, for the inductively coupled plasma analyses conducted as part of the leaching studies in Chapter 6; Peggy Kimball, Metropolitan Water District of Southern California (MWDSC) for editing a large portion of the manuscript; Robert La Londe (MWDSC) for preparing many of the graphics; and Cordelia Hwang and Michael Sclimenti (MWDSC) for "pitching in" at the end to get the final report completed.

The authors also wish to express their appreciation for the advice and assistance of Jarmila Popovicova, the AwwaRF Project Manager, and of the Project Advisory Committee: Gary Burlingame, City of Philadelphia; David Holt, Thames Water Utilities (UK); and Diana Rashash, North Carolina Cooperative Extension.

EXECUTIVE SUMMARY

BACKGROUND

Water provided to the general public should not only be of sufficient quantity at adequate pressure, but should also maintain acceptable quality as it passes from the treatment plant to the consumer. It is generally recognized that consumers are affected by deterioration of water quality within the distribution system. Such deterioration can cause impairment of the organoleptic characteristics of the water, resulting in consumer complaints. If the water's taste is not good, consumers will wonder if it is safe to drink. The American Water Works Association (AWWA), in a 1989 survey of 300 water utilities in the United States (Suffet et al. 1993), found that 65% of the utilities surveyed reported the distribution system as the primary cause of their taste-and-odor problems.

Although the last 20 years have seen major advances in taste-and-odor control, this effort has been directed toward changes in quality that either originated in the source water or were produced during the treatment process. Very little is known concerning changes that develop once the water leaves the treatment plant for the distribution system. Burlingame and Anselme (1995) provide the most comprehensive review of distribution system tastes and odors published to date.

Distribution systems are very complex, and investigation of the causes of changing water quality in distribution systems is difficult. Reporting of distribution-related taste-and-odor cases has been primarily anecdotal; a central database should be established to record such cases and document their outcome. Often, the need to respond rapidly to an emergency does not allow the utility time to investigate and understand the cause of the problem, but merely requires the quickest possible solution. Solving these problems may be time-consuming and costly. In addition, distribution systems have usually not been designed to minimize tastes and odors.

STUDY OBJECTIVES

The objectives of this study were to identify the key parameters implicated in the production of tastes and odors in the distribution system, determine the chemical and microbiological causative agents, assess the potential solutions to specific water quality problems, and establish guidelines for water utilities to resolve these problems or ultimately prevent them from occurring.

APPROACH

This study consisted of a systematic investigation of taste-and-odor problems caused by:

- Water quality entering the distribution system, including surface and groundwater, various treatment processes (low- and high-dose coagulation and pre- and post-ozonation), and the type of disinfectant used (chlorine or chloramines)
- Distribution system conditions and operations
- Biological activity
- System materials
- Detention time
- Blending issues (e.g., issues involving treated waters of different origin and quality, or chlorinated and chloraminated waters)

Four principal tasks were identified. The first consisted of a literature review and gathering of information. The second task involved bench- and pilot-scale experiments to produce different types of treated water and to simulate distribution system conditions in the laboratory. The third task was to determine the organoleptic properties (taste and odor thresholds and dose-response relationships) of the chemicals causing tastes and odors. The fourth task consisted of an investigation of actual case studies and identification of the causes of the problems.

FINDINGS AND CONC LUSIONS

The principal findings and conclusions of the study fall into five categories of taste-and-odor causation or association: (1) biological activity, (2) disinfectants, (3) disinfection by-products (DBPs) and continuing reactions, (4) leaching of distribution system materials, and (5) blending factors.

Tastes and Odors Associated with Biological Activity

The following findings are based on results obtained from the pipe-loop experiments:

- In the absence of disinfectants, an increase in the intensity of earthy-musty flavors with increasing detention time was observed in the pipe-loop experiments. No linear correlation between the microbial counts (of fungi and actinomycetes) in bulk water and the earthy-musty flavor threshold number (FTN) could be found.
- The use of disinfectants (chlorine, chloramines) can induce a delay in observation of earthy-musty flavors and a reduction in FTN.
- The control of earthy-musty flavors by disinfectants is based on a chemical reaction or sensory masking action, as well as the biostatic effect with its resulting lower microbial activity.
- These pipe-loop experiments determined that chlorine can induce unpleasant flavors, even at very low concentrations. No significant increase in FTN was noticed with the use of monochloramine.

Generally, the pipe-loop and annular reactor studies show that they generate results similar to those observed in actual distribution systems. During a case-study investigation, 2,4,6-tribromoanisole (TBA) was identified as the cause of an earthy-musty odor. The odor threshold concentration (OTC) of TBA is approximately 0.03 ng/L at 25°C. Although the biological origin of TBA was not confirmed, it can be assumed that the mechanism leading to its formation is similar to that of 2,4,6-trichloroanisole (TCA). It has been reported that TCA is formed by the biomethylation of chlorophenols in the presence of fungi.

Tastes and Odors Caused by Disinfectants

In the United States, the background flavor of treated water is chlorinous. The U.S. Safe Drinking Water Act requires that public water supplies maintain a chlorine residual of 0.2–4.0 mg/L. In some countries (e.g., France and England), however, the chlorinous flavor is considered objectionable, and measures are taken to lessen the chlorine residual in the distribution system.

The pipe-loop experiments conducted in France led to additional findings related to the taste-and-odor behavior of disinfectants:

- Jar tests with flavor-free water (Evian) allowed the determination of flavor threshold concentration (FTCs) of 0.025 mg /L for chlorine as Cl_2 and 0.28 mg /L for monochloramine as Cl_2. (The FTC is given by the amount of chemical in the greatest dilution volume in which a flavor can be perceived by a panelist and the dilution ratio is the FTN). These FTCs were determined at 25°C.
- The chlorine FTC is influenced by the initial FTN and flavor description of the water. No significant effect is observed for monochloramine.
- Except for chlorine decay, the distribution system has no specific effect on the chlorinous flavor induced by disinfectants.
- Chlorine flavors became unnoticeable at concentrations below 0.07 mg /L for chlorine as Cl_2 and 0.55 mg /L for chloramine as Cl_2 and were replaced by earthy-musty flavors.

The results of the field and bench-scale studies yielded similar findings:

- Field studies showed that chlorinous flavors were perceived down to 0.05 mg /L for chlorine as Cl_2 and that earthy-musty flavors appeared at concentrations below 0.1 mg /L for chlorine as Cl_2.
- Bench-scale experiments performed under various treatment conditions confirmed the sensory masking effect of disinfectants on various waters with different origins and different water-quality characteristics. It was observed that longer detention times and subsequent chlorine decay support the appearance of other taste-and-odor characteristics. When untreated water was compared with treated water at various post-treatment detention times, it was seen that a masking effect of odors (originally present in the untreated water or produced during ozonation) occurred in the presence of a chlorine or chloramine residual. Sometimes the masking effect diminished after an extended detention time and the original odor reappeared, but under other conditions, the original odor was permanently eliminated. In most cases, the initial quality of the water (discounting the chlorine residual) was the determinant factor in

the sensory characteristics of the water received at the consumer's tap, especially in the more remote areas of the distribution system.

- Bench-scale experiments on three groundwaters, using disinfection only, showed that in general, decay of the disinfectant residual after longer detention times appeared to improve the organoleptic properties of the water. An exception was the appearance of a medicinal odor in one water.
- Thus, the shelf life of water with respect to tastes and odors is largely related to the persistence of a total chlorine residual.

When chloramines are used for the disinfectant residual in a distribution system a free ammonia residual is often carried. In this study, the OTC at 25°C for ammonia was estimated at 0.9 mg/L as nitrogen, which is at a level greater than would normally be present in the distribution system.

Tastes and Odors Caused by DBPs and Continuing Reactions

The formation of DBPs is most likely to continue in the distribution system in the presence of a disinfectant residual. The odor characteristics of a large number of DBPs were determined through the course of this study. Descriptors, OTCs at 45°C, and dose-response plots were also determined (see Appendix A).

Iodinated trihalomethanes (ITHMs) exhibited medicinal characteristics that became more distinct with the number of iodine atoms in the structure. Tests performed in this study indicated that ITHMs form slowly in the distribution system. It was observed that the chlorine-to-nitrogen ratio during chloramination was an important factor in the formation of ITHMs. Higher concentrations of the compounds are generally formed when the 5:1 ratio is used at the low residual (e.g., 0.5 mg/L monochloramine as Cl_2). Higher concentrations are generally formed when the 3:1 ratio is used at the high residuals (e.g., 4 mg/L monochloramine as Cl_2). An increase in dichloroiodomethane and bromochloroiodomethane concentrations was seen with longer detention times.

Tastes and Odors Caused by Leaching of Distribution System Materials

Bench-scale and case studies indicated that leaching of distribution system materials may cause the appearance of tastes and odors, either directly from the compounds leached or through the reactions of the disinfectant residual with these leached chemicals. Every material that is in contact with water has some potential to affect the flavor of the water, given the right conditions.

Tastes and Odors Associated with Blending Factors

It was determined that when blending two chlorinated waters of different origins and different water quality, differences in odors appeared before the differences in flavors; differences in odors in the chlorinated waters appeared before differences in the dechlorinated waters; and differences in odors were greater when tested at 45°C than at 25°C. There were no significant differences between blends prepared in the laboratory and tested at zero time and blends collected from the actual distribution system at unknown detention times.

The trends observed for blending chlorinated groundwater with chloraminated surface water generally followed the trends observed by Barrett et al. (1985). The changes in the organoleptic properties of the blended water were very subtle. The organoleptic characteristics of the water changed according to the contact times, with the "swimming pool" and "sweet" characteristics becoming more prevalent at 24 and 120 hours and the "straw" odor decreasing at longer detention times.

PRACTICAL APPLICATIONS FOR UTILITIES

This study highlighted certain factors that need to be considered in order to minimize tastes and odors in the distribution system and reduce consumer complaints. As with tastes and odors originating in source waters or treatment plants, there are actions that can be taken to correct or prevent distribution-generated tastes and odors, thus increasing customer satisfaction. The first step is to understand where potential changes may occur and how they can be avoided.

TASTE-AND-ODOR INVESTIGATION PROCEDURES

Although it is not possible for water utilities to avoid all tastes and odors in their systems, many of them can be anticipated, and preventive or mitigative measures can be taken. When quality changes cannot be anticipated, and taste-and-odor complaints are received by a water utility, it is important to conduct a thorough investigation to identify the origins and chemical causes of taste- and odor-related consumer complaints. Tracking down the problem is often simple, and experienced water utility personnel may correct it easily. However, because of the complexity of distribution systems and the intricacy of their pipe networks, it is sometimes difficult to find the source of the change and decide on a simple solution. Only a systematic investigative procedure can resolve this type of situation.

An investigative procedure was developed to help water utilities understand and respond efficiently to taste- and odor-related consumer complaints. Appendix B presents the field investigation form and questionnaire.

SOURCES OF TASTES AND ODORS IN DISTRIBUTION SYSTEMS

Numerous causes and associations of potential tastes and odors in the distribution system have been described. They can be summarized as follows:

- Replacement, repair, or coating of pipes or reservoirs
- Disinfectant residuals that are too high or too low
- Biological activity in the water or on surfaces contacted by the drinking water
- Changes in water quality
- DBP formation and continuing reactions
- Blending factors

Replacement, Repair, or Coating of Pipes and Reservoirs

Some of the bench-scale and case studies demonstrated the potential of chemical leaching from distribution system materials as a causative factor in taste-and-odor production. It is generally assumed that if the material has been used in the past without creating water quality problems, it will then be safe to use under all conditions. However, case studies have shown that a defective material, or even a slight change in the product formulation, can lead to the leaching of chemicals associated with tastes and odors. In other cases, it was seen that leached compounds can react further with the residual disinfectant, producing additional taste-and-odor changes at the consumer's tap.

Migration of additives from newly installed synthetic material is sometimes accompanied by the appearance of intense tastes and odors. When coatings have not been adequately cured, their application solvents have a tendency to leach taste- and odor-causing compounds such as alkyl benzenes.

The bench-scale studies and a number of case studies highlighted the need to include a taste-and-odor approval procedure before new pipes and other materials are used and before pipes or recently coated reservoirs are released for use. The product should be tested and approved for use by a recognized authority before the new material is installed or coating applied. The recognized authorities should include taste and odor testing in their approval procedures. Ideally, the same conditions that exist in the specific system should be used. This means that water of similar quality, with the same disinfectant at a similar concentration and a similar contact time, should be used. These tests, when applied systematically, should minimize the consequences of the potential water quality changes and reduce consumer complaints.

Disinfectant Residuals

Disinfectant residuals should be evaluated regularly throughout the entire distribution system, and the system should be adjusted for optimal customer satisfaction while still maintaining adequate microbial control. Even small changes in residual concentrations (free chlorine) can provoke consumer complaints. A regular flushing program may be necessary to maintain good water quality. The system should be designed, maintained, and operated so as to carry the lowest total chlorine residual possible, at a consistent level throughout the system, at all times.

Biological Activity

Biological activity in the distribution system is inversely related to disinfectant level. As the residual decreases, biological activity increases. Removing dead ends, if possible, and instituting a regular flushing program may help minimize biological activity. Use of a different disinfectant and corrosion control may be necessary in certain situations.

Water Quality

It is very important for water utilities to be aware of changes in the water quality they provide to their consumers. If the water is obtained from a wholesaler, the latter should notify the water utility when blends are changed. Results should be provided for flavor-profile analysis (FPA) of current and projected blends. When the use of a new well is anticipated, FPA should be performed on the product water prior to distribution. The water tested should be representative of the water that will ultimately be delivered to customers. This means that the water from the well should be of consistent quality, the untreated water should be tested before and at the end of bench- or pilot-scale treatment, and the treated water should be tested at appropriate detention times. Analyses for bromide, iodide, ammonia, and sulfur at low detection limits, together with analyses for the usual regulated chemicals, should be conducted on the untreated and treated water.

DBP Formation and Continuing Reactions

The sensory characteristics of a large number of DBPs are listed in the present report. This compilation could be used to estimate potential taste-and-odor changes. By comparing the results of the chemical analyses with the OTCs of these compounds, it should be possible to anticipate the taste-and-odor potential of a water sample. It may be possible to optimize treatment to minimize DBP formation while maintaining adequate disinfection for microbial control.

Blending

When it is necessary to use blends of water of different qualities, it is recommended that bench-scale experiments, such as the ones presented in Chapter 5, be performed. These experiments will help determine the optimum blend that can be used without impairment of the sensory characteristics of the water distributed.

DISTRIBUTION SYSTEM TASTE-AND-ODOR WHEEL

Figure A presents the distribution system taste-and-odor wheel. This wheel stems from the original flavor wheel of Mallevialle and Suffet (1987) but has been updated and modified for the specific taste-and-odor compounds associated with distribution systems. The inner circle shows the four categories of basic tastes that pertain to water, one category of mouthfeel and nosefeel, and eight categories of odors. The middle circle from the center shows the common descriptors for each category, and the outer circle lists the chemical compounds that can be used as reference standards for the specific descriptors. All compounds designated by an asterisk have been actually confirmed to cause tastes or odors in the distribution system. Compounds such as methyl butanal and the iodomethanes have actually been identified in some of the case studies presented in this report.

Table A complements the taste-and-odor wheel by identifying the possible sources (column 2) of the different taste- and odor-causing agents and by suggesting appropriate corrective actions (column 3) to be taken to eliminate or prevent tastes and odors. The taste-and-odor wheel is intended for use as a guideline by water practitioners to recognize the types of quality changes that may be or are being experienced.

Record-Keeping and Databases

It is recommended that water utilities keep records of taste-and-odor complaints, even when they are easily addressed. A documented and properly compiled database could help water utilities predict taste-and-odor changes before they occur. Information on location, frequency of occurrence, and type of complaint, along with other relevant water quality parameters, could

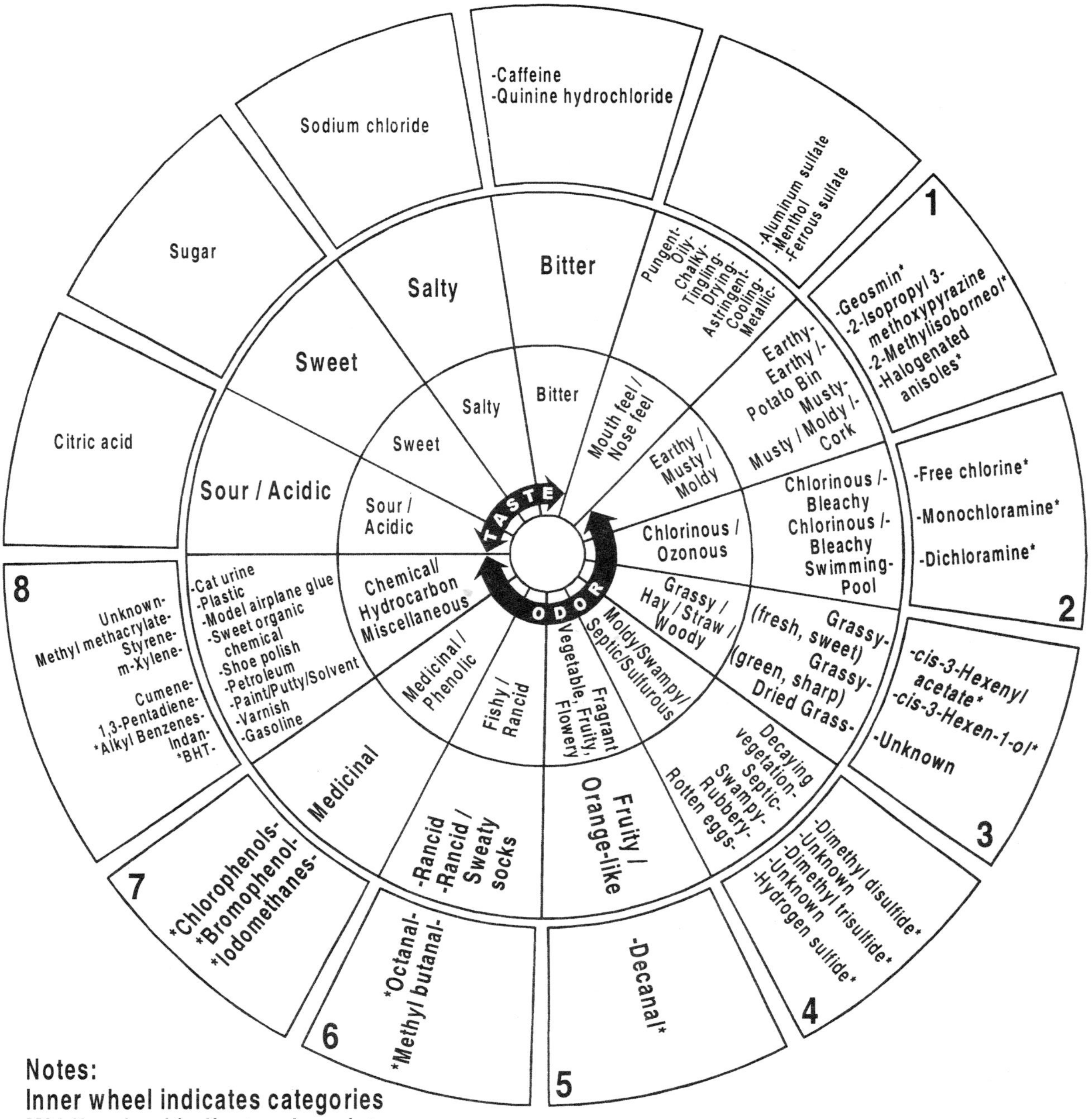

Notes:
Inner wheel indicates categories
Middle wheel indicates descriptors
Outer wheel indicates reference standards
Distribution system has not been positively confirmed to be the source of the compounds in italics
***Presence confirmed in water**

Figure ES.1 Distribution-system taste-and-odor wheel

Table ES.1

Sources and Possible Corrective Actions

Category	Possible Source	Possible Corrective Actions
Sweet	• Open reservoirs, biological activity	• *Cover reservoir. Breakpoint-chlorinate. Clean regularly*
Salty	• Cross-connections • Treatment breakage at reverse osmosis membrane • Well contamination	• *Eliminate cross-connection. Survey for potential cross-connections and install appropriate backflow prevention devices.*
Sour/Acidic	• Hot water systems	• *Raise water temperature in water heaters. Flush water heaters annually. Replace sacrificial anodes in water heaters.*
	• Poorly maintained point-of-use (POU) device	• *Replace POU device filters more regularly.*
	• Cross-connections	• *Eliminate cross-connection. Survey for potential cross-connections and install appropriate backflow prevention devices.*
Bitter	• High pH from corrosion of cement lining	• *Fill, hold, and flush to waste newly installed or relined cement-lined pipelines left to stand. Periodically flush.*
	• Metal leaching	• *Stabilize water by adding polyphosphates or adjust pH.*
	• Cross-connections (e.g., backflow from carbonation)	• *Eliminate cross-connection. Survey for potential cross-connections and install appropriate backflow prevention devices.*
Mouthfeel/Nosefeel	• Galvanized or copper pipes, corrosion, rust, and other corrosion by-products	• *Add stabilizing agents such as polyphosphates, or adjust pH.*
	• New plumbing and water coolers	• *Fill, hold, and flush new plumbing.*
	• Cross-connections or backflow from carbonation	• *Eliminate cross-connection. Survey for potential cross-connections and install appropriate backflow prevention devices.*
Earthy/Musty/Moldy	• Open reservoirs, biological activity	• *Cover open treated-water reservoirs. Chlorinate open treated-water reservoirs. Install aeration systems.*
	• Dead ends or low-flow sections of system	• *Perform routine flushing of distribution system dead ends. Eliminate dead ends by looping mains.*
	• Loss of chlorine and un-masking of background odors	• *Maintain detectable residual at all times.*
	• Biomethylation of halogenated phenols	• *Treat with chlorine dioxide.*
	• Cross-connections	• *Eliminate cross-connection. Survey for potential cross-connections and install appropriate backflow prevention devices.*

(continued)

Table ES.1 (continued)

Category	Possible Source	Possible Corrective Actions
Chlorinous	• New main disinfection	• *For super-chlorinated mains, dechlorinate and discharge.*
	• Residual boosting	• *Optimize weight ratio of chlorine to ammonia-nitrogen for chloraminated water. (Make sure the water is where it should be with respect to the breakpoint curve.)*
	• Blending	• *Eliminate breakpoint-chlorination occurrence in blended water. Reduce blend percentage or residuals of chlorinated water compared to chloraminated water.*
	• Continuing reactions	• *Chloraminate distribution system.*
	• DBP formation	• *Optimize treatment.*
Grassy/Hay/Straw/ Woody	• Open reservoirs, biological activity	• *Cover reservoirs. Treat reservoirs with chlorine. Utilize selective withdrawal from reservoirs. Blend with other sources.*
Marshy/Swampy/ Septic/Sulfurous	• Poorly circulated reservoir	• *Aerate (or aerate hypolimnon of) source-water reservoir. Change withdrawal levels from reservoirs.*
	• Poorly maintained POU devices	• *Replace POU device filters more often.*
	• Poorly maintained hot water systems	• *Flush hot water systems annually.*
	• Poorly maintained sedimentation basins and/or dirty or stagnant filters in treatment plants	• *Clean sedimentation basins frequently. Backwash filters before returning to service.*
	• Dead-end mains	• *Loop or periodically flush dead-end mains.*
	• Oxidation of polysulfides	• *Breakpoint-chlorinate, followed by chloramination.*
	• Cross-connections or backflow	• *Eliminate cross-connection. Survey for potential cross-connections and install appropriate backflow prevention devices.*
Fragrant/Vegetable/ Fruity/Flowery	• Open reservoirs, biological activity	• *Aerate source-water reservoirs. Breakpoint-chlorinate, followed by chloramination of distribution system.*
	• Continuing reactions	• *Adjust chloramine:ammonia ratios.*
Fishy	• Open reservoirs, biological activity	• *Aerate source-water reservoirs. Pre-oxidate source waters prior to treatment.*

(continued)

Table ES.1 (continued)

Category	Possible Source	Possible Corrective Actions
Medicinal/Phenolic	• Post-chlorination of biofilms	• *Convert to chloramines. Maintain high chloramine residuals.*
	• Slow kinetics and DBP formation	• *Optimize treatment.*
	• Reservoir and tank linings, chlorination of phenolic linings	• *Use coatings approved by a recognized authority. Allow proper curing times on new coatings. Fill and hold, followed by flushing newly coated tanks and pipelines.*
	• Cross-connections or backflow	• *Eliminate cross-connection. Survey for potential cross-connections and install appropriate backflow prevention devices.*
Chemical/Hydrocarbon/ Miscellaneous	• Chlorine dioxide/carpet, volatilization in indoor air	• *Add ferrous salts or reduced sulfur compounds to remove chlorite ion at the treatment plant or substitute chloramines for chlorine in the distribution system*
	• New plumbing and pipe materials	• *Use plumbing materials and fixtures approved by a recognized authority. Fill, hold, and flush plumbing.*
	• Plastic pipes, lubricants, new mains, asphaltic coatings, paint, epoxy/solvent-based linings	• *Use lubricants approved by a recognized authority. Allow proper cure times for new linings. Fill, hold, and flush newly lined mains.*
	• Permeation of buried plastic pipes	• *Replace plastic pipelines or remove contaminated soils surrounding pipelines.*
	• Old coal-tar linings on cast-iron mains	• *Gently flush to refresh water quality in mains. (Aggressive flushing can damage the coal tar lining the mains).*

provide insights into the underlying causes of taste-and-odor changes and help to document trends over time or follow changes in treatment or operations.

Although this report includes representative case histories and examples of potential taste-and-odor changes, new situations will inevitably arise that have not previously been experienced or reported. It is hoped that this report, including appendixes, will assist water utility personnel in providing the best-tasting water possible on a cost-effective basis and at a globally competitive price.

RECOMMENDATIONS

1. A "taste-and-odor formation potential test" similar to the one used for trihalomethanes, but based on sensory evaluation, should be established. Parameters such as detention time, pH, disinfectant residual, and temperature should be taken into consideration. This test should be performed with the specific water matrix and the specific treatment in use by the utility doing the evaluation (site-specific water), and the test should be regularly used in evaluating water quality changes.
2. Certification procedures for materials to be used in the distribution system should require sensory evaluation prior to approval of the materials. In addition, all new facilities including pipes should be tested for off-odors prior to their release for use.
3. Utilities should use only approved materials in their distribution systems and certification should be by the recognized authorities (e.g., NSF International or AWWA). The recognized authorities should include taste and odor testing in their approval procedures.
4. Methods of taste-and-odor evaluation should be standardized. It is only by means of standardization that the results of a taste-and-odor investigation can be used for comparison with other occurrences.
5. Water utilities should record and compile their taste-and-odor complaints and subsequent investigations in a database that can be accessed for future reference. Such a database could provide insights into the origins and causes of recurrent quality changes. Ideally, these databases should be made available to other researchers and water practitioners for optimal understanding and solution of problems.
6. Additional work is necessary to identify the causes of chlorinous odors in distribution systems in which the disinfectant residual is very low.
7. Careful flushing of the pipes and minimum use of lubricants will help to minimize impacts of pipes and lubricants on tastes and odors.
8. New techniques are needed to measure lower concentrations of compounds that can cause taste and odors in the distribution system.

9. More work should be performed to identify the sulfur odors, their mechanisms and kinetics of formation and possibly their stability in treated waters.
10. More work is necessary to confirm actual relationships between biological activity and tastes and odors generated in the distribution system.

CHAPTER 1
INTRODUCTION AND BACKGROUND

BACKGROUND

Taste-and-odor problems are a major concern to the water industry. It is extremely important for a water utility to provide water that is free of objectionable tastes and odors because customers judge the quality of the water at their tap mainly by its aesthetic properties such as color, taste, and odor. Water provided to the general public should not only be of sufficient quantity at adequate pressure, but must also not be objectionably degraded as it passes from the treatment plant to the customers. This water quality degradation includes the production of undesirable tastes and odors.

The last 20 years have seen major advances in the taste-and-odor field. Research has covered multiple areas ranging from chemistry and microbiology to engineering and treatment processes. Various microorganisms that cause tastes and odors in source water have been identified and treatment processes have been devised for the sole purpose of removing taste-and-odor compounds. Source-water reservoir management practices (Izaguirre and Devall 1995) have been developed and are successfully used to prevent or minimize major taste-and-odor episodes. This progress has been made possible by the use of sophisticated analytical techniques, such as closed-loop stripping analysis (CLSA) with gas chromatography/mass spectrometry (GC/MS) (Bruchet, Khiari, and Suffet 1995) and sophisticated sensory testing with flavor-profile analysis (FPA) (Krasner et al. 1985, Bartels et al. 1986, APHA et al. 1998).

Most of this effort has been directed at problems originating in source water or produced during the treatment process (Mallevialle and Suffet 1987, Suffet et al. 1995). Despite these major developments in the taste-and-odor field, one area that has not been thoroughly investigated is the distribution system. It is commonly believed that when drinking water smells and tastes bad year-round, the cause may be the distribution system (Blyth and Johnston 1997). However, tastes and odors from the distribution system, like the ones originating in source water, can be chronic when the problem originates from the quality of the water that enters the distribution system. They may be episodic when cross-contamination, leaching material, or maintenance events are involved. Burlingame and Anselme (1995) published the most comprehensive review of tastes and odors in distribution systems to date. Distribution systems are very complex, and the task of tracking down the problems is extremely difficult. Reports of cases of distribution-related taste-and-odor problems are mostly anecdotal and a central database to record them in distribution systems is lacking.

In many cases, the need to respond rapidly to the emergency does not allow utilities to investigate the problem properly or to understand the cause of the problem; instead, they must seek the quickest possible solution and minimize the number of complaints. This encourages anecdotal literature.

The American Water Works Association (AWWA) Taste and Odor Committee conducted a nationwide survey of water utilities in 1989 to assess the extent of taste-and-odor problems in the United States (Suffet et al. 1993). Figure 1.1 summarizes the responses received, showing that 65% of the 388 utilities surveyed indicated the distribution system as the primary cause of their problems. The second highest cause, as reported by 60% of the utilities, was disinfection. The major odor problems reported were chlorinous, earthy, fishy, and medicinal odors (Figure 1.2). The major taste problems cited were sour, metallic, and chlorinous flavors (Figure 1.3). Utilities using groundwater also reported complaints of sulfide odors (Figure 1.4).

Earthy and musty odors are produced by microorganisms and sometimes appear in the distribution system, even when they have not been reported in the treatment plant effluent or linked to the presence of geosmin or 2-methylisoborneol (MIB). One explanation for this phenomenon is that geosmin and/or MIB were actually present in the effluent water, but their odor could not be detected because of the masking effect exerted by high concentrations of disinfectant residual. Once the disinfectant has dissipated, then the earthy-musty–smelling compounds, such as trichloroanisoles, can be produced in distribution systems. Marshy, swampy, septic, or sulfurous odors are generally associated with anaerobic conditions and may occur in the distribution system. Chemical, hydrocarbon, and other miscellaneous odors usually result from spills or industrial contaminants. The sour taste reported was a major complaint but has not received much attention in the literature.

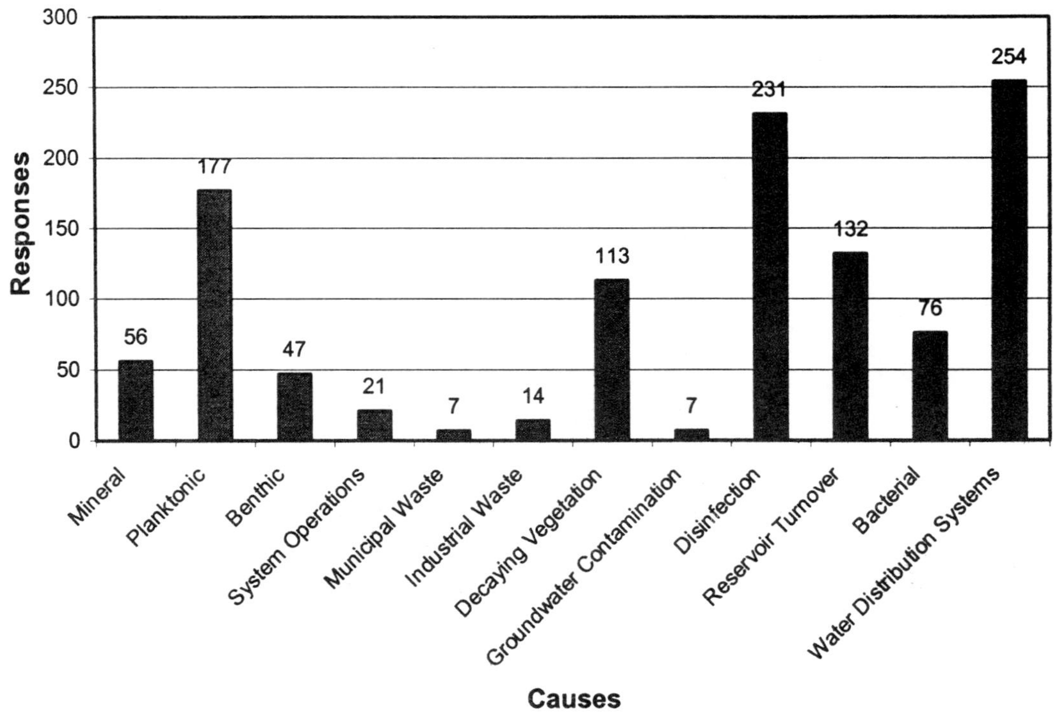

Figure 1.1 Reported causes of taste-and-odor problems in U.S. utilities (Suffet et al. 1993)

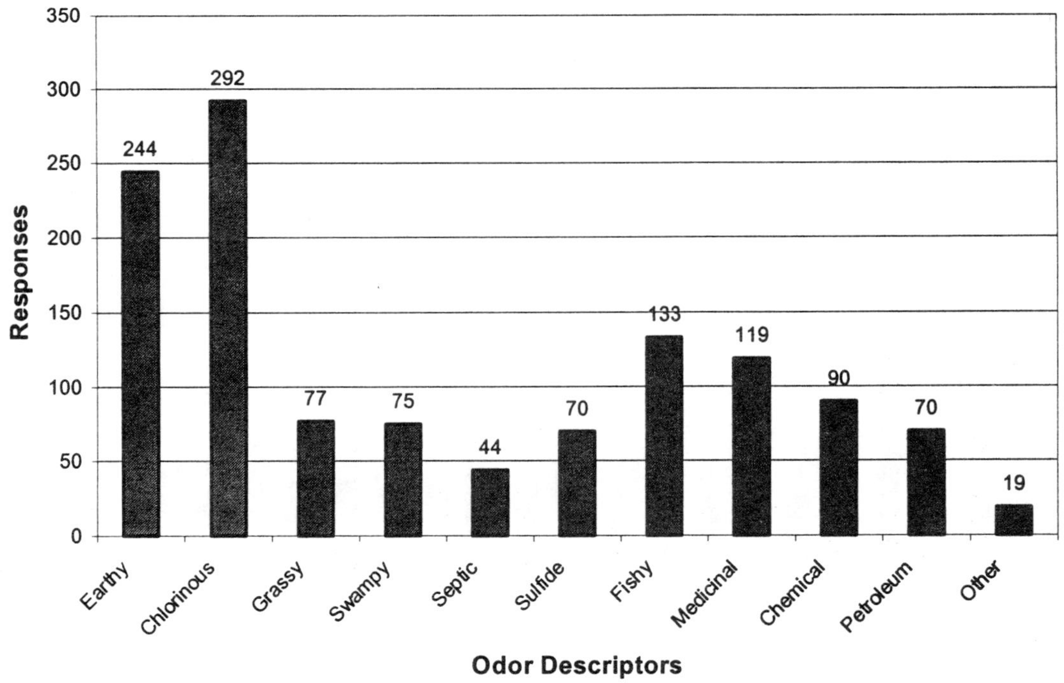

Figure 1.2 Types of odor problems experienced by U.S. utilities (Suffet et al. 1993)

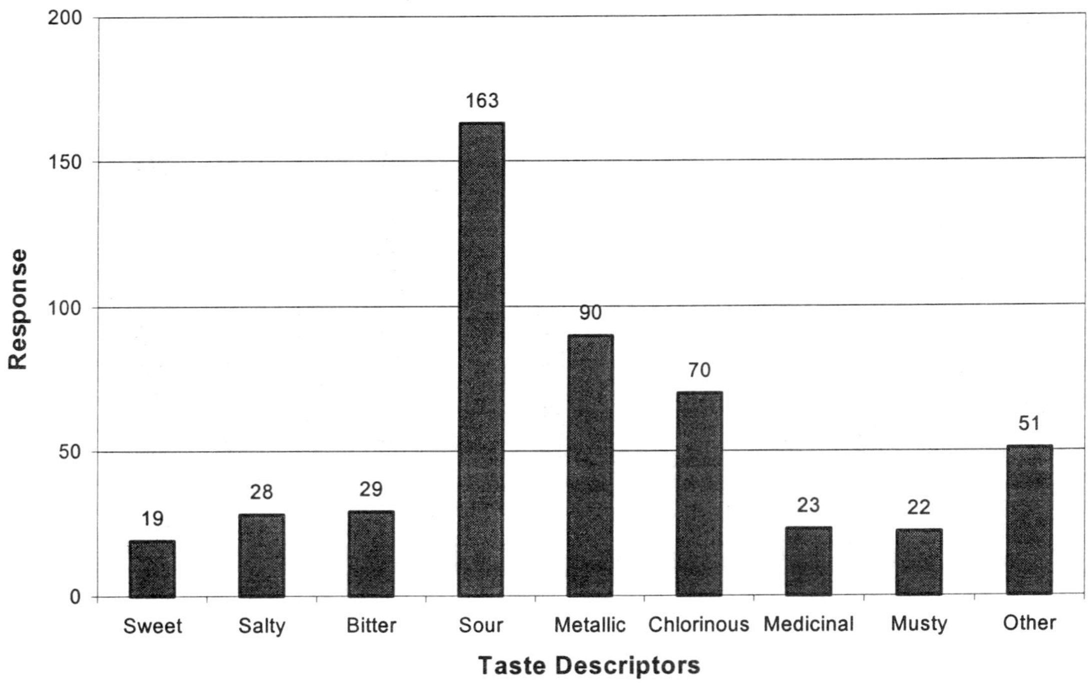

Figure 1.3 Types of taste problems experienced by U.S. utilities (Suffet et al. 1993)

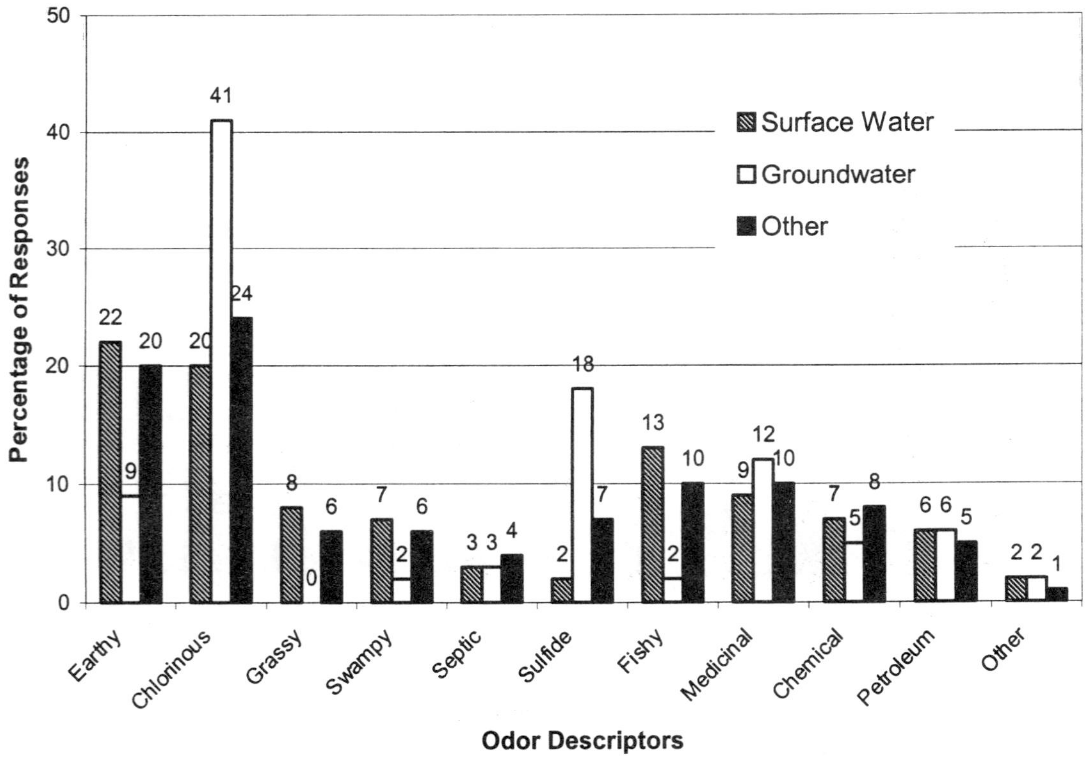

Figure 1.4 Types of taste-and-odor problems shown by water source (Suffet et al. 1993)

Figure 1.5 presents a breakdown of the tastes and odors that were attributed to the distribution system. Approximately 49% of the utilities indicated chlorine residual as the cause of taste-and-odor problems in their distribution systems. Domestic plumbing or home treatment devices were the second most important cause noted, followed by corrosion by-products.

An informal survey conducted in 1994 by Societé Lyonnaise des Eaux on the number and types of drinking water complaints in western Paris (France) showed that tastes and odors ranked second (32%) after complaints about color (39%) (Figure 1.6). "Chlorinous" problems accounted for 48% of the taste-and-odor complaints and "musty" for 21%. In 1995, a random survey in southern Paris showed that 33.5% of the samples collected exhibited some type of sensory impairment with only 1.8% rating at strong intensity. Similarly to the 1994 survey, "chlorinous" reports accounted for the majority of the problems (77%), followed by "earthy" (9%). (Figure 1.7). It is interesting to note the relatively high percentage of "medicinal" descriptors perceived.

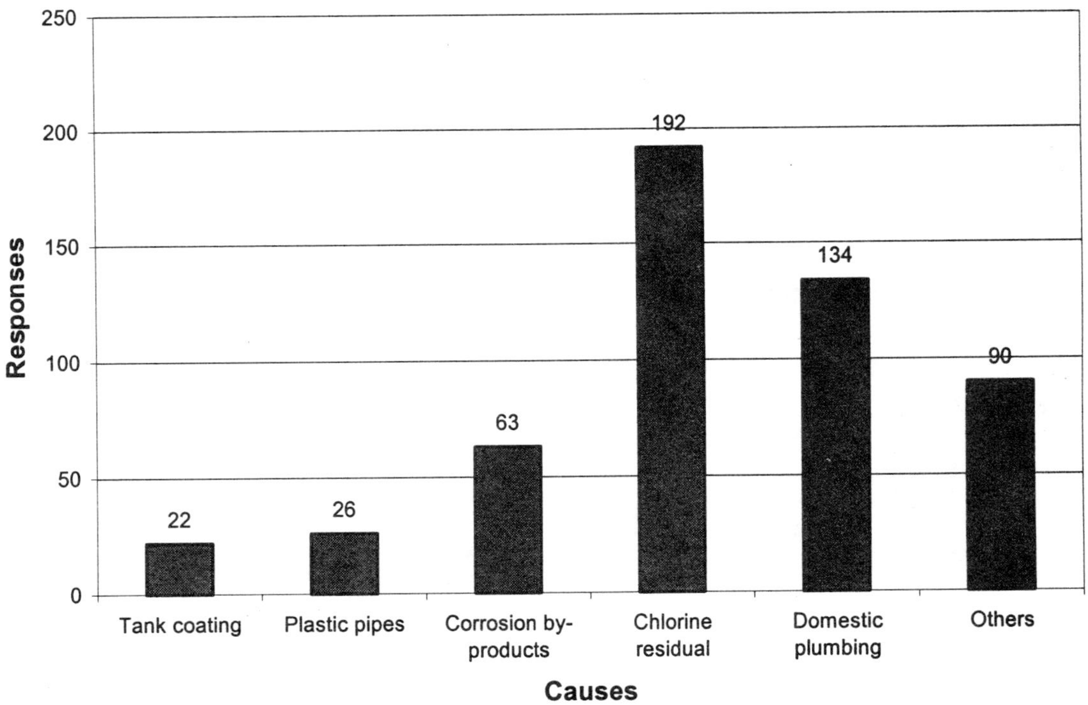

Figure 1.5 Probable causes of taste-and-odor problems in distribution systems (Suffet et al. 1993)

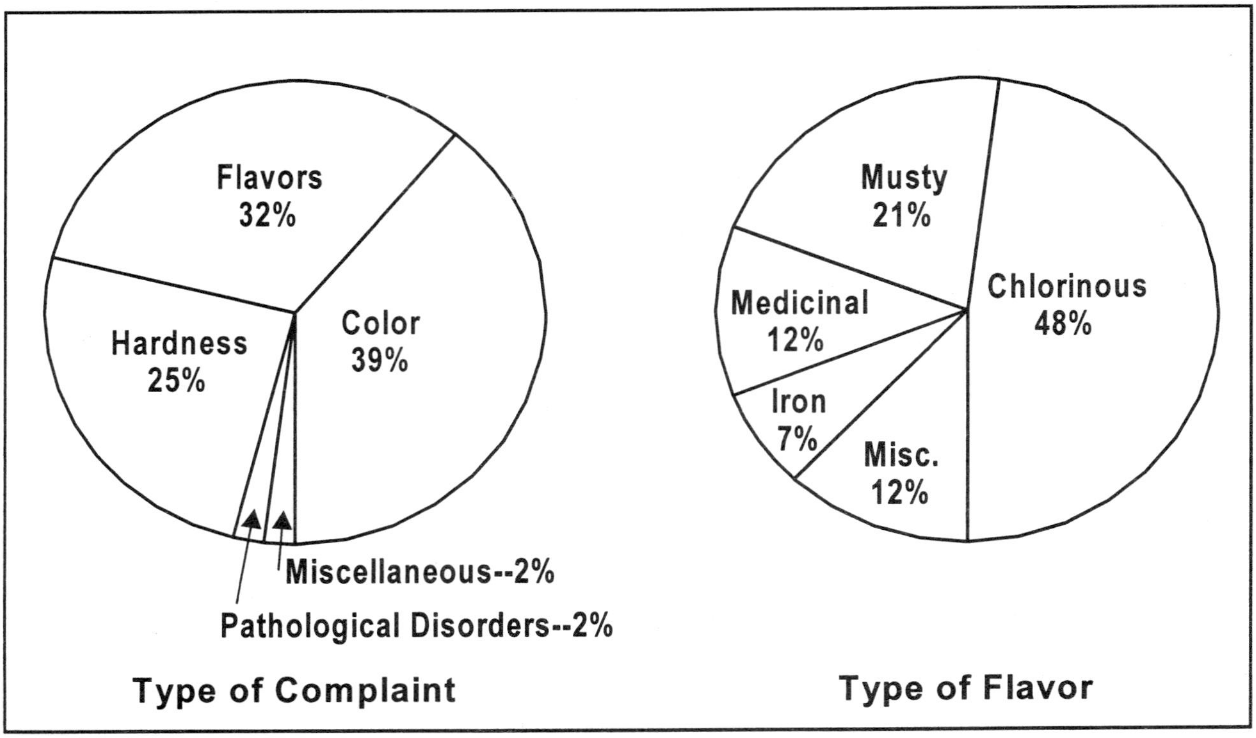

Figure 1.6 Survey results in the western Paris regional center—1994

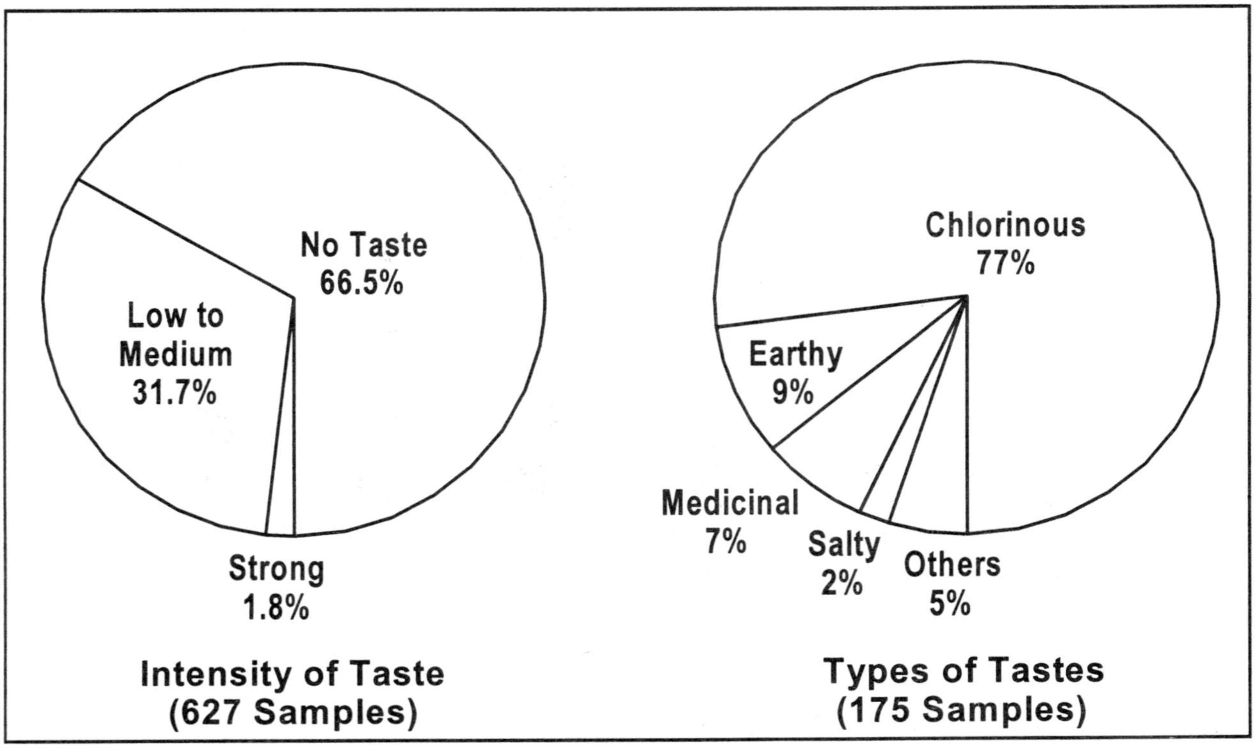

Figure 1.7 Survey results in the southern Paris regional center—1995

CAUSES OF TASTES AND ODORS IN DISTRIBUTION SYSTEMS

Tastes and odors in distribution systems may arise from biological sources (e.g., crustaceans, fungi, or bacteria) or from chemical sources (e.g., disinfectant residuals, disinfection by-products [DBPs], or chemicals leaching from reservoir cover material). Some problems are created by the type of disinfectant used (e.g., the reaction of chlorine dioxide with volatile organics from new carpets [Hoehn et al. 1990, Gallagher et al. 1994, Ellenberger et al 1998]), whereas others result from the system design (e.g., dead ends, or the hydraulic design of a storage reservoir) or the way the system is operated (e.g., blending chloraminated water with chlorinated water). Table 1.1 summarizes the various taste-and-odor problems in distribution systems.

Several factors can be responsible for the changes occurring in the distribution system. These changes can be biological, such as bacterial growth or biofilm formation; chemical, such as disinfectant residual decay, continuing reaction of disinfectant with remaining dissolved organic carbon (DOC), and subsequent formation of DBPs; or physical, such as leaching of chemicals from materials used and cross-connection contamination. Residence time and blending may also affect water quality.

Tastes and Odors Produced by Biological Activity

MacKenzie (1938) reported that 15 out of 17 (88%) of the fungi isolated from water that exhibited taste and odor were able to produce musty flavors. It was noted that the intensity of these flavors increased synergistically by the addition of chlorine. Burnham (1970) reported the isolation of various species of fungi directly in consumer complaint samples.

The most studied case of biological activity in distribution systems relates to the biomethylation of trichlorophenols to trichloroanisoles (Montiel et al. 1991, Sävenhed et al. 1991, Nyström et al. 1992). Bruchet (1999) reported a number of instances where 2,4,6-trichloroanisole was identified by chemical analysis as the cause of earthy-moldy types of odors. In areas with low velocity and no disinfectant residual (e.g., dead ends or when the water has not been used for an extended period), swampy or rotten vegetable odors can appear. Krasner et al.

Table 1.1

Known taste-and-odor problems in distribution systems

	Problem
Biological	Swampy and decaying vegetation odors in dead ends and poorly circulated storage facilities
	Chlorophenols formed from chlorine reacting with phenol and transformed to trichloroanisoles by fungi
Chemical	Precursors leaving the treatment plant continue to react with disinfectants to form odorants
	Brominated and iodinated trihalomethanes (THMs) producing strong medicinal odors
	Residual chlorine taste and odor is desirable from a disinfection perspective, but consumers may object to the tastes and odors of chlorine species
	Chlorine dioxide reacts with volatile organics from new carpet to produce odors that resemble kerosene and cat urine
Physical Characteristics and Materials	New mains have wet paper, putty, or chalky odor from lubricants used during installation
	Polyethylene pipes can release phenolic compounds
	Service connections and building materials contribute to metallic tastes
	Polymer pipes can allow permeation of hydrocarbons
	Cross-connections during pesticide applications can yield solvent-like tastes and odors
	Temperature (The warming of water increases total odor)

(1989) reported the presence of septic-swampy odors in finished drinking water in southern California. Chemical analysis identified organic sulfides as the cause of these odors. The source of the problem was traced back to the bottom of the reservoir, where anoxic conditions had developed, possibly causing bacterial activity and the production of sulfides.

A recurring problem in the Perth (Western Australia) distribution network was described by Wajon et al. (1985, 1986). Swampy odors were found in some parts of the distribution network, where anaerobic conditions and low chlorine residuals prevailed. Sulfur compounds such as hydrogen sulfide, methylmercaptans, and dimethylpolysulfides were identified in samples taken from various locations. The authors suggest that the decomposition of microorganisms (bacteria, algae, fungi) under these conditions leads to the production of odorous compounds.

Water heaters (especially when the hot water has not been used for a long period of time or when the heater thermostat is set too low) are another area where bacterial growth and subsequent production of organic sulfur compounds are favored. Finally, poorly maintained point-of-use devices have been found to favor anaerobic conditions, leading to formation of odorous sulfur compounds.

Tastes and Odors Caused by Disinfectants

One of the most frequently reported taste-and-odor complaints is directly associated with the use of chlorine as a disinfectant. The species involved are hypochlorous acid (HOCl) and the hypochlorite ion (OCl^-). "Bleach" is the descriptor generally used to describe both HOCl and OCl^-. HOCl has an odor threshold concentration (OTC) at 25°C of about 0.28 mg/L as Cl_2 and a flavor threshold concentration (FTC) at 25°C of about 0.24 mg/L as Cl_2, whereas OCl^- has an OTC of about 0.36 mg/L as Cl_2 and an FTC of about 0.30 mg/L as Cl_2 (Krasner and Barrett 1984).

When ammonia is used in combination with chlorine, several species of chloramines are formed. Monochloramine, dichloramine, and trichloramine are the three major species of inorganic chloramines. As with free chlorine, monochloramine has a "bleach" descriptor. Monochloramine's OTC and FTC at 25°C are approximately 0.65 and 0.48 mg/L as Cl_2, respectively, as observed by Krasner and Barrett (1984). The same authors report a "swimming pool" descriptor for the dichloramine species and an intensity level of 4 (slight) to 8 (moderate) for concentrations ranging from 0.1 to 0.5 mg/L as Cl_2.

White (1999) states that tastes and odors are not likely to form from the chlorine compounds themselves when the concentrations of these compounds are less than 20.0 mg/L as

Cl_2 for HOCl, 5.0 mg/L as Cl_2 for monochloramine, 0.8 mg/L as Cl_2 for dichloramine, and 0.02 mg/L as Cl_2 for trichloramine. Pertaining to the palatability of water, White (1999) states that the best ratio of monochloramine to dichloramine is site-specific. Williams (1949) found that trichloramine can form in alkaline waters that are high in nitrogenous matter where free-chlorine residuals are maintained. It is generally known that trichloramine will form in water with a pH of 4 or less. Williams (1949) documents trichloramine formation in drinking water and describes the odor as a very pungent "geranium" type of odor. Williams (1966) refers to the odor of trichloramine as "geraniums" or "rotting grapefruit." The combination of chlorine species found depends on factors such as the pH and the ratio of chlorine to ammonia-nitrogen (Palin 1975, Snoeyink and Jenkins 1980, Wei and Morris 1975, Whittle and Lapteff 1975).

In addition to the direct effect of the disinfectant on the taste and odor of the drinking water, the number of complaints increased when utilities changed disinfectant during different times of the year. This observation can be generally explained by the fact that customers are adapted to certain types of odors, and the appearance of unfamiliar odors may trigger those complaints. Consumers tend to complain of any noticeable change, irrespective of its nature.

Blending can, in some instances, improve the taste-and-odor characteristics in certain areas of the distribution system by a simple dilution effect. In other instances, especially when chloraminated water that has excess ammonia is blended with chlorinated water, the production of di- and trichloramines may be favored (Barrett et al. 1985). Di- and trichloramines are more odorous than monochloramine.

Continuing Reactions and Disinfection By-Product (DBP) Formation

Although the loss of residual may improve organoleptic characteristics, when treated water leaves the treatment plant it has a certain amount of DOC and a disinfectant residual. As a result of slow reaction kinetics, the remaining DOC reacts with the disinfectant residual, and by-products absent in the treatment plant effluent appear in the distribution system. Phenols and halophenols (chlorinated, brominated, and, to a lesser extent, iodinated) are of particular interest in the taste-and-odor field because they are known to cause medicinal odors (Burttshell et al. 1959, Whitfield 1988, Dietrich et al. 1999). 2-Chlorophenol, 2,4-dichlorophenol, and 2,6-dichlorophenol have OTCs (at about 25°C) lower than 10 µg/L (Burttshell et al. 1959).

Although phenol is usually found in some source waters and is (in most cases) caused by industrial pollution, it is sometimes found in materials used in the distribution system. When phenol that has leached from the pipe material into the chlorinated water reacts with the residual disinfectant, it forms various chlorophenols, some of which are known to be potent taste- and odor-causing compounds. When trichlorophenols are formed, their biomethylation by fungi can lead to formation of trichloroanisoles, which may produce earthy-musty odors at low pg/L levels (Montiel et al. 1999). Most trichloroanisoles have extremely low OTCs, but 2,4,6-trichloroanisole is the most potent, with an OTC between 20 and 80 pg/L.

Another category of medicinal taste-and-odor compounds includes the iodinated trihalomethanes (ITHMs) (Gittelman and Yohe 1989, Bruchet et al. 1989). Of all the possible THMs formed in drinking water, iodinated THMs appear to have the greatest potential for imparting tastes and odors to treated drinking water. Iodinated THMs, with the exception of iodoform, have received only limited attention, in part because no analytical standards could be obtained commercially. Iodoform has a medicinal taste and an FTC of about 5 μg/L (Hanson 1987). Two iodinated THMs that are frequently found in drinking water are dichloroiodomethane and bromochloroiodomethane (Gould et al. 1985). Because of the low levels of iodide usually found in nature, the concentrations of these iodinated THMs are usually too low to cause any tastes and odors. There are some waters in which it is possible to find rather high concentrations of iodide and, consequently, high levels of iodinated THMs.

Amino acids are important precursors for the aldehydes and nitriles that result from oxidation. Amino acids are an important part of the natural organic matter (NOM) present in drinking water. Biofilms are the major source of amino acids found in the distribution system. The reaction of amino acids with residual chlorine or chloramines leads to the production of aldehydes and, ultimately, nitriles (LeCloirec and Martin 1985). A series of short-chain, branched aldehydes has been associated with objectionable odor episodes (Hrudey et al. 1988). Furthermore, the reaction between chloramines and aldehydes produces nitriles as a secondary product of the amino acids–chlorination reaction. Both aldehydes and nitriles can cause tastes and odors (Mallevialle and Suffet 1987). The taste-and-odor properties of aldehydes and nitriles are not very well understood.

Other Reactions Producing Tastes and Odors

It is known that chlorinous odors are generated during the disinfection process and are generally caused by the disinfectant itself. However, some cases of chlorinous odors can develop even in the absence of a chlorine residual. This type of problem occurs frequently in France, where disinfectant residuals are traditionally kept at very low levels because customers reportedly object to chlorinous odors in their tap water. Welté and Montiel (1999) investigated the possible causes of persistent chlorinous tastes and odors developing in the distribution system at low or absent chlorine residuals. The strength of the odor increased as the water residence time increased. It was assumed that the cause of the change was trichloramine produced by chlorination of some organo-nitrogen compounds with slow formation kinetics during the winter. This assumption could not be positively confirmed, and, to date, neither have the compounds responsible for these tastes been clearly identified nor the production mechanism elucidated.

Leaching of Distribution-System Material

Plastic pipe, organic coatings and membranes, and epoxy relining products are progressively replacing traditional metallic or cementitious materials. Despite existing controls, and for various reasons, additives, solvents, or monomers can be leached into drinking water. In a large number of cases, materials used in distribution systems have been reported as the cause of tastes and odors. Although the material used is generally approved and certified by recognized authorities (e.g., NSF International or AWWA) the certification processes do not involve organoleptic parameters. In some cases, only a specific batch of the material can be found to be defective.

Rigal and Danjou (1999) reported several cases of tastes and odors caused by chemicals leached from the material used. Amongst these chemicals, styrene used in the manufacture of fiberglass-reinforced polyester resin was found to cause resinous flavors in the water. The same authors reported another example of organoleptically impaired water caused by the leaching of styrene, acetone, and methylisobutylketone from material used in the coating of a water tower.

Rigal and Danjou (1999) also reported an incident that occurred in another location following the rehabilitation of a reservoir. A very strong flavor, described as "medicinal" by consumers, appeared in the drinking water. The reservoir had been rebuilt using cement containing organic additives. This material had been approved by the health authorities and did not give any flavor to the water, and no leaching of organics was observed during the migration test. It was found that a vinyl polychloride net had been added to the material used for the reservoir without additional testing. The authors hypothesized that chlorophenol was formed from the reaction of chlorine with triphenylphosphate leached from the vinyl polychloride additive. In all these cases, it was noted that the material, although certified, had not been tested under conditions specific to the treatment plant or the distributed water.

Polyethylene is another type of pipe material used in the distribution system that has been linked to taste-and-odor reports. The change in quality is usually caused by leaching of large quantities of additives used as antioxidizing agents (Anselme et al. 1985a, 1985b, 1986). The release of these additives (2,6-di-*tert*-butyl-4-methylphenol, or BHT) and aldehydes (from the oxidation of polyethylene during extrusion) can cause strong tastes and odors.

In addition to the factors described above, many other contributors to tastes and odors remain misunderstood. Table 1.2 identifies some of the areas that need to be investigated to help understand the origin of tastes and odors in the distribution system.

STUDY OBJECTIVES

The purpose of this study was to identify tastes and odors that are specifically generated in the distribution system. As highlighted in Table 1.2, many questions still need to be answered to explain the specific causes of taste-and-odor problems generated in the distribution system and to identify the specific chemicals associated with them. More research is necessary to understand the biological and chemical causes and conditions that result in tastes and odors in distribution systems. More work is needed to move forward in this field and to increase public confidence in drinking water supplied at the tap. A key factor is to identify the chemicals causing tastes and odors in order to understand the production mechanisms within the distribution system. This knowledge will help to identify the appropriate solutions to specific water quality changes or to prevent the changes in the first place.

Table 1.2

Distribution-system taste-and-odor problems requiring additional investigation

	Problem	Information needed
Biological	Actinomycetes	What are the conditions that allow growth of actinomycetes?
	MIB and geosmin	Are they produced in the distribution system, or are they already present in the plant effluent?
	Heterotrophic bacteria found in suspension and attached to pipe surfaces	Do they produce tastes and odors? If so, which ones?
	Chlorophenols from disinfection transformed to trichloroanisoles by fungi	Which conditions favor this type of reaction?
Chemical	Di- and trichloramines have more pronounced characteristics than monochloramine	What is the effect of increased monochloramine concentration and ratio of chlorine to ammonia-nitrogen?
	DBPs are formed as a result of the reaction of disinfection with NOM	Do DBPs produce tastes and odors? If so, which ones have taste-and-odor characteristics? What are the conditions that favor their production?
Physical Characteristic and Materials	Plastic pipes can release tetrachlorethylene, phthalate esters, and other compounds	What is the correlation between these compounds and tastes and odors in a drinking water distribution system?
	Odors from gaskets	Which ones? Under which conditions?
	Tastes and odors related to corrosion	Which types? Under which conditions?
	Bitumastic coating can cause tastes and odors	Which ones? Under which conditions?
	Leaching of organic solvents from lining	Which ones? Which types?
	Water with earthy-musty descriptors in the absence of geosmin or MIB	What are the compounds implicated?
	Blending—chemical changes resulting from blends and change to a different source	What are the mechanisms?

The objectives of the study were to:

1. Bring together the information that already existed and develop it to explain the causes and sources of distribution-system tastes and odors (the information is divided into biological sources, chemical causes related to disinfection practices, and other distribution sources in general).
2. Simulate water treatment processes and distribution-system conditions, at the bench and pilot scale, that produce tastes and odors in order to determine their sensory and chemical characteristics.
3. Investigate case studies of taste-and-odor problems that occur in actual distribution systems.
4. Identify preventive actions and adequate solutions to assist water utilities in minimizing consumer complaints and maximizing customer satisfaction
5. Develop guidelines to help water practitioners understand the problems and identify practical solutions to their specific needs.

SIGNIFICANCE OF THE STUDY

Management and control of tastes and odors in distribution systems require an understanding of the mechanisms that produce them, so that measures may be taken to prevent or alleviate the water quality changes and satisfy the needs of the public. The rationale for control is twofold. First, it is generally more efficient to prevent tastes and odors (and other water quality problems) when their potential occurrence is taken into consideration during the design and planning phase rather than dealing with these changes later when they actually occur. Second, it is more cost-effective to avoid the problems altogether than to resort to expensive corrective measures such as switching to a new treatment or changing to alternative materials.

It is the premise of this study that a thorough understanding of the processes and conditions leading to the production of tastes and odors in the distribution system is essential. This study provides answers to a number of questions by following a systematic approach involving laboratory experiments and investigating case studies.

CHAPTER 2
ANALYTICAL PROCEDURES, ANALYTICAL METHODS, AND EXPERIMENTAL DESIGN

ANALYTICAL PROCEDURES

Bench-scale experiments were performed on several waters, which covered a wide range of water quality and origin, to evaluate the taste-and-odor formation potential of various treatment processes. Experimental parameters were selected to create worst-case conditions that would maximize the production of taste-and-odor compounds.

Conventional Coagulation Treatment Protocol

All coagulation tests were performed using a standard jar test apparatus equipped with six 2-L square Plexiglas™ containers. From 12 to 18 L of sample water (depending on taste-and-odor analytical requirements) was placed in the jars. Polymer and $FeCl_3$ were simultaneously added to the reactor at their specified dosages. Two coagulant doses were tested, 3 and 30 mg/L. The samples were then rapid-mixed for 1 min (velocity gradient [G] ~ 100 s^{-1}), flocculated for 15 min (G ~ 40 s^{-1}), and settled for 60 min. After settling, 10–12 L of sample was withdrawn from a depth of 10 cm below the water surface and placed into an organic-free glass container. Care was taken to minimize excessive pouring or volatilization that could result in a loss of taste-and-odor compounds.

Ozonation Protocol

Ozone was produced using a tube-type, water-cooled generator operating on a low-frequency (60 Hz), variable-voltage power supply (Trailigaz Labo 76; Praxair-Trailigaz Ozone Co., Cincinnati, Ohio). Evaporated USP-grade liquid oxygen was used as the feed gas in all experiments. The O_3/O_2 feed gas concentration, maintained typically at 2.0% (by weight), was monitored continuously (PCI Ozone Controls HC-12; PCI-Wedeco Environmental Technologies,

Caldwell, N.J.). Gas flow to the ozonation reactor was controlled with calibrated glass/sapphire rotameters, with corrections for temperature and diffuser pressure. Excess O_3/O_2 gas was generated and fed to destruct to ensure that periodic sampling by the monitor did not significantly affect O_3 production or flow to the reactor. Prior to initiating this study, all stainless steel (grade 316L SS) and perfluoromeric (PFM) materials in contact with O_3/O_2 gas were cleaned with trichloroethylene or methanol, air dried, and ozonated for an extended period.

O_3/O_2 gas from the generator was passed sequentially through two 1.5-L fritless washing bottles. The first bottle contained 1.0 L of 5% H_2PO_4 solution; the second bottle contained 1.0 L of deionized, distilled water. The gas washing procedure was used to ensure that the feed gas was free of readily soluble gas contaminants. O_3/O_2 gas was then introduced into a bench-scale, semi-batch, continuously stirred tank reactor (CSTR) equipped with a six-blade Rushton turbine impeller (LIGHTNIN, Rochester, N.Y.) and quadrant baffles (Yoshida and Miura 1963). All gas- and liquid-contacting surfaces of the 19-L CSTR were constructed of borosilicate glass, 316L SS, or PFM materials. Gas was dispersed into the reactor through a 1.6-mm horizontal orifice located 10 mm under the tip of the 100-mm-diameter turbine mixer. The turbine mixer was typically operated at 450 rpm (G ~ 1000 s^{-1} for 16 L water) to ensure both rapid liquid mixing and extensive entrainment of headspace gas (via vortexing) into bulk solution. A PFM microgear pump (model 415A, Idex Corp., Vancouver, Wa.) was used to withdraw ozonated water from the CSTR. All tests were conducted at ambient room temperature (22–25°C).

Applied ozone dosage was determined by the O_3/total organic carbon (TOC) ratio and was set at 2:1 for all influent waters. Ozone was applied at a rate of 0.5 mg/L-min. This dose rate produced peak residuals similar to those found in full-scale contactors. Ozonated water was then kept within the CSTR for approximately 1 h to ensure complete decomposition of the remaining O_3 residual. Dissolved O_3 concentrations in this study were analyzed by Standard Method 4500-O3(B), with some modifications (APHA et al. 1998) after incorporating gravimetric sampling modifications (Yates and Stenstrom 2000).

For all tests evaluating pre-ozonation, ozonated water was transferred directly from the CSTR to Plexiglas jars using a PFM microgear pump. Alternately, for tests evaluating post-coagulation ozonation, jar-test effluent was transferred directly from the organic-free collection bottles to the ozonation CSTR using the PFM pump.

Chlorination and Chloramination Protocols

The simulated distribution system (SDS) procedure was used to perform chlorination and chloramination experiments. The SDS method was developed to evaluate the formation of various disinfection by-products (DBPs) in a given water on a bench scale. A more detailed description of the SDS test can be found in Koch et al. (1991). The SDS procedure was used in this study to evaluate the effect of the type of disinfectant and disinfection parameters on the organoleptic properties of various water matrices. The disinfectants studied were chlorine and chloramines at different doses, chlorine-to-nitrogen ratios, and contact times. The SDS parameters used in specific sections will be described with the relevant experiments.

Distribution-System Simulators

Two types of distribution-system simulators were used: pipe loops and annular reactors.

Pipe-Loop Study

The following describes the experiments associated with the pipe loops and associated field studies.

Inlet Water Characteristics. All pipe-loop experiments were performed on chlorinated finished water obtained from a conventional treatment plant (Figure 2.1). The initial free and total chlorine residuals were 0.1 ± 0.05 mg/L and 0.17 ± 0.05 mg/L, respectively. The water quality underwent no significant changes, except for chlorine decay, during its travel through pipes before reaching the experimental site where the pipe loop was located. Before entering the pipe loop, the water was dechlorinated using a granular activated carbon (GAC) cartridge.

Table 2.1 shows the main characteristics of the water before entering the pipe loop. The water entering the pipe loop showed consistent organoleptic properties; however, musty flavors (flavor threshold number [FTN] 1–3) were sometimes detected in the dechlorinated water.

Table 2.1

Main characteristics of the feed water

Parameters	Values
Flavor threshold number (FTN)	1-3 (musty)
Nitrite (mg/L)	<0.05 ± 0.1 L
Nitrate (mg/L)	30 ± 2
pH (units)	7.5 ± 0.2
Ammonium ion (mg/L)	<0.1
Total organic carbon (mg/L)	1.5 ± 0.1
UVA_{254}* (m^{-1})	1.1 ± 0.1 m^{-1}
HPC**–R2A (CFU***.mL^{-1})	100-300

*UVA_{254}: Ultraviolet absorption at 254 nm

**HPC: Heterotrophic plate count

***CFU: Colony-forming units

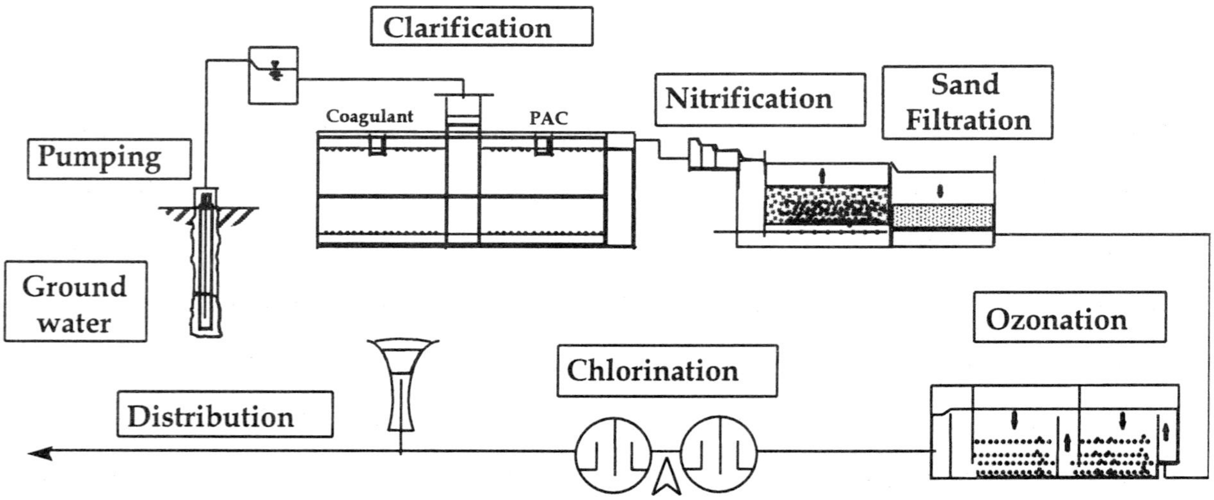

Figure 2.1. Description of the Aubergenville treatment plant

Pipe-Loop Design. The general design meets criteria relating to functional similarities with conditions encountered in actual networks (e.g., hydraulic conditions and types of material) (Piriou and Levi 1994). The pipe loop is composed of three loops (Figure 2.2) made of old cast-iron recovered on-site. The pipe loop is fed from a tank, and either chlorine or chloramine is injected before entering the pipe loop. The staging of residence times is obtained by placing several loops in series and enables the monitoring of water degradation kinetics for controlled flow conditions (recirculation velocity, residence time). Temperature control was performed for each loop, using a cooling system. The pipe loop offers the possibility of two types of sampling. Water samples can be taken at the outlet of each loop. A representative sampling of the inner surface of the pipes is also possible by the introduction of polyvinyl chloride (PVC) coupons (surface = 1 cm^2) at the surface of the pipes. Table 2.2 shows the main characteristics of the pipe loop and the operating conditions.

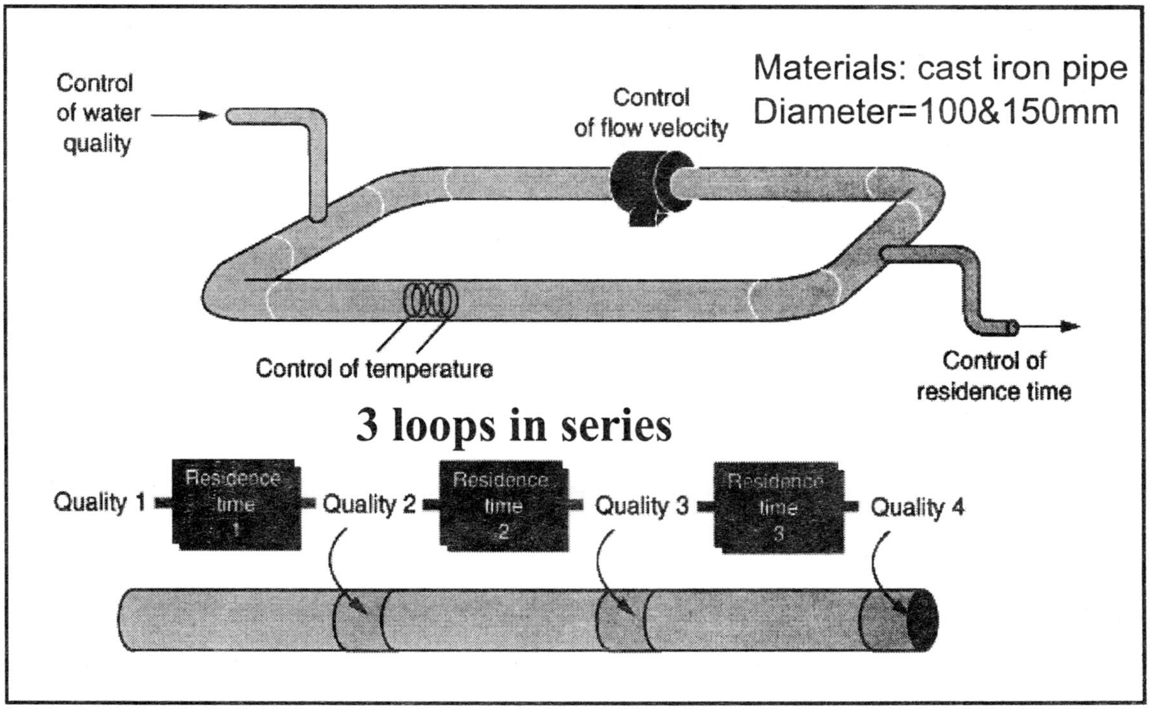

Figure 2.2 Schematic diagram of the pipe loop

Table 2.2

Pipe-loop characteristics

	Loop 1	Loop 2	Loop 3
Material	Cast iron	Cast iron	Cast iron
Diameter (mm)	150	100	100
Volume (L)	360	120	120
Length (m)	20	15	15
Residence time (h)	8	8	8
Reynolds number (Re)	45000	45000	45000
Velocity ($m.s^{-1}$)	0.3	0.45	0.45
Temperature (^oC)	20	20	20

The pipe loop (Figure 2.3) is used to study water quality evolution under steady-state conditions. Thus, each experiment was carried out until steady-state conditions were reached, based on water quality parameters such as chlorine residuals, taste-and-odor thresholds, and microbiological counts (2 to 3 weeks).

Laboratory Control Experiments (*Water Aging Experiments and Water Chlorine Decay Kinetics*). To study the aging of water with time, control experiments were performed in glass jars. A 1-L water sample was placed in a clean glass bottle. All glassware was previously soaked in sulfochromic solution and rinsed with high-purity water.

For the chlorine decay kinetic determination, at time t = 0, sufficient sodium hypochlorite solution was added in order to obtain the desired chlorine concentration (≈ 1 mg /L as Cl_2). The chlorine demand of the bottle was minimized by allowing the bottle to sit overnight in chlorinated water (≈ 10 mg /L as Cl_2) before the test. During the test, the bottle was covered with aluminum foil to avoid the effect of light. The chlorine depletion was measured as a function of time using the spectrophotometric *N,N*-diethyl-*p*-phenylenediamine (DPD) method. The long-term kinetics of chlorine depletion can be expressed by the decay rate (k'_w in units of min^{-1}), assuming first-order kinetics.

Figure 2.3 Picture of the pipe loop

Field Experiments. Sampling campaigns were carried out on two different distribution systems: the network of Vigneux (suburb south of Paris) and Cholet (western region of France). Cholet is an average-size town of about 60,000 inhabitants. The source water comes from an open reservoir and is treated conventionally (by prechlorination, coagulation, and sand filtration) with ozonation as a final step before final disinfection. Table 2.3 shows the general characteristics of the treated water. The production capacity is 30,000 m^3/day. The distribution network consists of over 250 km of pipe and two storage tanks. Either gray cast iron or cement-lined ductile iron constitutes the majority of the pipe material. Figure 2.4 shows the four sampling sites.

A conventional treatment plant (prechlorination, clarification, GAC filtration, ozonation and final chlorine disinfection) pumps raw water from the Seine River to supply the Vigneux distribution system. Table 2.4 shows the main characteristics of the raw and finished water. Table 2.5 gives a general description of the distribution system. Figure 2.5 shows the sampling site locations.

Table 2.3

Physico-chemical characteristics of Cholet plant effluent

Parameters	Measurement
Free chlorine (mg/L)	0.6 ± 0.3
pH (units)	8.4 ± 0.1
Total organic carbon (mg/L)	1.9 ± 0.4

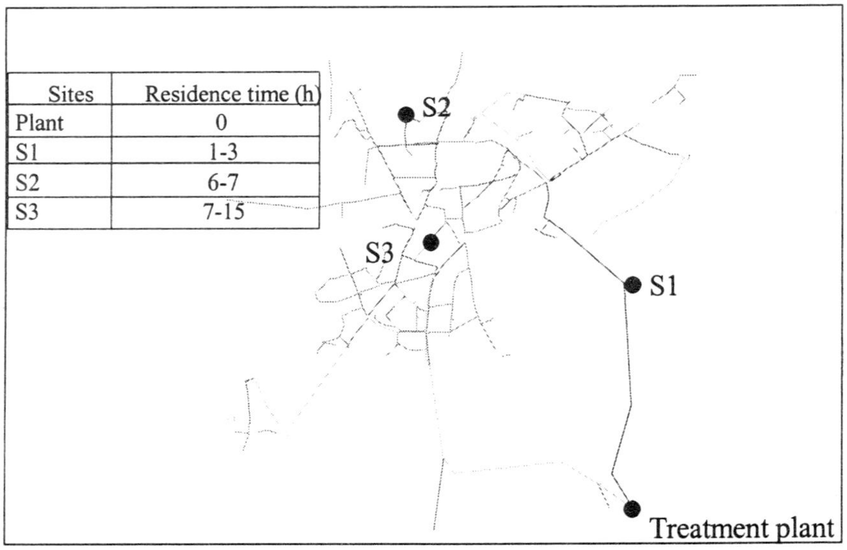

Figure 2.4. Location of sampling sites in the distribution system of the city of Cholet

Table 2.4

Physico-chemical characteristics at the Vigneux treatment plant inlet and outlet

Parameter	Raw water	Finished water
Particles > 5 μm (number/L)	10^5-10^6	10^5-4 10^6
Turbidity (NTU*)	5-50	0.1-0.2
Total organic carbon (TOC) (mg/L)	2.6-3.3	0.5-1.0
Biodegradable organic carbon (mg/L)	0.5-1.3	NR**
UV (m^{-1})	4.9-6.5	2.5-3.0
UV/TOC	1.7-2.5	3-5
pH	NR	7.5-7.7
HPC at 22°C (CFU/mL)	2 10^4-3 10^4	10^3-3 10^3
Total coliforms (CFU/100 mL)	10^3-10^4	10^2-10^3
Flavor threshold number (FTN)	5–12 mainly musty flavor	4–10 mainly chlorine flavor

*NTU: Nephelometric turbidity units

**NR: Not reported

Table 2.5

Vigneux distribution-system characteristics

Parameter	
Population served	200,000 inhabitants
Material	Unlined cast iron Cement-lined ductile iron High-density polyethylene
Diameter	63 to 600 mm
Storage tanks	One at 1,500 m^3 Four at 10,000 m^3** Four at 1,500 m^3**
Taste and dor complaints	Yes

**Water from Vigneux-sur-Seine (33%) and other resources (67%)

N° Sites	Residence time (h)
1	2
2	3
3	4
4	5.6
5	8
6	9
7	10.5
8	10.6
9	11.5
10	11.5

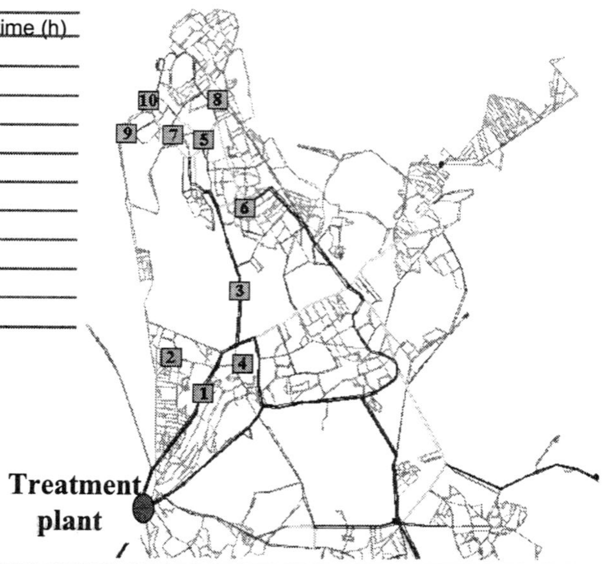

Figure 2.5 Location of sampling sites in the distribution system of Vigneux

Annular Reactor (AR) Study

ARs are devices that are used to simulate drinking water distribution systems. They were developed by the Center for Biofilm Engineering of Montana State University (Bozeman, Mont.) and consist of a rotor turning inside a stationary outer cylinder. Figure 2.6 shows a schematic diagram of the ARs used in this study. Hydraulic conditions inside the reactor (shear stress, water velocity) depend on the rotor's rotational speed. The Center for Biofilm Engineering recommends operating the ARs at a rotational speed of 60 rpm, which represents a water velocity in pipes of approximately 1 ft/s. Four draft tubes located inside the inner cylinder enhance liquid mixing. The fluid volume of a reactor is 1150 mL. The water residence time in the distribution system can be simulated by the AR's influent flow. For example, an influent flow of 1.6 mL/min simulates a distribution system residence time of 12 h. Removable wall segments (slides) are located on the outside wall of the inner cylinder; they permit the study of the biofilm fixed on the AR walls. Biofilm measurements can provide information on the number, species, and viability of bacteria fixed on AR walls, as well as the fixation capabilities of organisms and compounds responsible for tastes and odors. This unique characteristic of ARs (the possibility of collecting both liquid and biofilm samples) makes them cost-effective in simulating a distribution system. Because some taste-and-odor compounds are produced by bacterial activity, relating these compounds to biomass may provide novel and important information.

ARs can more accurately simulate distribution systems because of the presence of a biofilm on their walls, which is not present, for example, in batch incubation. However, the development of this biofilm requires a startup period (from 6 weeks to 3 months, depending on the water temperature and organic loading of the water) during which the ARs are fed on a continuous basis with the water to be analyzed. Once the biofilm is at a steady state (the rate of detachment of bacterial cells is equal to their rate of attachment), ARs can be considered to be representative of a distribution system. As soon as the water flow to the AR is stopped, the

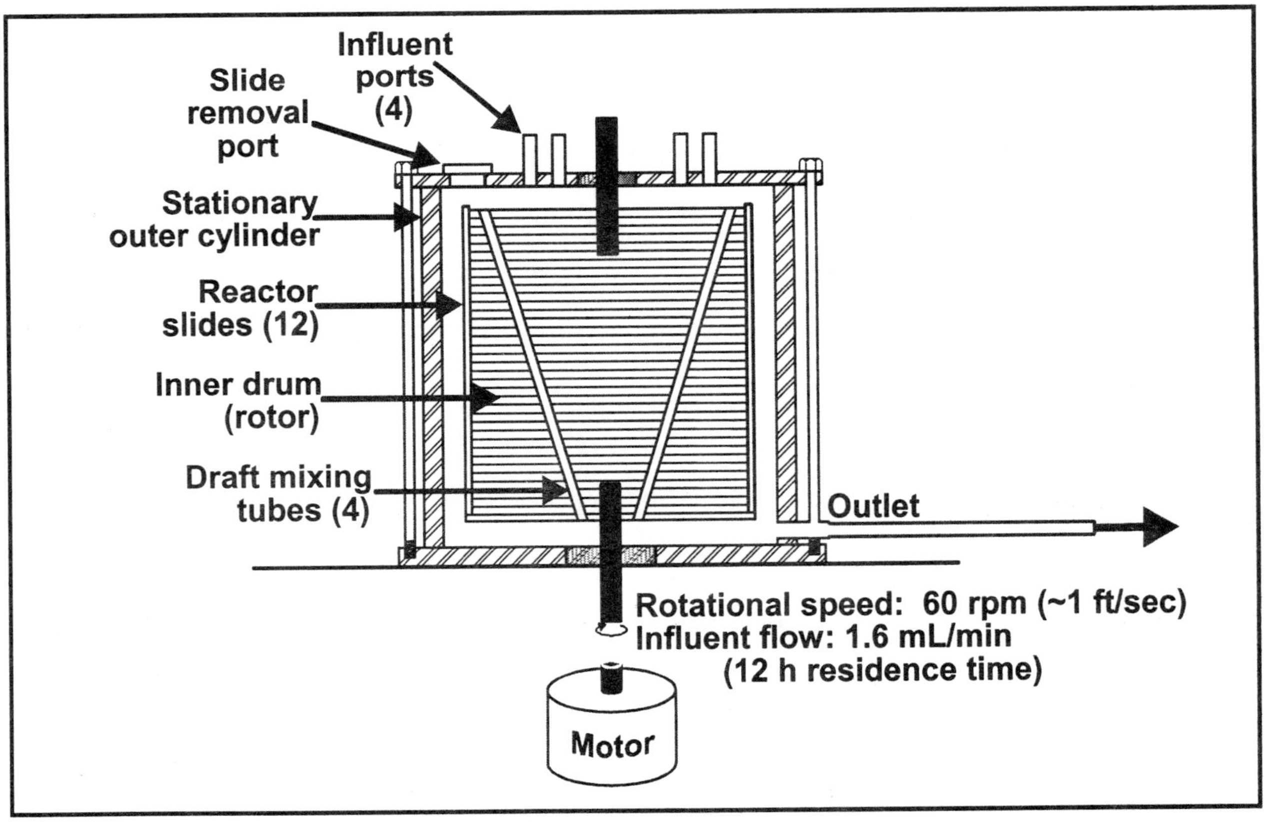

Figure 2.6 Schematic of an annular reactor

rotational speed is changed, or the AR is emptied, the biofilm will start to deteriorate. Moreover, it has been shown that any change in the characteristics (e.g., temperature, pH, chemical constituents) of the water entering an AR (or a distribution system) can lead to biofilm perturbation.

Four ARs were used in this study. Two of them received treated water from the F. E. Weymouth Filtration Plant of the Metropolitan Water District of Southern California (MWDSC). These two ARs were placed in parallel and received the same water, with the only difference being that one received chloraminated water and the other received chlorinated water (before ammonia addition). In an attempt to study the biomethylation of chlorophenols in distribution systems, spiking experiments with chlorophenol added to the influent water were performed on these two ARs. The third AR was placed in a system that used ozonated surface water. A fourth AR was placed in a system that used chlorinated groundwater. With the exception of the third AR, which was disconnected after only 6 months of service because of technical difficulties, all ARs were operated for a period of 2–3 years.

ARs Receiving Surface Water. Two ARs were installed at MWDSC's Weymouth plant, which has a capacity of approximately 400 MGD and receives water from the Colorado River and from northern California through the State Water Project. Treatment includes conventional treatment (alum coagulation, sedimentation, and filtration) with prechlorination followed by ammonia addition to form chloramines. In recent years—until 1998—chlorine was substituted (for one month during the summer) for chloramine disinfection to prevent nitrification in the distribution system.

Two ARs were installed in parallel. The first AR received chloraminated water with an average chlorine residual of 1.8 mg/L during the first year of operation and 2.5 mg/L after that. The second AR received only chlorinated water (before ammonia addition).

Every other day, chlorine residual readings were taken, and the rotational speed (60 rpm) and flow rate (1.6 mL/min) were checked. On June 28, 1997, the plant switched to free chlorine. In order to maintain the AR system's integrity and to prevent biofilm alteration, it was decided to add ammonia to the AR influent water reservoir to form chloramines. Each week a fresh solution of ammonia was prepared and fed into the reservoir to maintain an average 1.8 mg/L monochloramine residual as Cl_2 at a 5:1 weight ratio of chlorine to ammonia-nitrogen (all chloramines—no free chlorine).

A third AR was installed at a southern California water utility using surface water on August 6, 1997. From the month of October until April, the utility distributes water from MWDSC. For the remainder of the year (May through September), the utility treats water from a lake filled with water from MWDSC. The treatment consists of in-line flocculation followed by ozonation and chloramine disinfection. The water from the utility treatment plant is blended with MWDSC water before it reaches the first customer.

ARs Receiving Groundwater. A fourth AR was installed at the effluent of a water utility treating groundwater with the following characteristics:

Total organic carbon (TOC): 1 mg/L

Bromide: 200 µg/L

Alkalinity: 196 mg/L as $CaCO_3$

Conductivity: 666 mhos

ANALYTICAL METHODS

Extraction and Concentration Techniques

Closed-Loop Stripping Analysis (CLSA)

The CLSA technique was used for extraction because it selectively isolates the volatile nonpolar fraction of the organic compounds present in the water sample, which includes compounds with organoleptic properties. This technique has been shown to be effective for the analysis of geosmin and 2-methylisoborneol (MIB), which are known to cause taste-and-odor problems (Krasner et al. 1985). It is now the standard method for the analysis of MIB and geosmin (APHA et al. 1998). For the bench-scale studies of pipe materials in Chapter 6, the temperature of the water bath was increased from 25°C to 45°C, and the temperature of the carbon trap was raised from 45–50°C to 65°C. Also, 50 g of sodium sulfate was added to every 1-L sample of water in the salting-out process.

Simultaneous Distillation-Extraction (SDE)

The SDE technique has frequently been used for the separation of volatile organic compounds (VOCs) from food, beverages, and other agricultural products. SDE was developed by Likens and Nickerson (1964). High extraction recoveries have been reported for VOCs found in beer (Nickerson and Likens 1966). Later, other researchers have successfully used the method for the extraction of fatty acids, phenols, and other compounds found in water (Richard and Junk 1984). A detailed description of the SDE procedure is given by Mallevialle and Suffet (1987). The SDE method was used in this study to complement the CLSA technique for the extraction of organoleptic compounds of greater polarity and higher molecular weight than those extracted by CLSA. For the bench-scale studies of metal pipe in Chapter 6, a modification of the original SDE method (Khiari et al. 1992) was made to enhance recovery. A glass dewar filled with dry ice was placed on top of the SDE apparatus for more efficient cooling of the dichloromethane extract. The evaporative concentration system (EVACS) (Ibrahim, Suffet, and Sakla 1987), with nitrogen blowdown to 0.1 mL, was selected as the final concentration apparatus.

Liquid/Liquid Extraction (LLE)

Batch LLE is probably the most widely used method today for extracting and concentrating organic contaminants from liquid matrices. In some instances, LLE was proven superior to CLSA or SDE for specific odorous compounds such as iodinated trihalomethanes (ITHMs), with recoveries between 80% and 90% for diiodomethane and iodoform (Suffet et al. 1995).

Chemical Analysis

Gas Chromatography/Mass Spectrometry

Gas chromatography coupled with mass spectrometry (GC/MS) was used for the chemical identification of compounds of organoleptic significance. It was complemented by the use of the sensory GC technique described in the sensory method section.

Large-Volume Injection (LVI) GC/MS

Identification of trace organics involved in taste-and-odor episodes primarily relies on the use of GC/MS. Capillary GC has been limited to the injection of 1- or 2-µL aliquots of the sample extract. With the sensitivity of modern bench-top GC/MS instruments, this injection volume allows detection limits around 1 ng/L to be achieved when CLSA is used as the extraction-concentration step. Such a detection limit sometimes proves insufficient and does not allow, for example, the detection of trichloroanisoles at the sub-ng/L odor threshold concentrations (OTCs). Another limitation encountered by analytical chemists is the detection of intense odors during sensory GC, which cannot be correlated to any observable peak on the GC/MS chromatogram.

A new generation of GC analysis allows the injection of large volumes of solvent (up to 150 µL) in the capillary GC columns. Figure 2.7 shows a schematic diagram of the instrument used in this study. During LVI–GC/MS, the sample extract (up to 150 µL) is injected by a controlled-speed autosampler through a cold on-column injector into a 15-m precolumn consisting of a 12-m empty deactivated phase and a 3-m retaining column (UNCORET, proprietary design of Fisons Instruments, Milan, Italy). In a subsequent step, both the solvent and the most volatile solutes evaporate from the rear of the solvent injection band while heavier solutes remain throughout the precolumn. Under controlled pressure and temperature conditions, the solvent vapors are vented through the solvent vapor exit until a preselected amount of final solvent volume is left inside the precolumn. During this evaporation step, the carrier gas is maintained at a high pressure (140 kPa in this case). The remaining solvent is then flushed into the analytical column. At the beginning of the chromatographic run, the carrier gas pressure is reduced (84 kPa in this case) to ensure optimum separation conditions. Both solvent effect and a 3-m retaining column allow refocusing of the deposited analytes before the flushing step.

Initial investigations had shown that the injection of 100 µL of a 1-pg/L solution of geosmin could yield a signal-to-noise ratio higher than 300 for the characteristic ion at m/z 112. Repeated injections of the same dilute solution led to a coefficient of variation less than or equal to 7%.

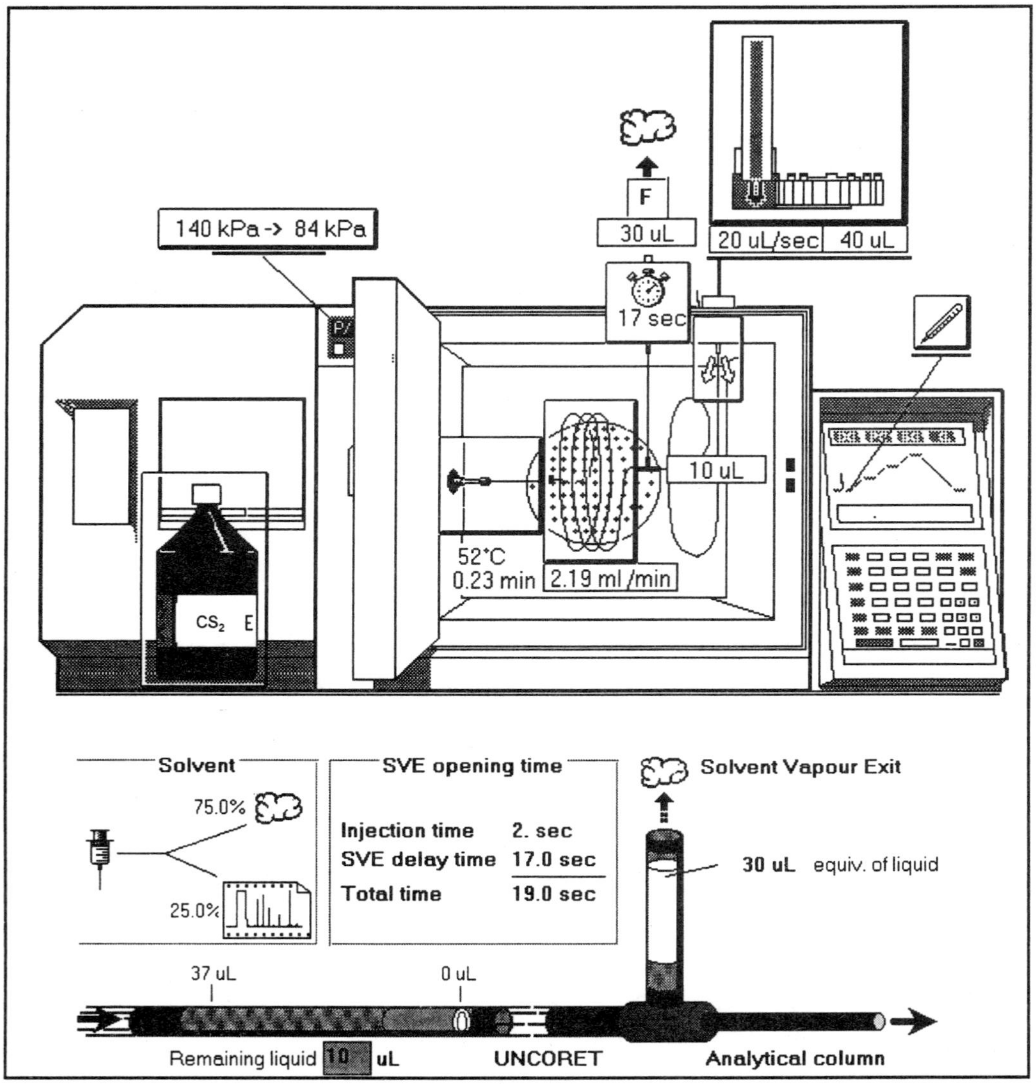

Figure 2.7 Large-volume injection–gas chromatography/mass spectrometry (LVI–GC/MS)

At the time of purchase, the software of the instrument for control of the parameters allowing the automatic injection of a wide range of commonly used solvents had not been programmed for the injection of carbon disulfide, the usual elution solvent used in CLSA. Hence, several months of empirical tests were necessary to find acceptable conditions for this solvent and to avoid flooding of the analytical column or complete evaporation of the solvent, with subsequent loss of analytes. The analytical conditions are shown in Table 2.6.

Ozone Residual

Ozone residual was determined in triplicate by Standard Method 4500-O3(B) with some modifications (APHA 1998). The modifications include using a smaller water sample volume

Table 2.6

Analytical conditions for LVI–GC/MS analysis with carbon disulfide as solvent

INJECTION	Retention gap: 12 m × 0.53 mm ID, deactivated fused silica Retaining precolumn: 3 m × 0.53 mm ID, SE54 Injection: volume 40 μL Injected speed: 20 μL/sec Inlet temperature: 52°C Inlet pressure: 140 kPA Solvent vapor exit delay time: 19 sec
CHROMATOGRAPHY	Column: 50 m × 0.32 mm ID (Chrompack CP-Sil 19-CB; Varian Analytical Instruments, Walnut Creek, Calif.) Initial oven temperature: 52°C (15 min) 3°C/min: up to 130°C 7°C/min: up to 250°C (10 min) Carrier gas: Helium Column head pressure: 84 kPa
DETECTION	Impact electronic ionization: (70 ev) Quadrupole mass spectrometer Mass range: from 80 to 400 0.45 seconds/scan

and a spectrophotometer cell with a shorter path length. In addition, equal volumes of distilled water and indigo reagent were added to equal volumes of ozonated and nonozonated water. The amount of ozone is proportional to the amount of indigo decolorized in the sample compared to a blank. The sample absorbance is measured at 600 nm using a spectrophotometer (Lambda 3-B model; PerkinElmer Instruments, Norwalk, Conn.).

Temperature, pH, Conductivity, Hardness, Alkalinity, Total Dissolved Solids (TDS), Dissolved Oxygen, Metals, and UVA_{254}

All disinfection experiments were performed at room temperature (20 ± 2°C) and at pH 8.0 ± 0.2 (prior to the addition of the disinfectants). If the ambient pH was outside this range, the pH was adjusted with 0.1 M HCl or 0.1 M NaOH. A pH meter with an accuracy of ±0.02 pH units (model 920A, Orion Research, Inc., Boston, Mass.) was used to determine the pH. This instrument was calibrated using pH 7.00 and pH 9.18 buffer solutions. Temperature was measured using a thermometer calibrated against a thermometer certified by the National Bureau of Standards. Analyses for conductivity, hardness, alkalinity, TDS, and UVA_{254} were performed according to Standard Methods 2510B, 2340, 2320, 2540C, and 5910, respectively (APHA 1998). Analysis for dissolved oxygen was conducted according to Standard Method 4500OC or 4500OD (APHA 1998), with the exception that liberated iodine was determined directly by absorption spectrophotometry, using a colorimeter (CHEMets kit K-7512; CHEMetrics, Inc., Calverton, Va.). Metals analyses were performed by flame atomic absorption spectrometry, inductively coupled plasma (ICP) spectroscopy, or ICP/mass spectrometry (ICP/MS).

TOC/Dissolved Organic Carbon (DOC)

TOC samples were analyzed by the UV/persulfate oxidation method (APHA 1998) using an organic carbon analyzer (Dohrmann model 80 or 180; Tekmar-Dohrmann, Cincinnati, Ohio). DOC is defined by a filtration step involving a pre-washed 0.45-μm nylon membrane filter.

Free Chlorine and Chloramines

Free chlorine and chloramines were determined using the DPD methods adapted from *Standard Methods* (APHA 1998) by the Hach Company (Loveland, Colo.) for use with a spectrophotometer (Hach model DR2000) and colorimeters (Hach models DR/700 and DR/1A). Where chloramine speciation was necessary, either amperometric titration method 4500-Cl D or DPD ferrous titrimetric method 94500-Cl F in *Standard Methods* (APHA 1998) was used for free chlorine, monochloramine, dichloramine, and trichloramine determinations.

Sensory Analysis

Sensory measurement is a subjective evaluation, and even OTCs may vary by orders of magnitude between individuals from day to day and with varying measurement procedures. Flavor-profile analysis (FPA) provides a mean value of the sensory characteristics by a panel (Krasner et al. 1985), but errors may be rather large (Meng 1990). Perception by an individual may be influenced by many factors, such as other panelists, synergism and antagonism between odorants, and personal habits such as smoking and fatigue. In addition, sensory analysis of mixtures of chemicals may produce a new odor that is different from the odors of the individual compounds (Khiari et al. 1992).

Flavor-Profile Analysis

FPA was used routinely in this study to describe the sensory characteristics of all water samples. This method was developed for the food industry by Arthur D. Little, Inc. (Caul 1957), and extended to water analysis by MWDSC (Krasner et al. 1985). In FPA, a trained panel evaluates the overall taste-and-odor qualities of a water. The panel uses reference standards to characterize odor qualities. The perceived organoleptic qualities are given a qualitative descriptor and a seven-point scale intensity rating. FPA is a sensory testing method by which a panel of trained analysts evaluates the taste-and-odor characteristics of a water sample (Krasner et al. 1985, APHA 1998). Characteristics that have been reported by more than 50% of the panel are called "descriptors." Their intensity is based on the average of the intensities reported by individual panelists. "Other notes" are characteristics that are reported by less than 50% of the panel. No intensity is assigned to "other notes." The following seven-point intensity scale was used in this study:

1 = Threshold
2 = Very slight
4 = Slight
6 = Slight to moderate
8 = Moderate
10 = Moderate to strong
12 = Strong

During these tests, the waters were not dechlorinated. The odor test temperature was 45°C (except at MWDSC and the University of California Los Angeles, where the test was done at 25°C), and the flavor test was done at 25°C.

FPA was used in this study to determine OTCs for all the chemicals under investigation.

Presentation of Odor Threshold Results. Results of odor thresholds are reported in µg/L (µg of compound per liter of odor-free water) at a 45°C temperature, unless stated otherwise. For each compound, the following are indicated:

- Number of panelists who perceived the compound
- Lowest concentration at which an odor was detected; these results will be referred to as "experimental OTC" (exp. OTC)
- Range of thresholds detected by panelists, with the number of panelists who detected these extreme concentration in brackets (terminal exp. OTCs)
- Best-estimate threshold (BET) concentration for each panelist (the geometric mean of the concentration at which the last was not detected and the next higher concentration was) with the standard log deviation also listed.
- Panel threshold concentration, i.e., the geometric mean of the BETs of the individual panelists (e.g., OTC)

- Constants m and b in the Weber-Fechner model, and the coefficient of determination, R^2

$$\text{Intensity} = m \log(\text{concentration}) + b$$

- Concentration where intensity = 1 in the Weber-Fechner model (OTC I = 1)
- Constants k and m in the Stevens Power model, and the coefficient of determination, R^2

$$\text{Intensity} = k\,(\text{concentration})^m$$

- Odor descriptors at low and high concentrations

Sensory GC Detection

The sensory GC technique consists of a gas-chromatographic separation of the aqueous sample extract followed by odor detection of the compounds at an olfactory port. Sensory GC is frequently used in the food industry for the detection of VOCs from fruits and vegetables (Acree 1976, Acree et al. 1984) and by the perfume industry (Parliment and Scarpellino 1977, Zurer 1987). The olfactory port system consists of a heated transfer section from the GC oven to the detection cone. Air drawn from the GC oven is used to heat the transfer tube to maintain the unit at a temperature high enough to transfer each sample component to the detection cone without losses due to condensation. The glass cone is purged with air that has been humidified by passing it through a bubbler containing water. This is to prevent nasal mucous membranes from drying out over a long testing period and thus maintain olfactory sensitivity. The flow rate of make-up air through the humidifier into the sample line can be adjusted to suit the olfactory sensitivity of each individual. Sensory GC and GC/MS have been shown to be capable of evaluating FPA data when the FPA intensity is above 4.0 on the seven-point FPA intensity scale of 1 (threshold), 2, 4, 6, 8, 10, 12 (Khiari, Barrett, and Suffet 1997).

Flavor Threshold Number

Although FPA was the sensory method preferentially used throughout the entire study, another method was used to monitor the organoleptic properties of the water during the pipe-loop experiments. This method, a modified protocol of the flavor threshold number (FTN) method, was deemed appropriate in this type of experiment. The analysis was performed by four panelists, and dilutions of the sample were carried out up to an FTN of 30 (Figure 2.8). The sample was incubated at 25°C for 30 minutes before the test. The flavor threshold concentration (FTC) of the chemical is given by the concentration in the greatest dilution volume in which a flavor can be perceived by a panelist and the dilution is reported as an FTN. Each panelist gives a description of the flavor associated with the FTN.

A statistical approach was used to express the panel results. Each panelist's response was taken into account, and a box-and-whisker representation (quartile) is used to express the test results. Table 2.7 shows an example of a panel's data, and Figure 2.9 shows the corresponding box-and-whisker graph.

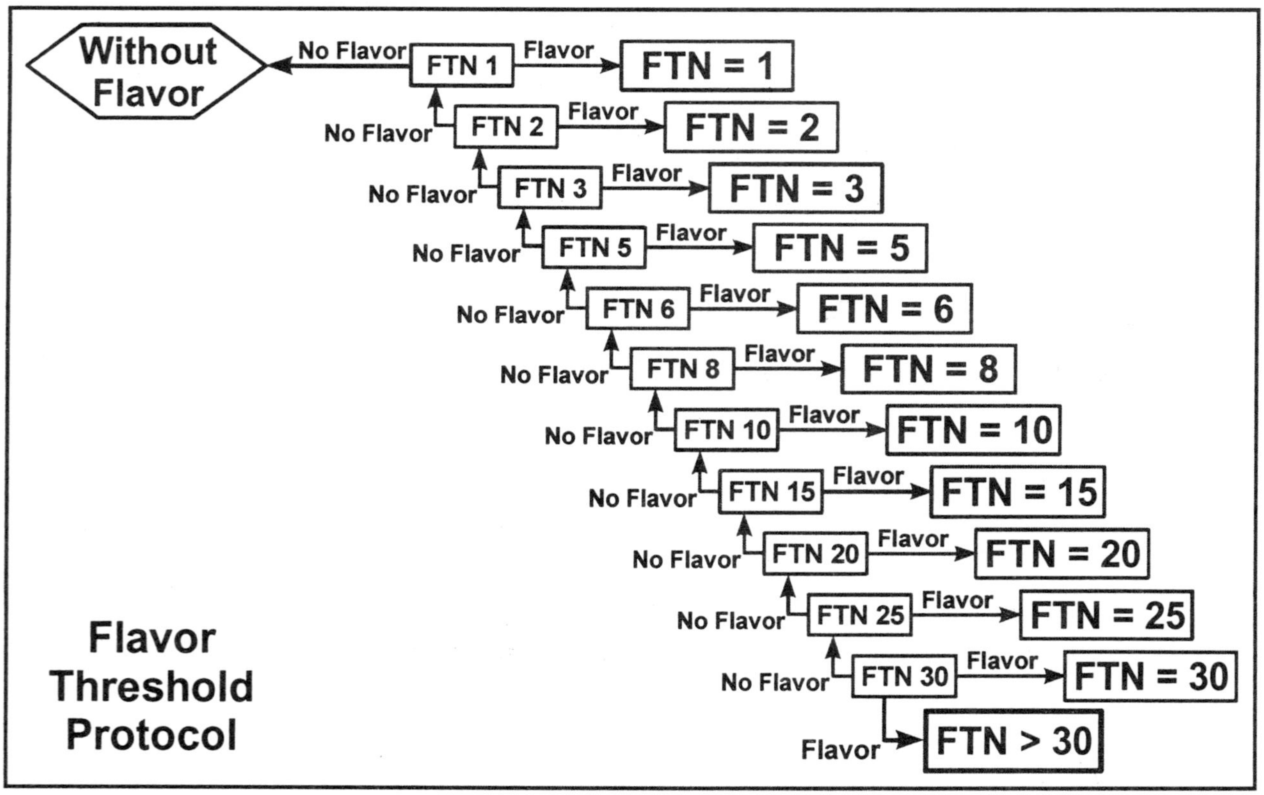

Figure 2.8 Schematic description of the adapted flavor threshold number (FTN) method

Table 2.7

Example of panel response

Panelist #	FTN
1	2
2	10
3	5
4	7

Typical FTN = 5

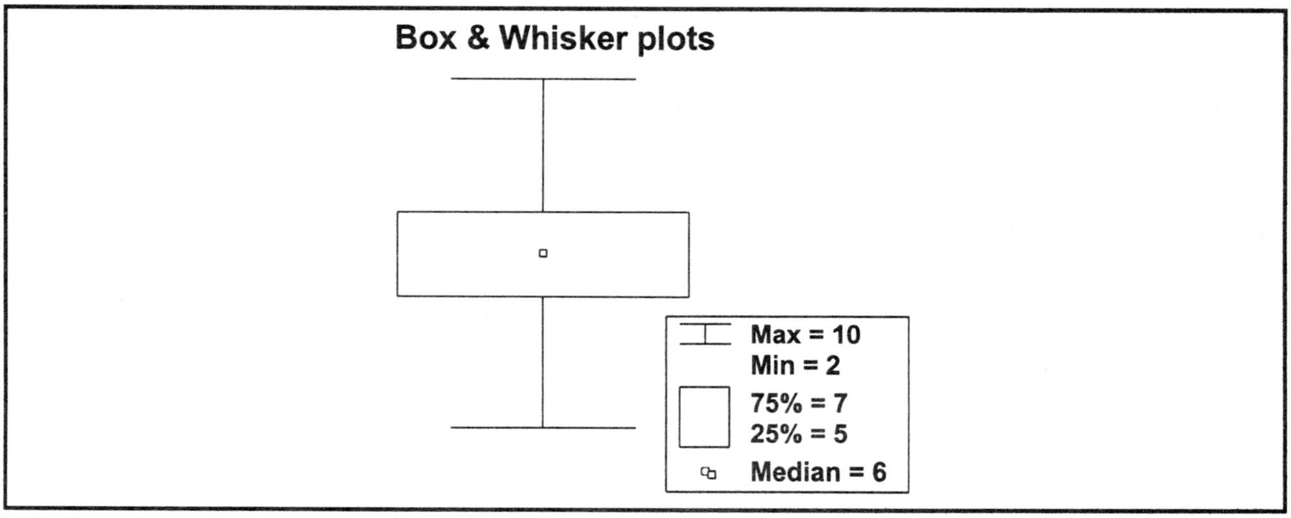

Figure 2.9 Box –and-whisker representation associated with Table 2.7

Microbiological Analysis

Biofilm measurements were performed to determine if a biofilm had reached steady state, which indicated whether the biofilm was representative of that found in distribution systems. Free bacteria at the AR effluents were enumerated by heterotrophic plate count (HPC) on selected samples. Cells fixed on AR slides were also enumerated by HPC after scraping the biofilm slides.

Biofilm Removal

Biofilms were removed from their substratum by a 400 W sonicator (Vibracell 72408; BioBlock Scientific, Cedex, France) fitted with a 3-mm stepped microtip and operating at 20 kHz. Experiments were carried out on 1-cm^2 PVC coupons, placed in sterile tubes, maintained in ice, and fed with 15 mL of normal saline water. The sonication characteristics are as follows: duty cycle = 25%, sonication time = 2 min, sonication power = 25%.

HPC–R2A

The culturable bacterial density of each sample was determined by pour-plating appropriate dilutions of water samples in R2A media (Reasoner and Geldreich 1985). Colony counts were enumerated after 7 days at 22°C.

Fungi Counts

Fungi were quantified after a 0.45-µm membrane filtration of 100 mL of the sample. Incubation was performed at 30°C for 3 to 10 days on specific medium (Sarbouraud) containing chloramphenicol to limit bacterial growth.

Actinomycete Counts

Actinomycetes were quantified after a 0.45-µm filtration of 100 mL of the sample. Incubation was performed during 2 weeks on a specific medium (Bacto Actinomycete Isolation).

EXPERIMENTAL DESIGN AND WATERS STUDIED

The following section describes the waters and the experimental conditions used for the bench-scale experiments performed in this study.

Effect of Increased Chloramine Residuals on Tastes and Odors

Untreated Colorado River water (CRW) and California state project water (SPW), as well as blends of the two waters, were conventionally treated in MWDSC's 10-gallon/minute pilot plant and then bench-scale chloraminated in the laboratory. Similarly treated select blends after preozonation and conventional treatment were also bench-scale chloraminated. The chlorine dose added to the pilot plant effluent was either 1.7 mg/L or 2.5 mg/L as Cl_2 at a 5:1 ratio of chlorine to ammonia-nitrogen. The contact times were chosen to simulate summer and winter conditions in MWDSC's system (summer, 48 h at 25°C; winter, 96 h at 10°C). Results were interpreted by looking for a change between the two residuals under the various conditions. It was assumed that the current residual being used was yielding acceptable results. Table 2.8 shows the general water quality of the raw waters used.

The waters studied were a blend of 50% SPW and CRW, 100% CRW, and 100% SPW. Nonozonated water and ozonated water were jar-treated and then bench-scale chloraminated at 3:1 and 5:1 ratios of chlorine to ammonia-nitrogen calculated as Cl_2. Samples were collected in glass bottles. One set was analyzed by FPA and for both free chlorine and total chlorine residual after 1 h. A different set was analyzed after 48 h. A final set was analyzed after 192 h. The target total chlorine residual after 1 h was 0.5 mg/L and 4.0 mg/L as Cl_2. The samples were held at 25°C.

Table 2.8

General water quality parameters

	Concentration (mg/L)	
Parameter	CRW	SPW
TDS	676	204
Alkalinity as $CaCO_3$	129	69
Total hardness as $CaCO_3$	316	88
TOC	3	3

CRW – Colorado River water

SPW – State project water

Tastes and Odors Produced by Disinfectants and DBPs

Effect of Full Treatment on the Production of Tastes and Odors

In addition to CRW and SPW described earlier, these experiments were performed on two other waters: Schuylkill River water and a groundwater from southern California. The Schuylkill River headwaters originate in the Appalachian Mountains of Pennsylvania. The water quality is somewhat compromised by acid mine drainage. Some tributaries have pH values reported as low as 3.5. Once this water mixes with two major tributaries—theTulpehocken and Maiden creeks—that drain valleys consisting of shales interbedded with sandstone and limestone, the acid mine drainage is neutralized. Run-off from agriculture and domestic waste from the towns of Reading, Pottsville, and Pottstown are factors affecting Schuylkill River water quality before it reaches the intake of the Philadelphia Suburban Water Company, near Phoenixville. Table 2.9 shows the water quality parameters for the four waters studied.

During these experiments, a large batch of source water was collected (50 L). The following parameters were used:

- Coagulant dose: Low–3 mg/L, high–30 mg/L
- Ozone dose to obtain an ozone: TOC ratio of 2
- Disinfectant dose: 2 mg/L
- Detention time: 1 and 48 h

The chlorinated and chloraminated samples obtained, as well as control samples collected at different steps during the treatment, were analyzed by FPA after 1 h and 48 h detention time. Figure 2.10 shows a schematic diagram of the treatment scenarios.

Groundwater Disinfection

Three groundwaters were used. The waters were obtained from different locations: a groundwater system in the northeastern United States (groundwater A) and from wells from

Table 2.9

General water quality parameters for the waters studied

	SPW	CRW	Sch. R.*	SoCal. Gr.**
TOC (mg/L)	3.0	2.6	1.5	1.0
Alkalinity as $CaCO_3$ (mg/L)	74	132	75	196
Hardness (mg/L)	105	327	150	
Conductivity (mhos)	547	1108	Not reported	666
Bromide (μg/L)	200	100	Not reported	200

*Sch. R.: Schuylkill River water
**SoCal. Gr.: A groundwater from southern California

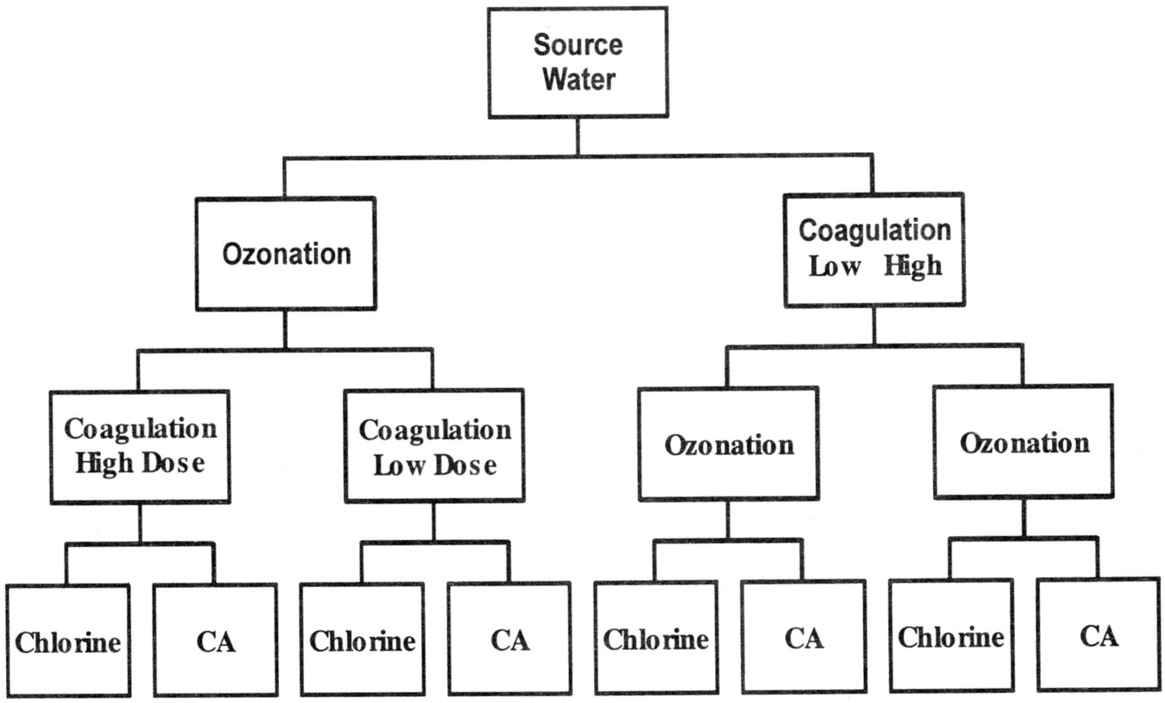

Chlorine: Chlorination
CA: Chloramination

Figure 2.10 Schematic diagram of the treatment scenarios

southern California (groundwaters B and C). Table 2.10 shows the water quality parameters for the three waters. Because it is common practice to use groundwater with no treatment except disinfection, the following experiments involved only limited treatment (membrane filtration through 0.45 μm followed by disinfection). Two disinfectants (chlorine and chloramines), and two detention times were used. The dose for both disinfectants was 2 mg/L.

Kinetics of Formation of ITHMs

The objective was to study the effect of treatment, chlorine-to-nitrogen ratio, disinfectant residual, and detention time on the formation of ITHMs. The parameters studied were as follows:

- Water matrices: CRW and SPW
- Treatments: conventional and preozonation/conventional
- Disinfectant: chloramines
- Ratios of chlorine to ammonia-nitrogen: 3:1 and 5:1
- Chloramine residuals as Cl_2: 0.5 and 4 mg/L
- Detention times: 1, 48, and 192 h.

Table 2.10

Water quality parameters for the three groundwaters studied

	Groundwater A	Groundwater B	Groundwater C
TOC (mg/L)	1.1	3.36	1.2
Ammonia (total)(mg/L)	Not reported	1.3	1.2
Bromide (mg/L)	Not reported	0.411	0.772
pH	Not reported	8.24	8.04

Effect of Blending Different Drinking Waters on the Organoleptic Profile

Barcelona's water supply is an example of a multiple-source system that blends two main types of water, which have very different compositions: water from the Llobregat River and water from the Ter River. The latter comes from a system of reservoirs in the middle course of the river that feeds the treatment plant and has, on average, a conductivity of 450 μS/cm. The water from the Llobregat River is taken in its lower course at the treatment plant of San Joan Despi, a few kilometers before reaching the sea. Its conductivity fluctuates between 600 and 1600 μS/cm. Despite the complete treatment applied at the San Joan Despi treatment plant, the organoleptic quality of the finished water is influenced by salinity, residual chlorine, and dissolved organic matter.

The automated management of the distribution system allows almost any supply area to receive water from either of the two main source waters or their blends. The consumers receive waters from different sources at different times, depending on the availability of the reserves, their management, and their distribution. When an area normally supplied with Ter water starts receiving Llobregat water, consumers perceive the change and complaints are often reported.

BLENDING OF CHLORAMINATED SURFACE WATER AND CHLORINATED GROUNDWATER

Waters Studied

Chloraminated Surface Water

The chloraminated water was prepared from SPW that had been jar-treated with alum at a 6-mg/L dose and ozonated with a 2.5-mg/L ozone dose. The water quality parameters of SPW were described earlier. Bench-scale chloramination was performed to obtain 3:1 and 5:1 ratios of chlorine to ammonia-nitrogen. The total chlorine residual was adjusted to 0.5 and 4.0 mg/L as Cl_2.

Table 2.11

General groundwater characteristics

Turbidity (NTU)	0.17
pH	7.87
Specific conductance (µmho/cm)	293
Ammonia (total) (mg/L)	0.036
Bromide (mg/L)	0.081
Total hardness as $CaCO_3$ (mg/L)	113
Alkalinity as $CaCO_3$ (mg/L)	164
TOC (mg/L)	0.36

Chlorinated Groundwater

Water from a local well was chlorinated using sodium hypochlorite to obtain free-chlorine residuals of 0.5 and 1 mg/L. Table 2.11 shows the groundwater's general water characteristics.

Experimental Design

The chloraminated water (at 3:1 and 5:1 ratio of chlorine to ammonia-nitrogen) and the chlorinated well water (0.5 and 1 mg/L free chlorine) were blended to obtain a target blend at chlorine-to-ammonia-nitrogen ratios of 5:1, 7:1, and 15:1. Water samples were analyzed by FPA and for chlorine species using the DPD–ferrous ammonium sulfate (DPD-FAS) titration method after 1, 24, and 120 h. Table 2.12 shows the experimental conditions used. Preliminary calculations done to obtain the percent of SPW water and well water needed to obtain the desired blend ratios showed that the 5:1 ratio could not be obtained when using water that had been chloraminated at a 5:1 ratio of chlorine to ammonia-nitrogen. Any chlorinated water would cause the ratio to be greater than 5:1.

Control samples of the untreated and treated waters under the different conditions were also analyzed by FPA and DPD-FAS titration.

Table 2.12

Experimental set-up for blending of chloraminated and chlorinated waters

	SPW		Well	Blend	% SPW	% Well
No.	Cl_2:N ratio	Chlorine residual (mg/L)	Cl_2 residual (mg/L)	Cl_2:N ratio	in blend	in blend
A-1	3:1	0.5	0.5	5:1	60	40
A-2			1		75	25
A-3		4	0.5		16	84
A-4			1		27	73
	5:1	0.5	0.5		100	0
			1		100	0
		4	0.5		100	0
			1		100	0
B-1	3:1	0.5	0.5	7:1	43	57
B-2			1		60	40
B-3		4	0.5		9	91
B-4			1		16	84
B-5	5:1	0.5	0.5		71	29
B-6			1		83	17
B-7		4	0.5		24	76
B-8			1		38	62
C-1	3:1	0.5	0.5	15:1	20	80
C-2			1		33	67
C-3		4	0.5		3	97
C-4			1		6	94
C-5	5:1	0.5	0.5		33	67
C-6			1		50	50
C-7		4	0.5		6	94
C-8			1		11	89
Control 1	3:1	0.5			100	0
Control 2		4			100	0
Control 3	5:1	0.5			100	0
Control 4		4			100	0
Control 5			0.5		0	100
Control 6			1		0	100

CHAPTER 3

TASTES AND ODORS ASSOCIATED WITH BIOLOGICAL ACTIVITY

PIPE-LOOP EXPERIMENTS

The following experiments were conducted in the pipe loop described in Chapter 2 to study tastes and odors thought to be produced by biological activity subsequent to disinfectant decay at long detention times. In order to dissociate the effect of detention time from that of the biofilm and iron pipe, parallel control experiments were conducted in glass jars. The flavor threshold number (FTN) method was used for the pipe-loop study.

Comparison of Aging of Water in Jars and in the Pipe Loop

Water collected from the Aubergenville treatment plant was used to compare chlorine decay kinetics and chlorine flavor loss in jars and in the pipe loops described in Chapter 2. The initial chlorine concentration was about 1 mg/L in both cases. Figures 3.1 and 3.2 show the loss of the chlorinous descriptor associated with the disinfectant in jars and in the pipe loops versus residence time. The residence times in the pipe loops are as follows: loop 1 = 8 h; loop 2 = 16 h; and loop 3 = 24 h.

Figure 3.3 shows the loss of the chlorinous descriptor and chlorine residual with time for jar and pipe-loop experiments. It can be observed that the chlorinous descriptor and residuals decreased in parallel in both instances. The rate of chlorine decay was greater in the pipe loops (first order: $k = 1 \times 10^{-3}\ min^{-1}$) than in jars ($k = 4 \times 10^{-4}\ min^{-1}$) and led to a faster decrease of the chlorinous descriptor with time.

Figure 3.4 compares the change in flavors in relation to the corresponding chlorine residuals for jar and pipe-loop experiments. It can be observed that no significant difference was found between the two types of experiments when the flavor induced by a chlorine residual was considered. This fact was confirmed by a nonparametric analysis of variance (ANOVA) test (Table 3.1). Small differences were noticed in the absence of chlorine because of generation of an earthy-musty flavor in the pipe-loop water.

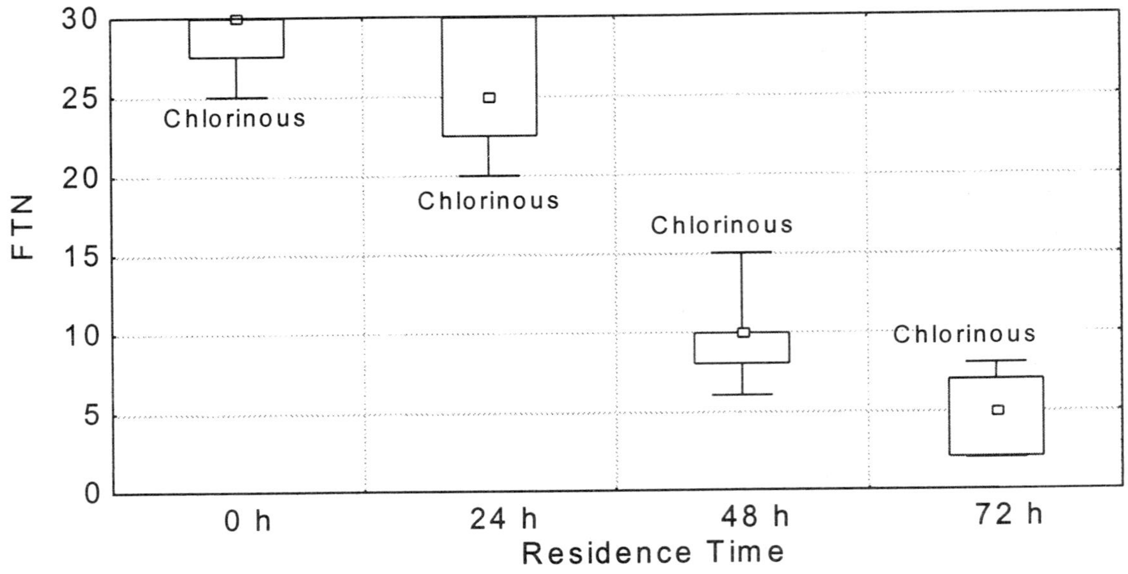

Figure 3.1 Loss of the chlorinous descriptor with time: jar-test experiments

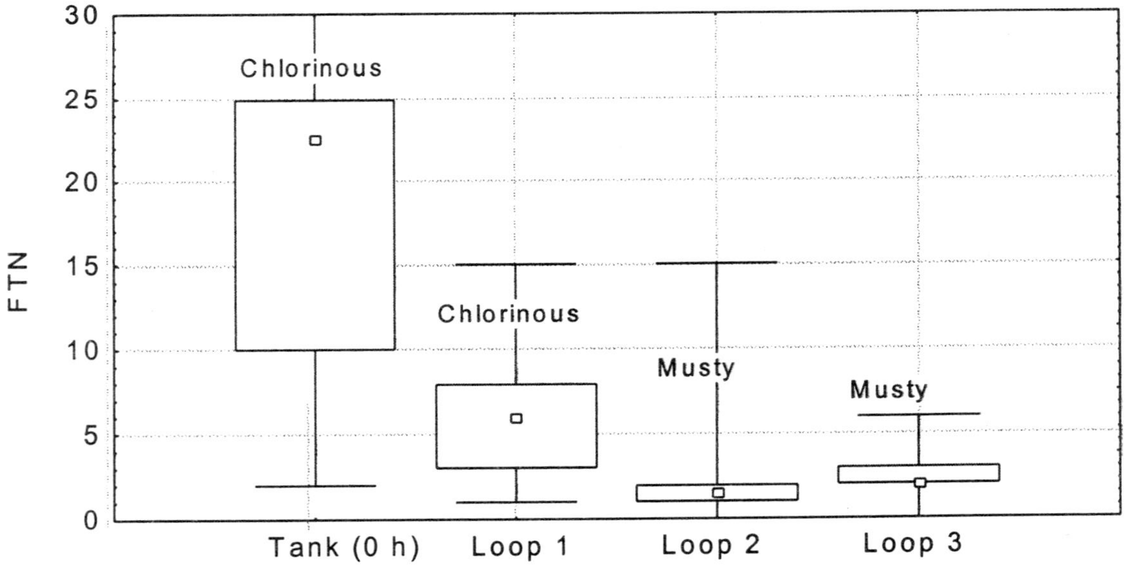

Figure 3.2 Loss of the chlorinous descriptor with time: pipe-loop experiments

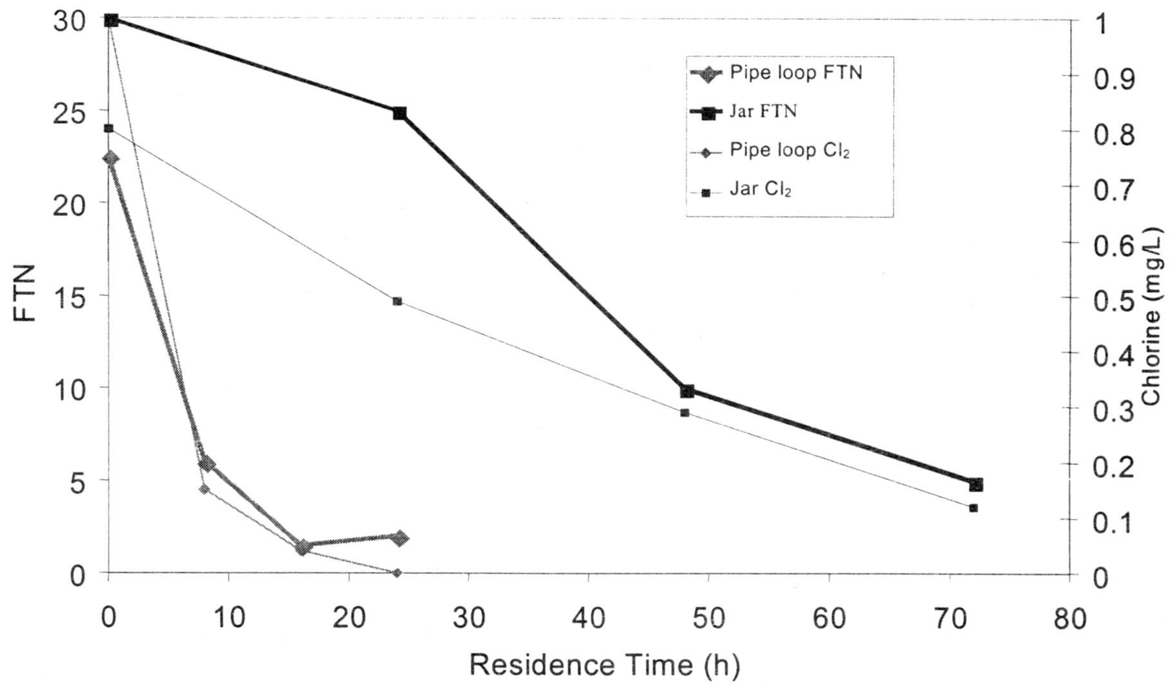

Figure 3.3 Jar and pipe-loop experiments: FTN and chlorine residual loss with time

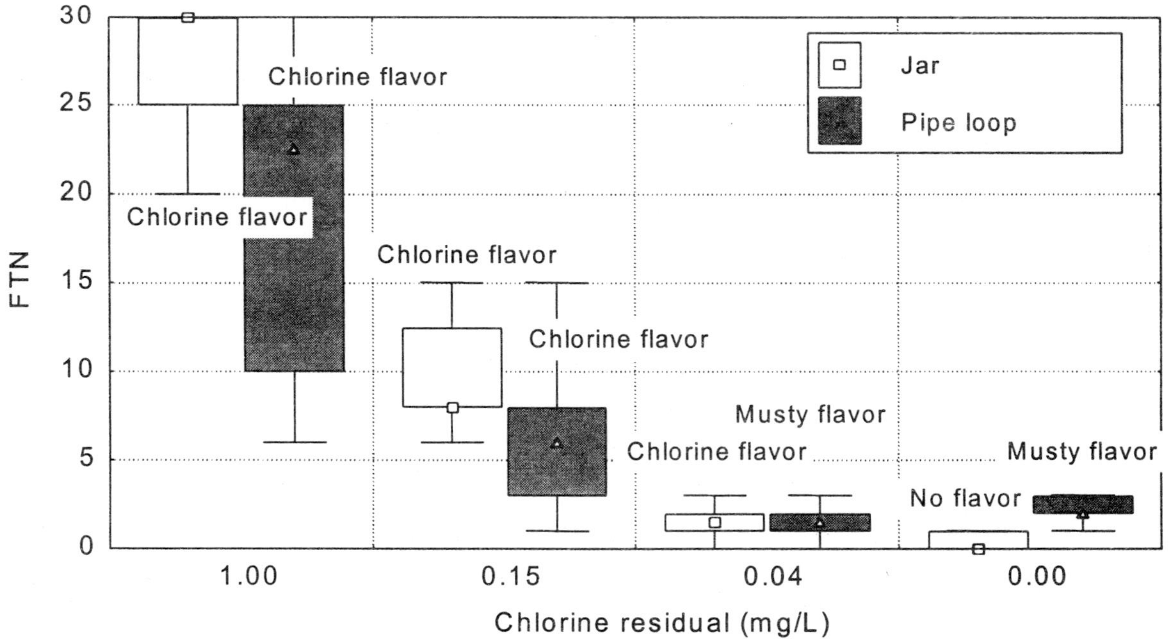

Figure 3.4 Flavor versus chlorine residual in jar and pipe-loop experiments

Table 3.1

Comparison of jar and pipe loop: Kruskal Wallis rank test

Experiment	n	Σ Ranks	Probability (null hypothesis)
Pipe-loop	80	5196.5	P=0.8564
Jar	48	3059.5	

Evaluation of the data using the Kruskal Wallis rank test showed that the chlorinous flavors observed in these two experiments were mainly caused by chlorine residuals and that a strong relationship existed between chlorine residual and flavor. In the presence of chlorine levels greater than or equal to 0.1 mg/L, no significant difference between controls and the distribution network was found, indicating no specific effect of the distribution network when the chlorinous descriptor was considered.

Chlorine residual seemed to induce higher flavor intensities than those reported in the literature. According to Krasner et al. (1984), the odor threshold concentration (OTC) and flavor threshold concentration (FTC), as determined by flavor-profile analysis (FPA) of hypochlorous acid, were 0.28 mg/L as Cl_2 and 0.24 mg/L as Cl_2, and those for the hypochlorite ion were 0.36 mg/L as Cl_2 and 0.30 mg/L as Cl_2, respectively. In this study, when the FTN method was used, the FTC for chlorine was found to be 0.03–0.05 mg/L as Cl_2 at a pH of about 7.

Comparison of Chlorination and Chloramination in the Pipe Loop

For these experiments, the Aubergenville treatment plant water used in the previous experiment was chlorinated or chloraminated at a level of 1 mg/L as Cl_2. Preformed monochloramine and sodium hypochlorite were used. Figure 3.5 shows chlorine flavor loss in both experiments.

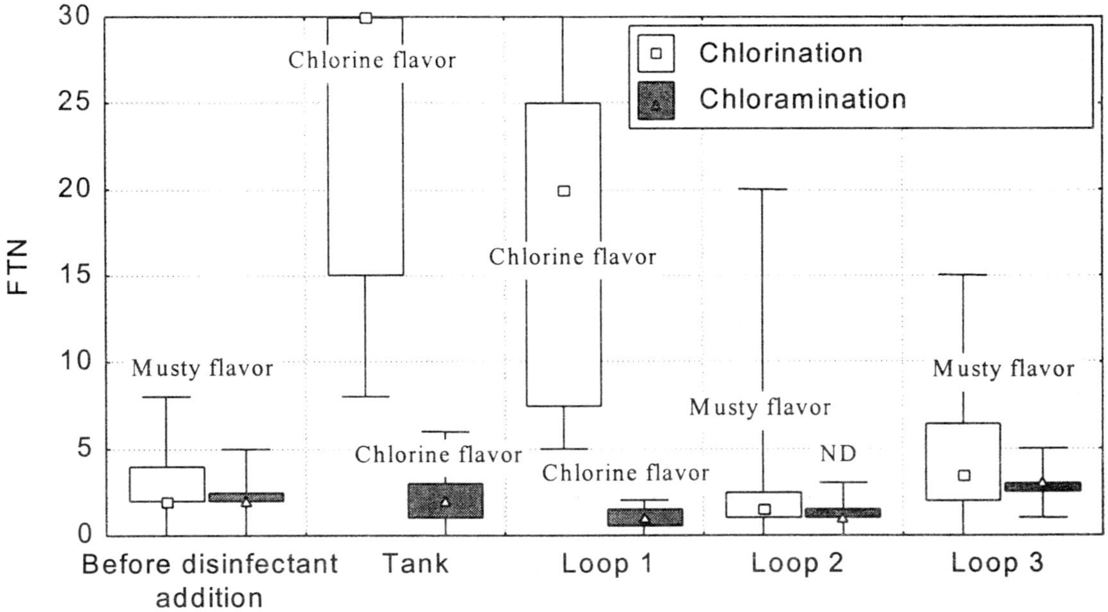

Figure 3.5 Pipe-loop experiments: flavor comparison of chlorination and chloramination

Figure 3.6 and Table 3.2 illustrate the chlorine and chloramine decay kinetics in the pipe loop. The chloramine decay rate was almost 10 times lower than that of chlorine. In the case of the chlorinated water, chlorine flavors decreased in parallel with chlorine residuals. Earthy-musty flavors were observed when chlorine residuals fell below 0.07 mg/L as Cl_2. These earthy-musty flavors were sometimes present at the inlet of the pipe loop, but experiments performed without any disinfectant indicated that these flavors were also generated in the pipe loop. In the case of the chloraminated water, no significant increase in FTN was noticed. For chloramine concentrations higher than 0.55 mg/L as Cl_2 (loop 3), chlorinous flavors were always detected. Earthy-musty flavors were noticed below this concentration.

Determination of FTC for Chlorine and Monochloramine

As mentioned in the previous section, the FTCs determined in this study were lower than those reported by Krasner et al. (1984). Bench-scale experiments were performed in order to investigate the discrepancy between the two results.

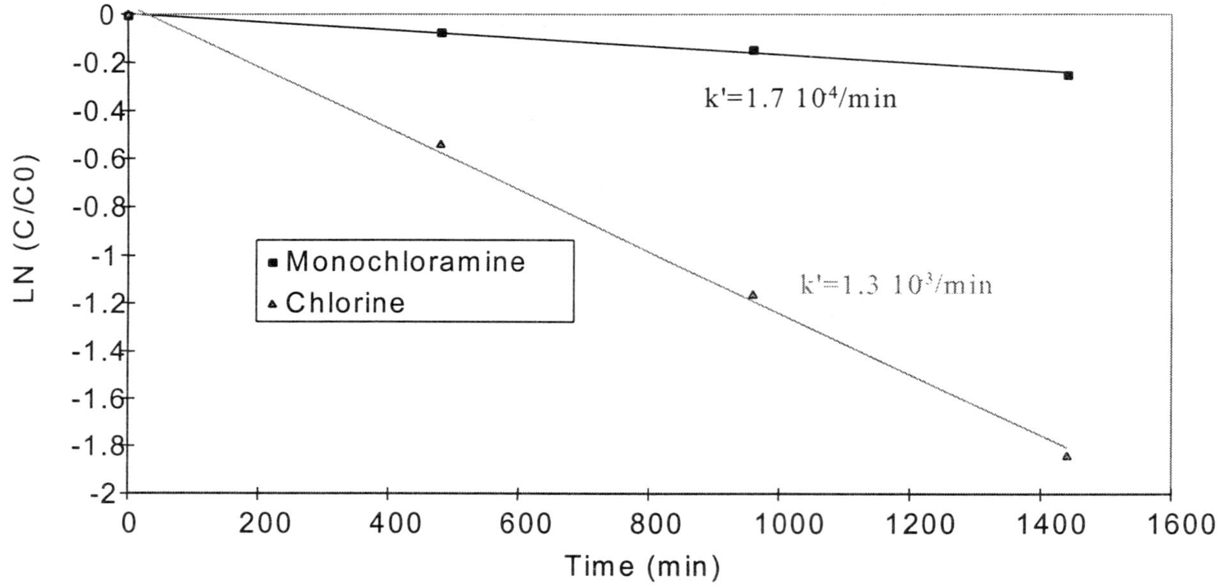

Figure 3.6 Pipe-loop experiments: comparison of decay kinetics for chlorine and chloramines

Table 3.2

Chlorine and chloramine residual along the pipe-loop

		Tank (0 h)	Loop 1 (8 h)	Loop 2 (16 h)	Loop 3 (24 h)
Chloramines (mg/L) as Cl_2	(average)	0.99	0.84	0.71	0.56
	(std dev.)	0.07	0.13	0.07	0.09
Chlorine (mg/L) as Cl_2	(average)	1.00	0.29	0.07	0.01
	(std dev.)	0.14	0.16	0.06	0.02

For both chlorine and monochloramine, flavor analysis was performed with Evian water (pH = 7.2) spiked with sodium hypochlorite or preformed monochloramine. Tests for the different disinfectant concentration solutions were performed at least 10–15 min after the addition of the disinfectant.

Figures 3.7a and b show the results obtained from the four panelists (positive answers) to chlorinous flavor in relation to disinfectant concentration. The FTC was determined for a 50% positive response level corresponding to an FTN of 1, in accordance with the FTN method. From the panel responses, an FTC of 0.025 mg/L as Cl_2 was determined for chlorine and an FTC of 0.28 mg/L was determined for monochloramine as Cl_2. These FTCs were 10 times lower for chlorine (0.025 versus 0.24 mg/L) and about two times lower for monochloramine (0.28 versus 0.48 mg/L) than those reported by Krasner et al. (1984). Three factors may explain the discrepancies between the results reported by Krasner and coworkers and those obtained in this study: the method of analysis (FTN versus FPA), the water matrix (Evian™ versus Sparkletts™), and the pH (7.2 versus 8).

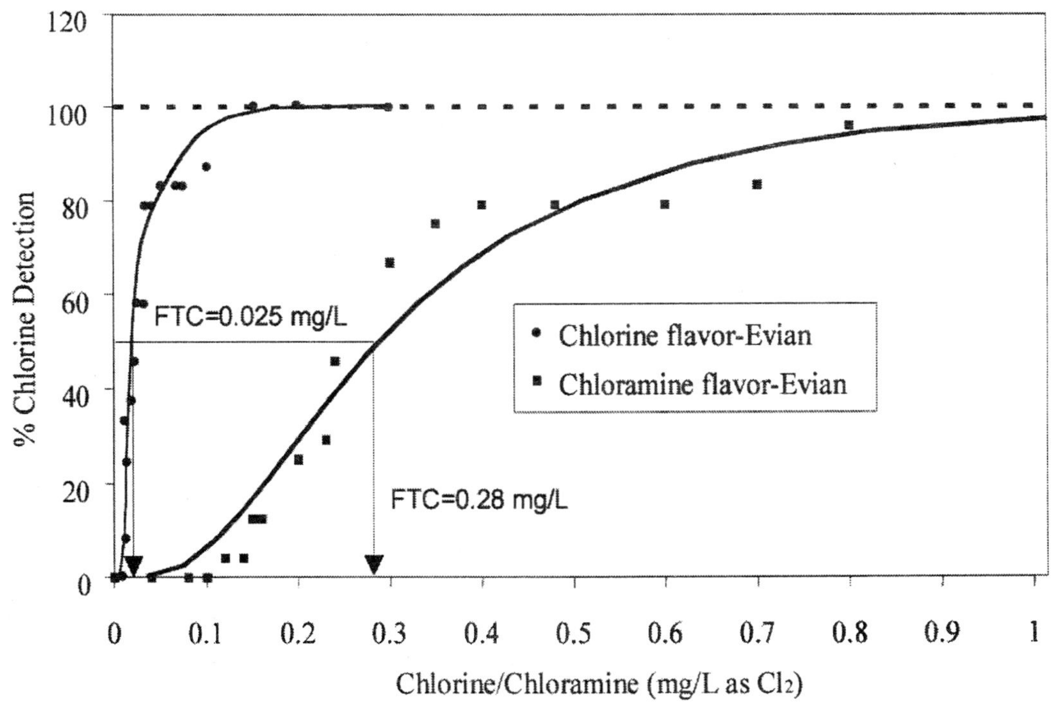

Figure 3.7a Flavor threshold concentration (FTC) determination for chlorine and chloramines in Evian water: panel results

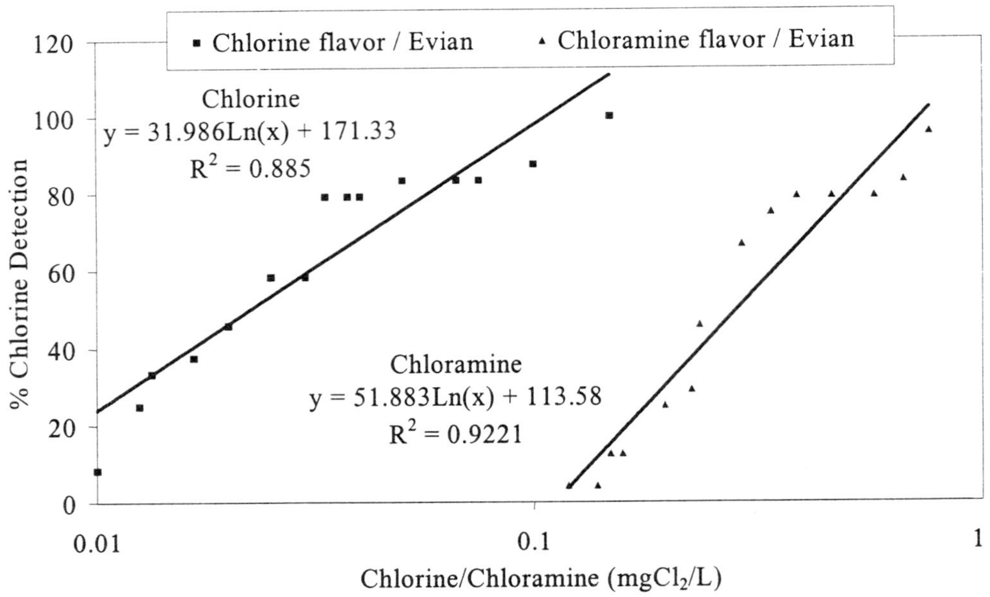

Figure 3.7b Log-scale plot of FTC determination for chlorine and chloramines in Evian water: panel results

Figures 3.8a and b compare the threshold concentration of chlorine for trained panelists and untrained individuals (consumer panel). Trained panelists were two times more sensitive than untrained panelists.

Chlorine and Chloramine FTCs in a Water with an Earthy-Musty Flavor

For these experiments, water collected at the outlet of the pipe loop was used. As described earlier, flavor analysis was performed on water that had been spiked with sodium hypochlorite or preformed monochloramine. The various levels of disinfectant solutions were analyzed at least 10-15 min following the addition of the disinfectant.

The pipe-loop effluent water had a slight earthy-musty flavor (FTN = 3) that was believed to be caused by the presence of trichloroanisoles at low-ng/L concentrations. Figures 3.9a and b and Figures 3.10a and b show the comparison of water from the pipe loop and Evian water. Although chlorine is often considered a masking agent, it was noticed that a slight

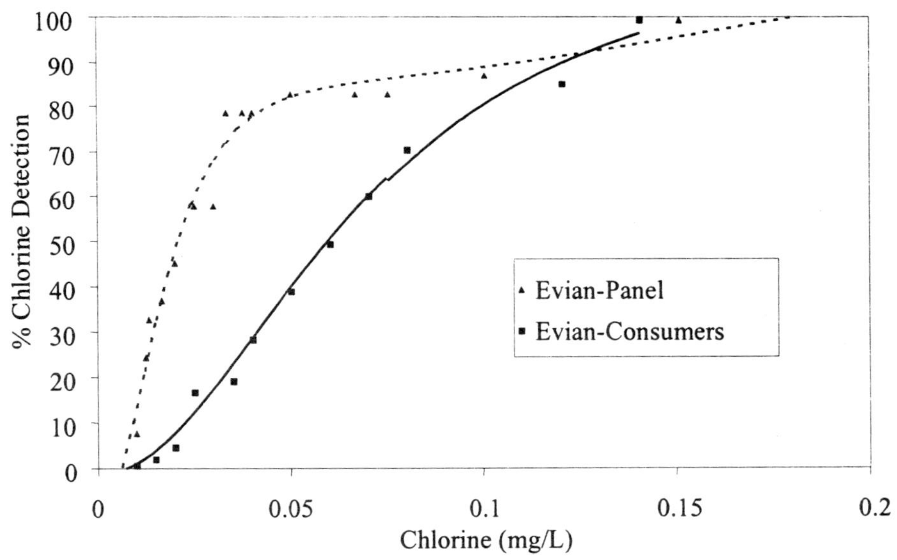

Figure 3.8a Comparison of chlorine threshold concentration: panel response/consumers

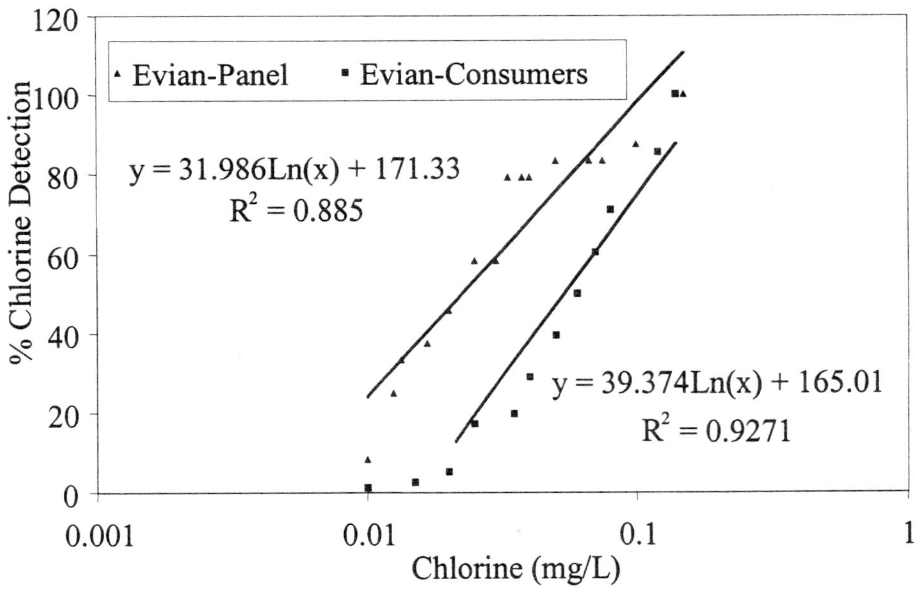

Figure 3.8b Log-scale plot of comparison of chlorine threshold concentration: panel response/consumers

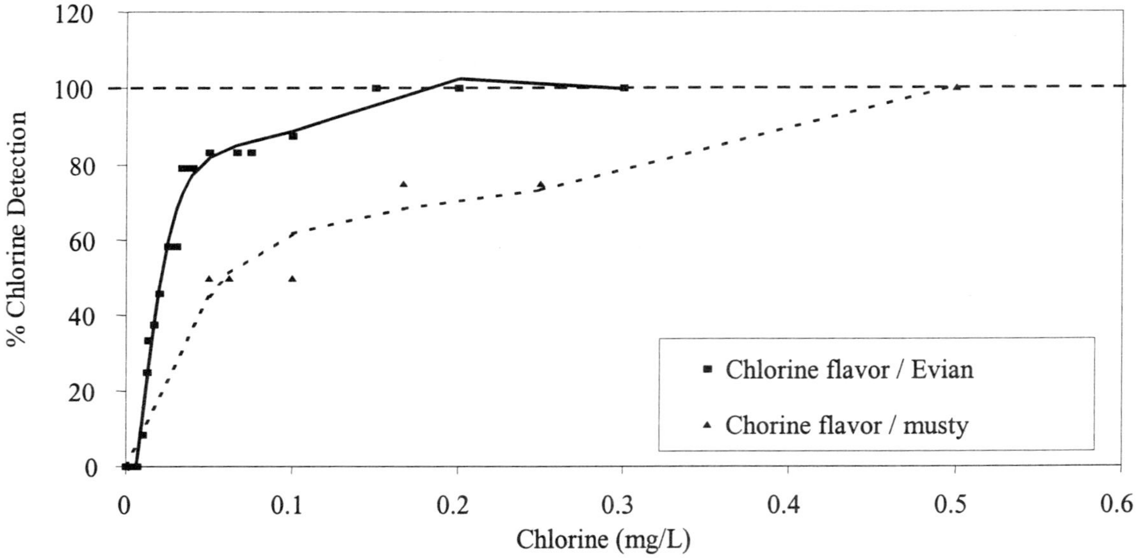

Figure 3.9a. Comparison of chlorine threshold concentration: Evian versus musty water

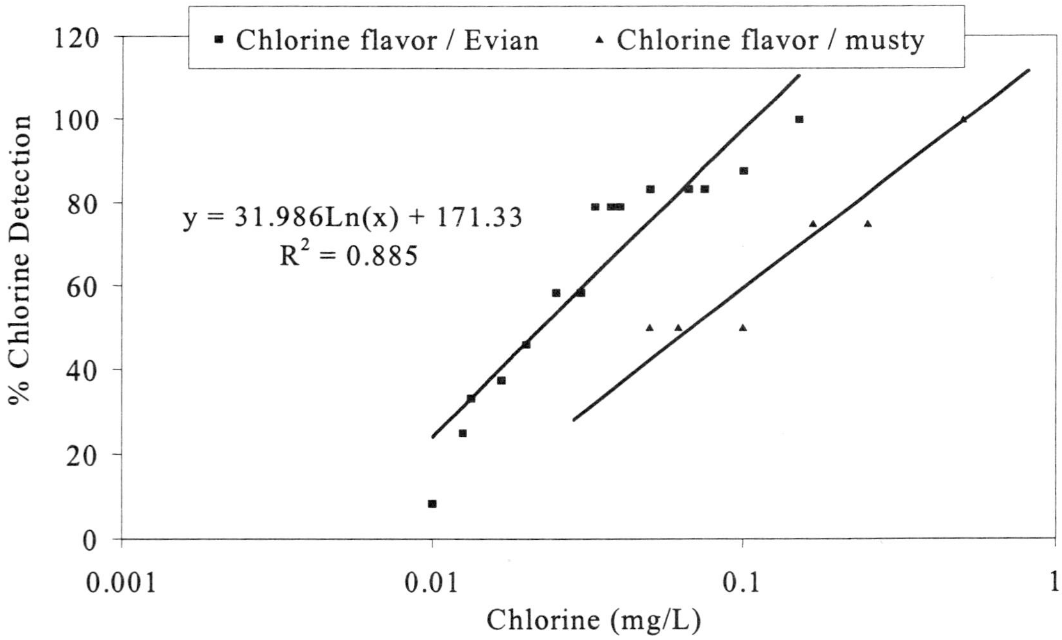

Figure 3.9b. Log-scale plot of comparison of chlorine threshold concentration: Evian versus musty water

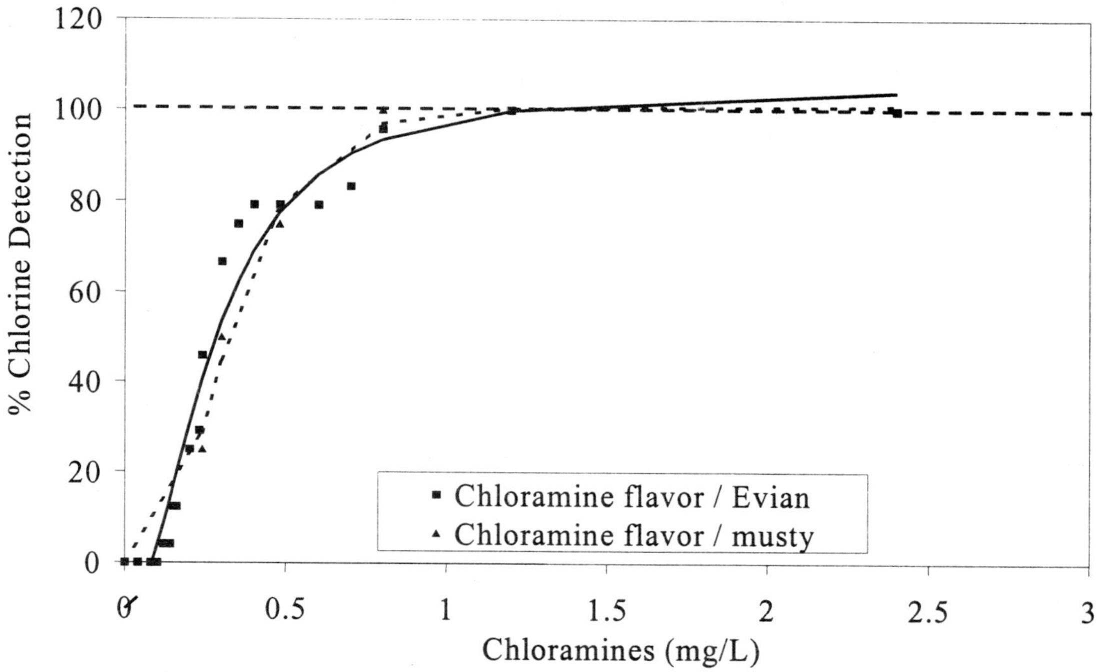

Figure 3.10a Comparison of chloramine threshold concentration: Evian versus musty water

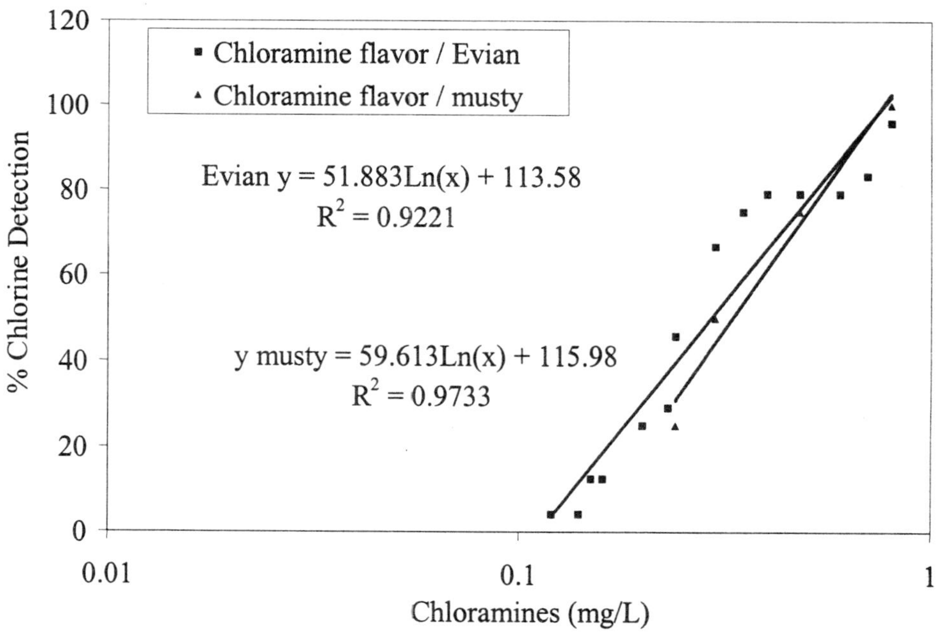

Figure 3.10b Log-scale plot of comparison of chloramine threshold concentration: Evian versus musty water

earthy-musty flavor (FTN = 3) could induce a twofold increase in the chlorine FTC. As shown in Figure 3.10, the initial FTN of the water had no effect on the FTC when monochloramine was present. No effect was noticed on the FTC for chloramine in the presence of an initial musty flavor.

Control of Musty Flavors by Chlorine and Monochloramine

To investigate the possible control of musty flavors by chlorination or chloramination, three different experiments were performed with the Aubergenville treatment plant effluent water. The parameters used were as follows:

- no disinfectant
- 1 mg/L chlorine as Cl_2
- 1 mg/L monochloramine as Cl_2

Musty Flavor Generation in the Absence of Disinfectant

Figure 3.11 shows the change of musty flavors with increased residence time in the pipe loop with no disinfectant. No significant change in FTN could be observed for up to 16 h, and

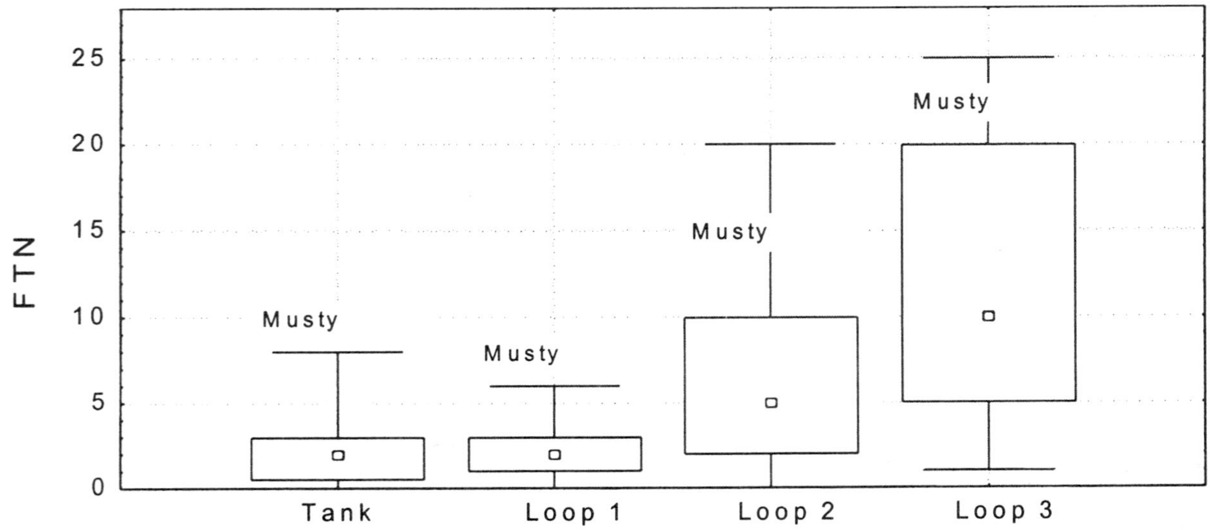

Figure 3.11 Musty flavor versus residence time in the pipe loop: no disinfectant added

an FTN of 10 was obtained after a 24-hour-residence time, indicating the formation of musty flavors. Additional experiments indicated that trichloroanisoles were the probable cause of the musty flavor. According to the literature (Montiel et al. 1991, Sävenhed et al. 1991, Nyström et al. 1992), similar odors relate to the biomethylation of trichlorophenols to trichloroanisoles.

Figure 3.12 shows suspended heterotrophic bacteria, actinomycetes, and fungi counts in relation to the corresponding musty FTN obtained in the pipe-loop effluent data. No relationship could be observed between biological counts and musty FTN in the water. Thus, in the presence of 1–2 logs of fungi and actinomycetes, the musty flavor kinetics seemed mainly driven by residence time. Fungi and actinomycetes were also quantified at the inner surface of pipes using coupons. Table 3.3 shows the results of the four sampling experiments. Compared to fungi, almost no actinomycetes were found. Fungi counts generally increased with residence time, but no significant difference between loops was determined because of the high standard deviation in the analytical data that is common for such measurements. A problem with fungi enumeration determined by plate count procedures is that the counts can be affected by vegetative cells or spores. So the counts are actually an estimate of fungi population.

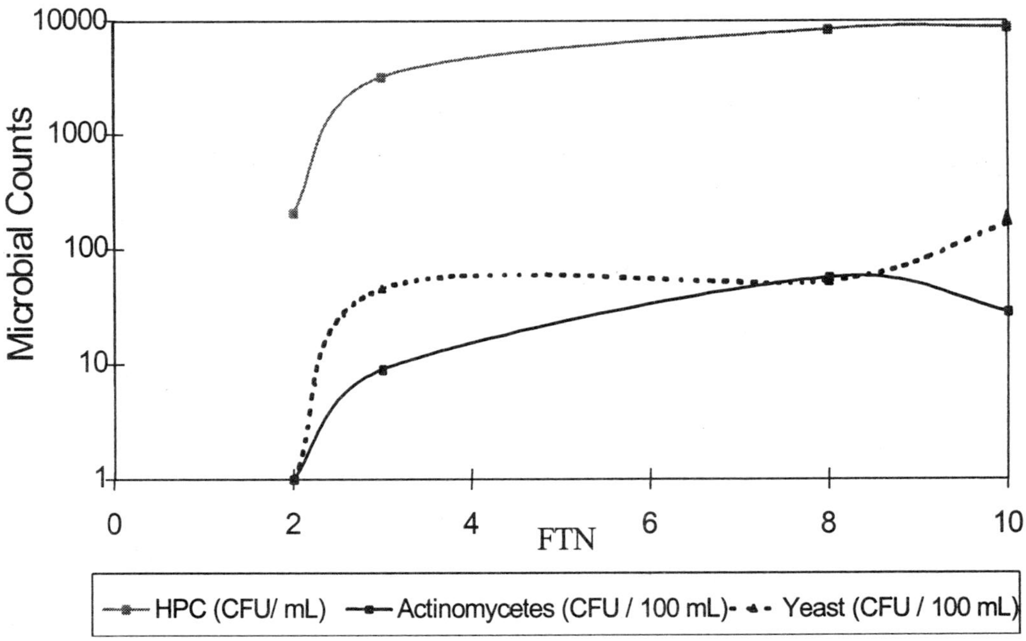

Figure 3.12 Relationship between biological counts (heterotrophic bacteria, actinomycetes, and fungi) and corresponding musty FTN in the bulk water: no disinfectant

Table 3.3

Biological surface colonization in the pipe-loop: with no disinfectant

CFU/cm^2	Loop 1	Loop 2	Loop 3
HPC– R2A	3300	3600	27300
HPC– R2A	17850	8100	30000
HPC– R2A	8850	87000	19500
HPC– R2A	32310	36675	5190
Geometric mean	11392	17465	16968
Median	13350	22387.5	23400
Actinomycetes	0	0	0
Actinomycetes	0	0	0
Actinomycetes	0	1	0
Actinomycetes	2	0	0
Geometric mean			
Median	0	0	0
Fungi	2	2	16
Fungi	6	2	2
Fungi	1	6	15
Fungi	50	20	850
Geometric mean	5	5	25
Median	4	4	15.5

HPC = Heteroptrophic plate count

Musty Flavor Control by Disinfectants

In the presence of disinfectants, musty flavors were detected only after a 16-h residence time for chlorine once the residual fell below 0.07 mg/L as Cl_2 and after a 24-h residence time for monochloramine after the residual decreased to 0.55 mg/L as Cl_2 (Figure 3.13). For shorter residence times, only chlorinous flavors were detected in either the chlorine or the chloramine experiments. In both cases (chlorine and chloramines), there was a three-to-four-fold reduction (from an FTN of 10 to an FTN of 2–3), indicating a masking of musty flavor in the presence of disinfectants or an effect of the disinfectant on the biological activity.

Figures 3.14, 3.15, and 3.16 show the comparison of biological counts according to the different operating conditions (no disinfectant, 1 mg/L chlorine as Cl_2, 1 mg/L monochloramine as Cl_2) for suspended heterotrophic plate count (HPC) bacteria, actinomycetes, and fungi, respectively.

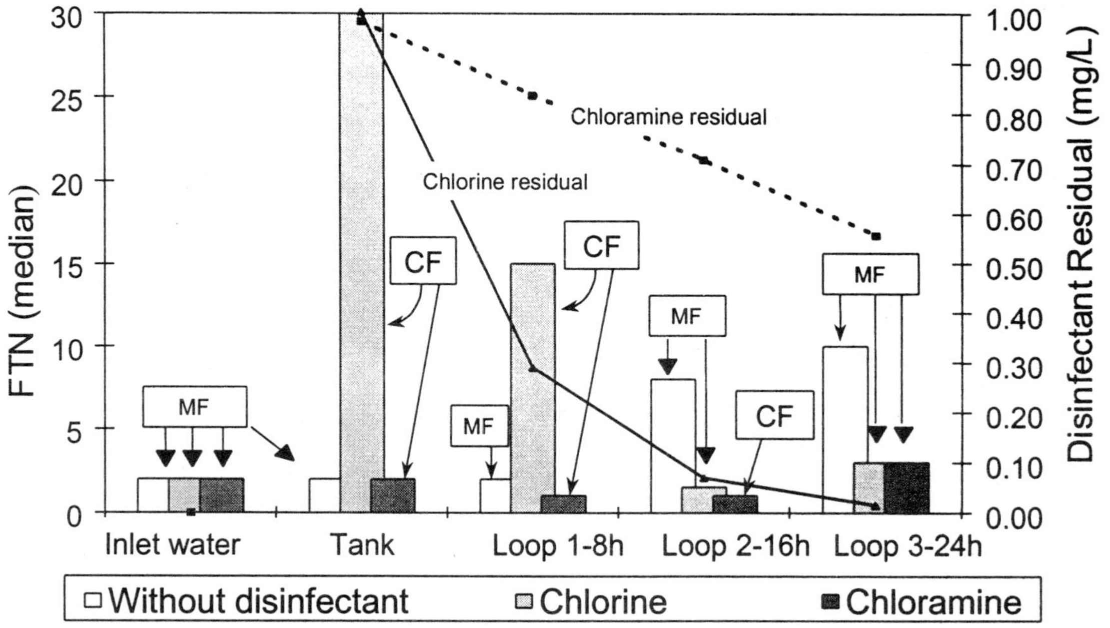

(CF: chlorinous; MF: musty)

Figure 3.13 Comparison of FTN change with time: no disinfectant, chlorine, and chloramines

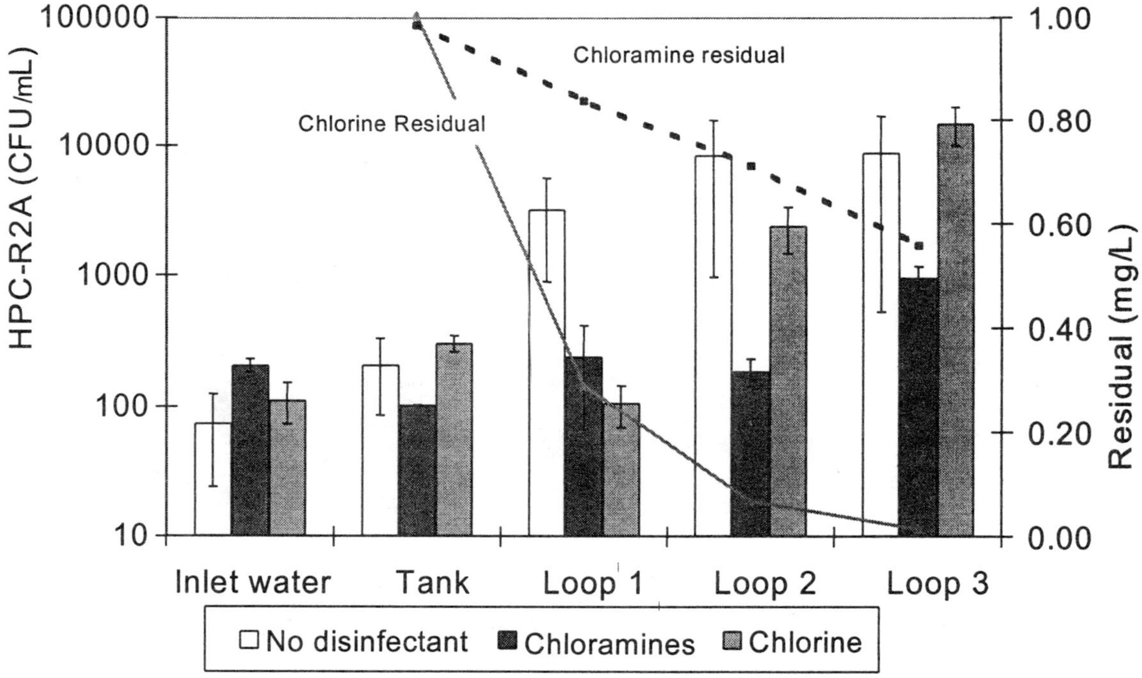

Figure 3.14 Suspended R2A heterotrophic bacteria count (HPC) evolution with time in the pipe loop: comparison of no disinfectant, 1 mg/L chlorine as Cl_2, and 1 mg/L monochloramine as Cl_2

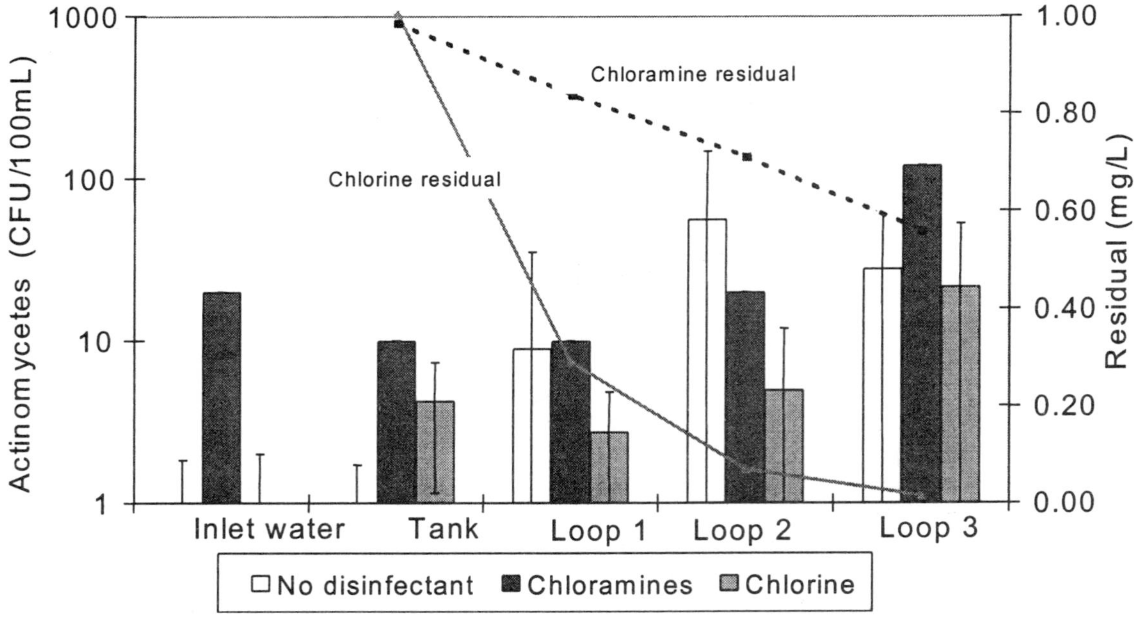

Figure 3.15 Suspended actinomycete evolution with time in the pipe loop: comparison of no disinfectant, 1 mg/L chlorine as Cl_2, and 1 mg/L monochloramine as Cl_2

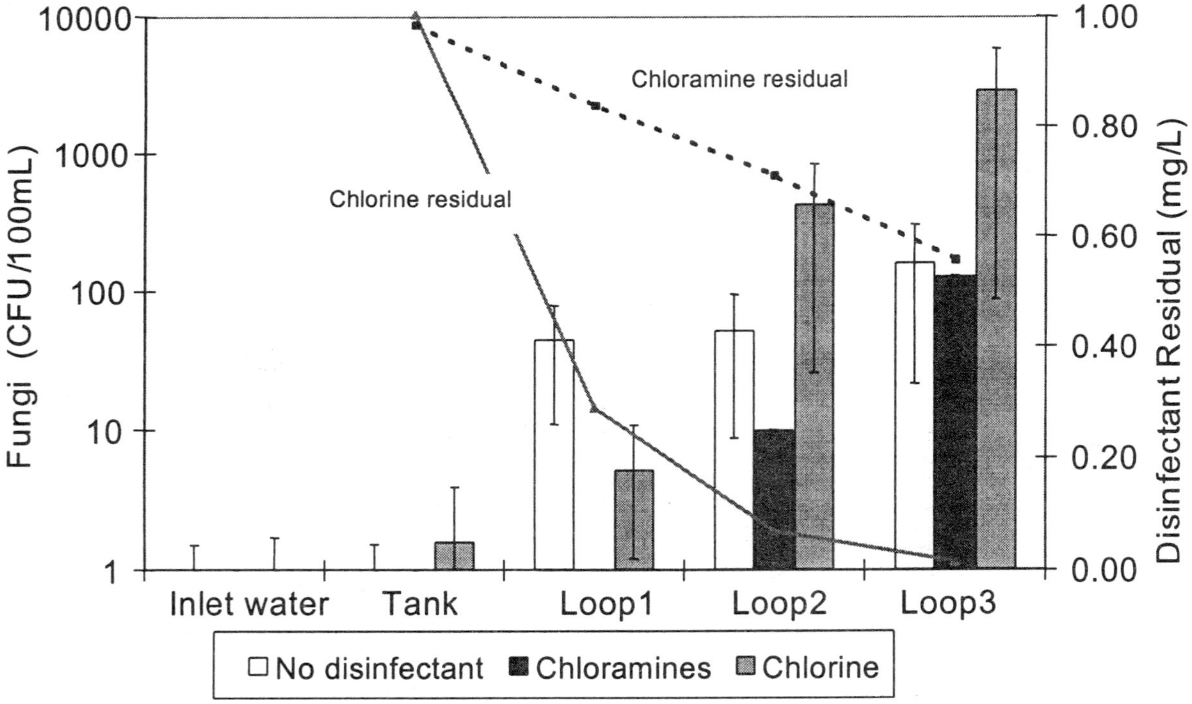

Figure 3.16 Suspended fungi counts with time in the pipe loop: comparison of no disinfectant, 1 mg/L chlorine as Cl_2, and 1 mg/L monochloramine as Cl_2

With no disinfectant, suspended HPCs increased with residence time. Bacterial counts of 10^4 colony-forming units (CFUs)/mL were reached after a 16-h residence time (a 2-log increase). In the presence of disinfectants, a decrease of approximately 90% was noticed in loop 1 (8-h residence time), attesting to the biostatic efficiency of both chlorine and chloramines when compared to values obtained without chlorine. These experiments were conducted in a pipe loop, so the increase of counts with residence time results primarily from release of bacteria from the biofilm rather than the growth of suspended bacteria. When the chlorine residual decreased to below 0.07 mg/L as Cl_2 (loop 2), the HPC increased again to reach values similar to those obtained with no disinfectant. The same trend was noticed for a chloramine residual below 0.55 mg/L as Cl_2. In this case, however, HPCs in loop 3 were 1 log lower because of the increased persistence of chloramines in the pipe loop.

With no disinfectant, suspended actinomycete counts increased with residence time. A 1–2 log increase was noticed for a 24-h residence time. The presence of chlorine led to lower counts in relation to the experiment with no disinfectant. However, almost no difference was noticed in the case of monochloramine. Because of changes in the inlet water quality with respect to suspended actinomycetes during disinfectant experiments, no real comparison could be made in relation to the experiment without disinfectant.

With no disinfectant, suspended fungi counts increased with residence time. A 2-log increase was noticed for a 24-h residence time. The presence of disinfectants led to lower counts in relation to the experiment without any disinfectant. A biostatic effect was noticed for chlorine residuals greater than 0.07 mg/L as Cl_2 and chloramine residuals greater than 0.55 mg/L as Cl_2. As a result, the use of disinfectants led to lower biological activity (significant effect on HPC and fungi counts), which may be linked to lower musty flavor generation. Because experiments performed without disinfectant had shown no linear relationship between the increase of biological counts and musty FTN, it was difficult to conclude that the musty flavor control by disinfectants was a result of their inhibiting action on microbiological activity.

For the chlorine experiment (Table 3.4), biological pipe surface colonization did not show a significant difference in relation to the experiment without any disinfectant (Table 3.3).

Table 3.4

Biological surface colonization in the pipe loop: chlorine experiment

CFU/cm^2	Loop1	Loop2	Loop3
HPC–R2A	81000	5475	15000
HPC–R2A	11175	33150	54000
Actinomycetes	723	8	125
Actinomycetes	1	4	3
Fungi	20	50	500
Fungi	2	1	1

HPC = heterotrophic plate count

Masking Effects of Disinfectants on Musty Flavor

Masking experiments with chlorine and chloramines were performed on water leaving the pipe loop (musty FTN = 3). Table 3.5 shows the results for these tests. First, it was noted that chlorine was more efficient than chloramines as a chemical masking compound. A concentration of monochloramine five times higher was necessary to achieve the same effect. A chlorine concentration of about 0.17 mg/L as Cl_2 was necessary to achieve complete masking of the musty flavor (FTN = 3). The same effect was obtained for a chloramine concentration of about 0.8 mg/L as Cl_2. Thus, the disinfectant masking effect did not explain the control of musty flavors described earlier. In pipe-loop experiments, musty flavors were detected for chlorine residuals less than or equal to 0.07 mg/L as Cl_2 and chloramine residuals less than or equal to 0.55 mg/L as Cl_2.

Field Experiments

Measurements were performed on two distribution systems. For the Vigneux distribution system, flavor analysis was performed according to the regular FTN method (dilutions: 1, 2, 3, and 5). For Cholet's network, the modified flavor analysis (up to FTN = 30) was used. Figures 3.17 and 3.18 show the relationship between chlorine residuals and associated FTN. Two types of flavor descriptors were monitored: chlorinous and earthy-musty.

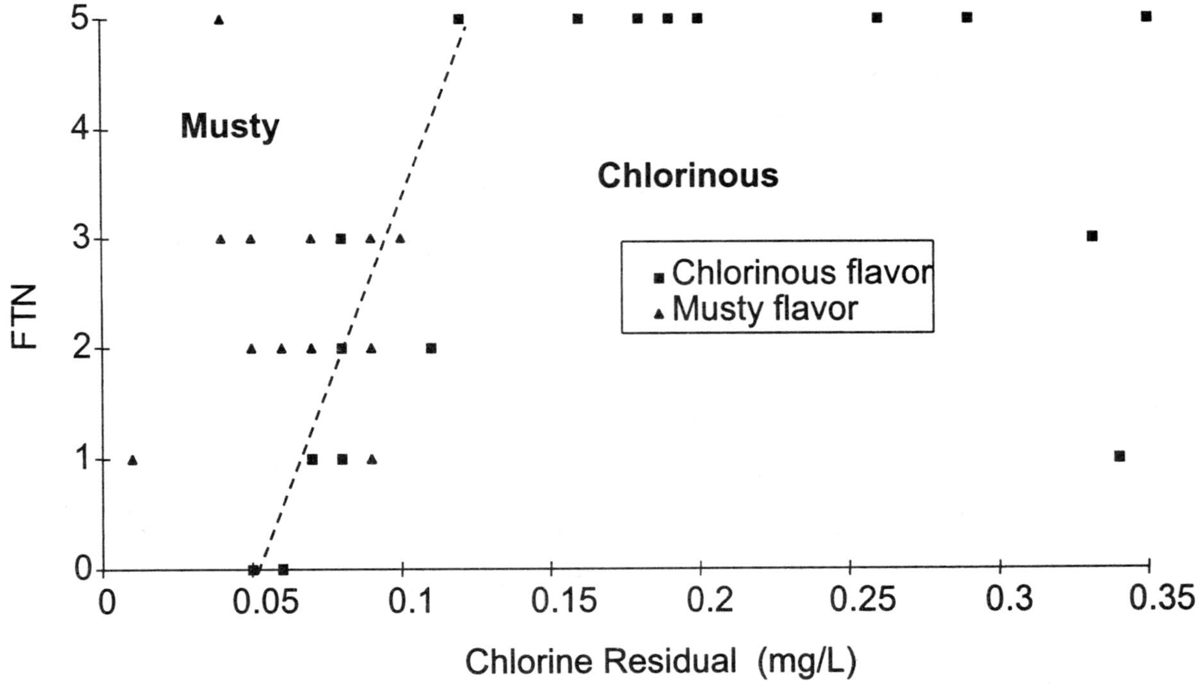

Figure 3.17 Vigneux network: FTN in relation to free-chlorine residual

Table 3.5

Masking effect of disinfectants on musty flavor leaving the pipe loop

Chlorine: Chlorine residual as Cl_2 (mg/L)	Chlorine: Musty flavor	Chlorine: Chlorinous flavor	Monochloramine: Chloramine residual as Cl_2 (mg/L)	Monochloramine: Musty Flavor	Monochloramine: Chlorinous flavor
0.500	no	yes	2.40	no	yes
0.250	no	yes	1.20	no	yes
0.167	yes	yes	0.80	yes	yes
0.100	yes	no	0.48	yes	no
0.062	yes	no	0.30	yes	no
0.050	yes	no	0.24	yes	no

For the Vigneux network, three areas in the graph can be defined based on Figure 3.17:

- Musty flavors were always detected for free-chlorine residuals less than or equal to 0.05 mg/L as Cl_2.
- Musty and chlorinous flavors were both detected for free-chlorine residuals greater than 0.05 mg/L and less than 0.1 mg/L as Cl_2.
- Chlorinous flavors were always detected for chlorine residuals greater than 0.1 mg/L.

These results were in agreement with those obtained with the pipe loop.

For Cholet's network, the same trend was noticed, but for total chlorine residuals instead of free-chlorine residuals. Because of its organic matter (total organic carbon level of about 2 mg/L), the water has a high chlorine demand. The high chlorine demand and the 24-h delay between sampling and flavor analysis may explain why no relationship was found between free-chlorine residual and FTN, probably because of chlorine decay during sample shipping. The city of Cholet is located 500 km from the laboratory conducting the flavor analysis. It was not possible to perform the flavor analysis on site.

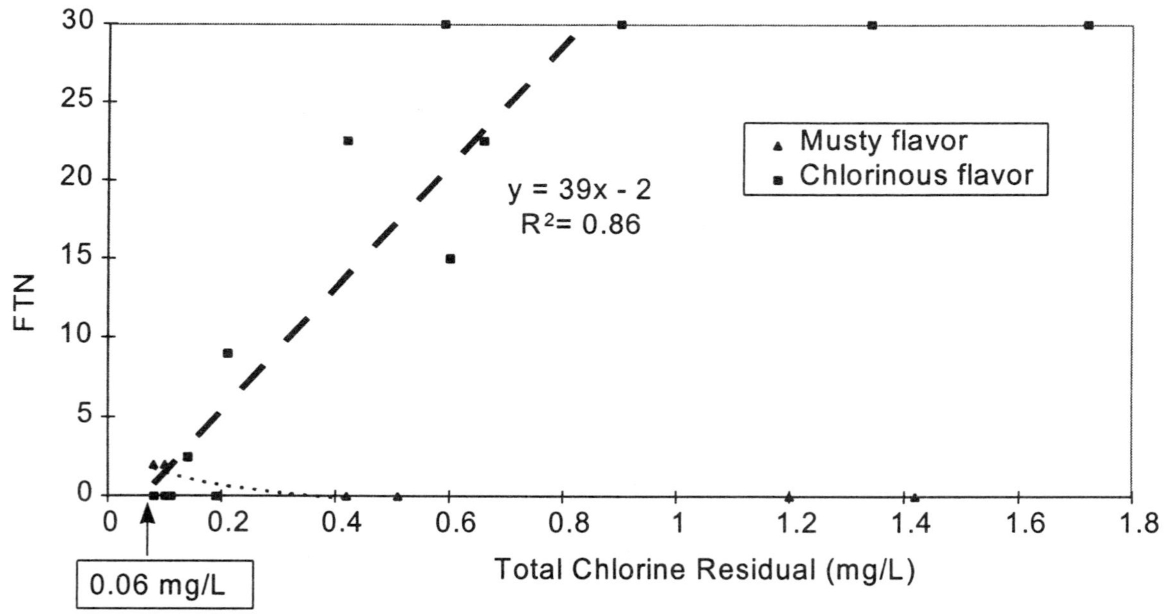

Figure 3.18 Cholet network: FTN in relation to total chlorine residual

For Cholet's network, water leaving the treatment plant was placed in jars, and water aging experiments were performed. Ozonated water (before final disinfection) was collected and chlorinated to achieve the same chlorine residual as at the outlet of the treatment plant. Flavors and chlorine residuals were monitored with respect to time and compared to those obtained in the network. Figure 3.19 shows the results of this comparison based on four sampling experiments.

Because panelists reported more than one descriptor for the same sample, the sum of the percentages exceeded 100%. For chlorine residuals greater than 0.45 mg/L as Cl_2 and residence times lower than 3 h, no significant difference between the two aging procedures (network or jar) was observed. For chlorine residuals lower than 0.1 mg/L (residence time longer than 9 h), jar-aging experiments led to FTN and flavor descriptors similar to those obtained with the pipe-loop ozonated water. However, in this case, network and jar samples showed entirely different organoleptic profiles. Fruity or bitter descriptors caused by aldehydes seemingly produced during ozonation had disappeared in the network, and musty flavors had appeared (biological activity by-products). In the case of the jar experiments, bitter-fruity descriptors that were first noticed before the disinfectant was added and reappeared when the disinfectant residual disappeared. This shows that the bitter-fruity descriptors were being masked by chlorine. These experiments demonstrate a clear contribution of the network to taste-and-odor phenomena.

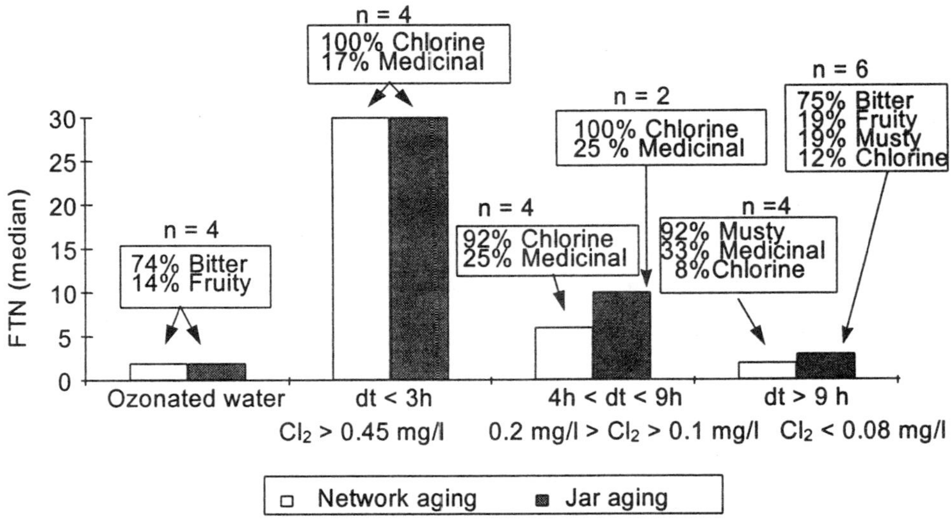

Note: Total exceeded 100% because more than one descriptor for the samples was reported by the panelists

Figure 3.19 Comparison of water aging in jars and in the network: Cholet distribution system

Summary and Conclusions

The main findings of this section are as follows:

- Jar tests with a taste-free water (Evian) resulted in FTC determinations of 0.025 mg/L as Cl_2 for chlorine and of 0.28 mg/L as Cl_2 for monochloramine (flavor perceived by at least 50% of the panelists).
- The same jar tests performed by individuals with no training resulted in an FTC for chlorine of 0.06 mg /L as Cl_2.
- Chlorine detection threshold was influenced by the initial FTN and flavor description of the water. No significant interference effect was observed during monochloramine detection threshold determinations.
- Except for a faster chlorine decay, which resulted in a reduced chlorinous flavor, the distribution system had no specific effect on chlorinous flavor induced by disinfectants.
- Pipe-loop experiments showed that chlorine flavors disappeared at concentrations below 0.07 mg /L for chlorine as Cl_2 and 0.55 mg /L for chloramines as Cl_2 and were replaced by earthy-musty flavors.
- Field experiments showed that chlorine flavors were perceived at concentrations as low as 0.05 mg /L for chlorine as Cl_2, and earthy-musty flavors appeared at chlorine concentrations below 0.1 mg /L as Cl_2.
- In the absence of disinfectants, an increase in the earthy-musty flavors with residence time was noticed in the pipe loop (an FTN of 1–2 increased to 10 with 24-h residence time). No linear relationship was found between the musty FTN and biological counts of suspended biomass. This means that the biomass concentration in the bulk water was not linked directly to the musty FPN, so musty flavors are not caused by the biological product contained in cells or released by the cells. This indicates that the musty flavor generation is a kinetic process, as in the case of the biotransformation of chlorophenols to chloroanisoles. For that process, it is more relevant to take into account all the biomass contained in the pipe loop, including free and suspended biomass.
- The use of disinfectants (chlorine or chloramines) induced a delay in the appearance of the earthy-musty flavor and a reduction of its FTN. A slight decrease of fungi and actinomycete counts in the bulk water was also noticed.

- Masking experiments performed in jars that contained water with an earthy-musty flavor collected at the outlet of the pipe loop (FTN = 3) showed that a chlorine concentration of 0.17 mg /L as Cl_2 or a monochloramine concentration of 0.8 mg/L as Cl_2 was needed to completely mask the earthy-musty flavors. These results, compared to those obtained in the pipe loop, indicate that the control of the earthy-musty flavor by disinfectants results not only from a chemical masking action but also from a biostatic effect which induces a lower microbial activity.

This study indicated that chlorine may induce unpleasant flavors, even at very low concentrations. Distribution systems can generate earthy-musty flavors that cannot be linearly related to biological activity and, thus, are mainly dependent on residence-time. These flavors can be partially controlled by the use of disinfectants.

ANNULAR REACTORS (ARS)

Details of laboratory and experimental procedures used in these studies are described in Chapter 2.

ARs Receiving Surface Water

AR-1

The first AR (AR-1) received conventionally treated chloraminated water with an average chloramine residual of 1.8 mg/L as Cl_2 during the first year of operation and 2.5 mg/L as Cl_2 starting June 6, 1998. The AR was started on June 19, 1997. Every other day, chlorine residual readings were taken and the rotational speed (60 rpm) and flow rate (1.6 ml/min) were checked. Figure 3-20 shows the results. The odor characteristics were monitored monthly, with little change observed. On June 28, 1997, the plant switched to free chlorine for one month. In order to maintain the AR system integrity and prevent biofilm alteration, ammonia was added to the AR influent water reservoir to form chloramines. Each week for a month, a fresh solution of ammonia was prepared and fed into the reservoir to maintain an average 1.8 mg/L

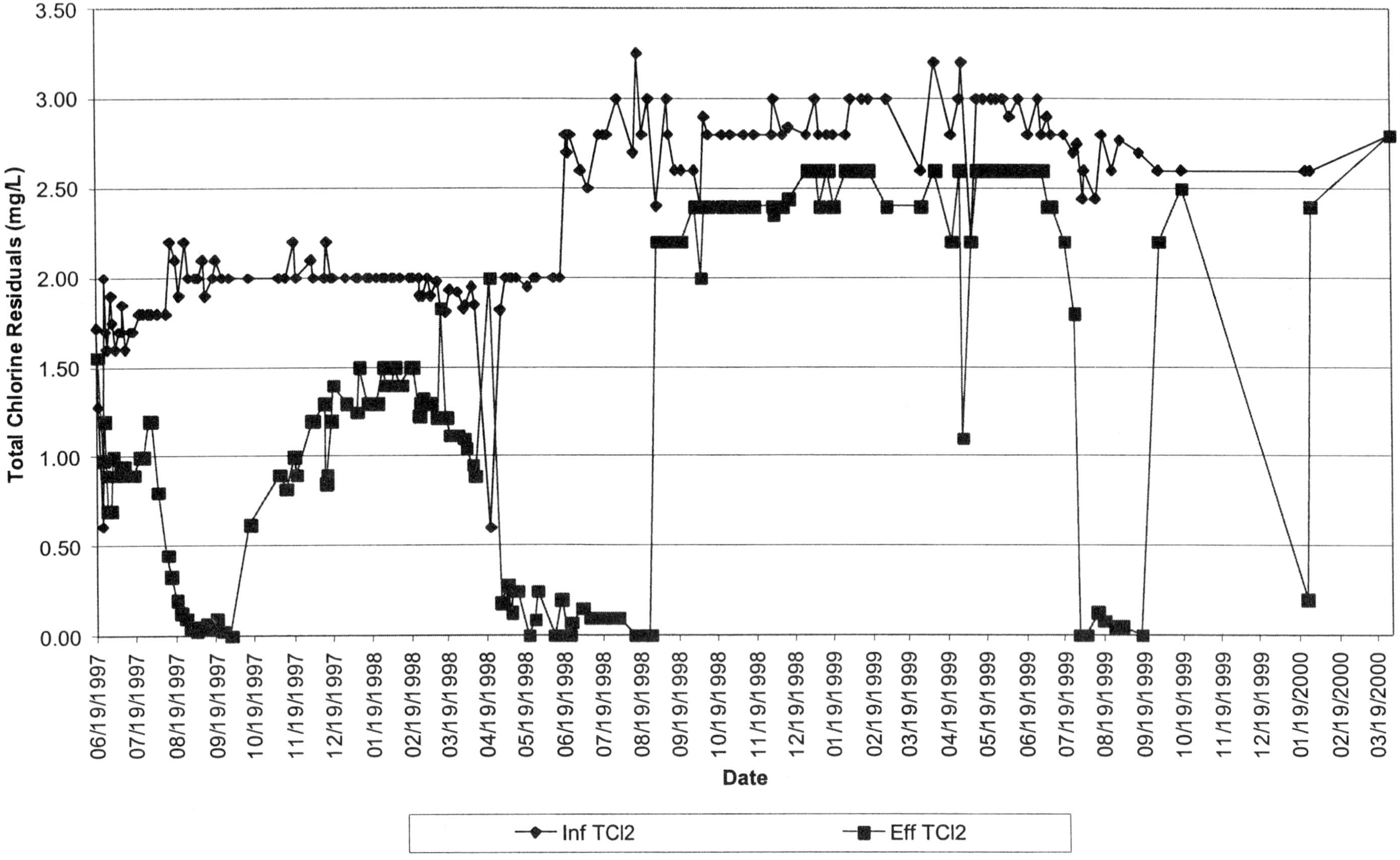

Figure 3.20 Chlorine residual profile for AR-1

monochloramine residual as Cl_2 at a 5:1 weight ratio of chlorine to ammonia-nitrogen (all chloramines–no free chlorine). On July 25, 1997, samples were taken from both the influent and effluent of the AR and analyzed by FPA (Tables 3.6 and 3.7), closed-loop stripping analysis (CLSA), and gas chromatography/mass spectrometry (GC/MS).

Both influent and effluent samples were extracted and analyzed by GC/MS. No major differences were found between the two extracts except for the large peak in the effluent that was identified by GC/MS as a branched hydrocarbon. This compound may be associated with the rubbery descriptor perceived in the effluent. It was suspected that the tubing used to feed water to the AR and the pump was the cause of the problem. A different type of tubing was used for the remainder of the experiment, and the rubbery-plastic odor subsided.

Table 3.6

Flavor-profile analysis (FPA) of AR-1 influent (July 25, 1997)

Odor		Flavor	
Descriptor	Intensity	Descriptor	Intensity
Chlorine	4	Chlorine	*
		Grassy	*
		Metallic-bitter	*
		Salty	*
		Musty-stale	*

*Other note (the descriptor was reported by less than 50 percent of the panel)

Table 3.7

FPA of AR-1 effluent (July 25, 1997)

Odor		Flavor	
Descriptor	Intensity	Descriptor	Intensity
Earthy	2	Astringent	2
Ozone	*	Rubbery	*
		Chalky	*
		Salty	*

*Other note (the descriptor was reported by less than 50 percent of the panel)

During the last week of August 1997, a sharp decrease in the chlorine residual was noticed in the effluent (Figure 3-20). Nitrite concentrations were 0.005 mg/L as nitrogen (N) in the influent and increased to 0.055 mg/L as N in the effluent. The reported detection limit (RDL) is 0.005 mg/L as N.

A second set of samples was taken on August 28, 1997. FPA, CLSA, and GC/MS analyses were again performed. The FPA results are shown in Tables 3.8 and 3.9. Because of the low FPA intensities, no chemical could be clearly identified for the cause of the sensory characteristics found in the water.

In an attempt to initiate the formation of trichloroanisoles, spiking experiments with 2,4,6-trichlorophenol were performed for three weeks, but no change of sensory characteristics was noted.

Table 3.8

FPA of AR-1 influent (August 28, 1997)

Odor		Flavor	
Descriptor	Intensity	Descriptor	Intensity
Chlorine	4	Chlorine	*
Floral/Ivory soap	*	Earthy	*
Suntan lotion (Coppertone)	*	Chalky	*
		Drying	*

*other note (the descriptor was reported by less than 50% of the panel)

Table 3.9

FPA of AR-1 effluent (August 28, 1997)

Odor		Flavor	
Descriptor	Intensity	Descriptor	Intensity
Earthy	2	Astringent	2
Sweet	2	Rubbery	*
Suntan lotion	*	Chalky	*
Cucumber	*	Salty	*

*other note (the descriptor was reported by less than 50% of the panel).

AR-2

The second AR (AR-2) was placed in parallel to AR-1. This reactor received similar water to AR-1 but was only chlorinated (prior to ammonia addition in the treatment process). It was monitored by FPA in the same manner as AR-1. Figure 3.21 shows the profile of the chlorine residual for the time of operation. Although the free chlorine at the effluent fluctuated between the detection limit of 0.05 mg/L as Cl_2 and 2.0 mg/L as Cl_2, the taste and odor remained relatively constant for the entire time of the study. The free chlorine at the influent fluctuated between 1.7 mg/L as Cl_2 and 3.6 mg/L as Cl_2 and was always perceived by FPA to be more intense than the effluent. Table 3.10 shows a typical sensory profile.

Spiking experiments similar to the ones for AR-1 were performed. It was believed that the high chlorine residual would hinder the microbial growth that is necessary for the methylation of trichlorophenols. To study this effect, the chlorine residual was decreased to approximately 0.1 mg/L by addition of sodium sulfite. Table 3.11 shows a typical sensory profile obtained. Weekly FPA monitoring performed on effluent samples showed that the sensory characteristics did not change.

No positive chemical identification could be found by GC/MS. Sensory GC was performed on the effluent extract. A moldy odor appeared at approximately 24 min of retention time. This retention time corresponded to the retention time of 2,4,6-trichloroanisole. This experiment could not be repeated because nitrification in the AR started to occur soon after that. Nitrite levels of up to 0.5 mg/L as N were detected. The HPC levels also increased significantly, to the levels shown in Table 3.12.

Table 3.10

Typical sensory profile of AR-2

Influent		Effluent	
Descriptor	Intensity	Descriptor	Intensity
Chlorine	2 - 4	Chlorine	1 - 2

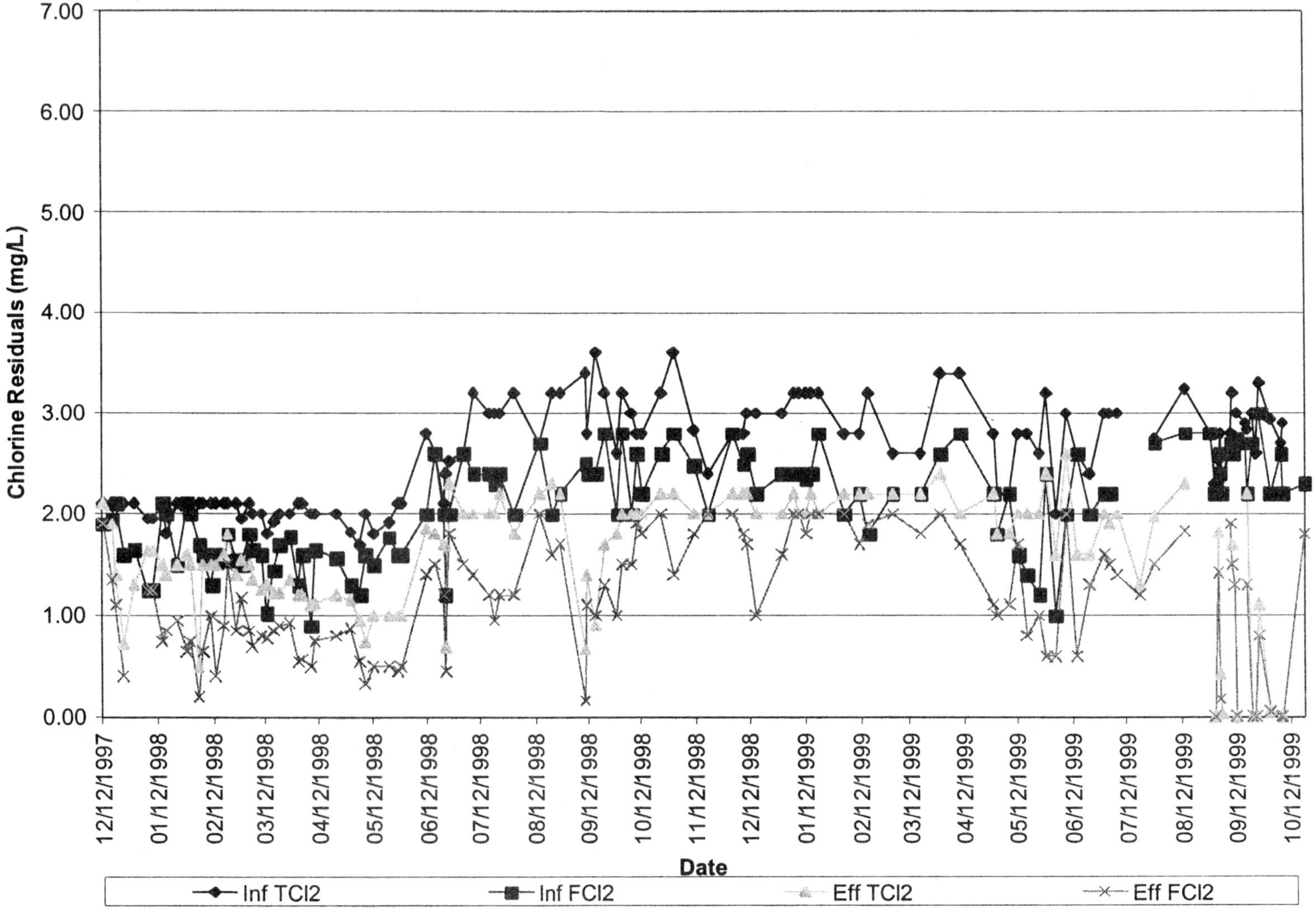

Figure 3.21 Chlorine residual profile for AR-2

Table 3.11

Typical sensory profile of AR-2 (with sodium sulfite addition)

Influent		Effluent	
Descriptor	Intensity	Descriptor	Intensity
Decaying vegetation	8	Grassy	8
Earthy	1	Moldy	1

Table 3.12

Heterorophic plate count (HPC)–R2A results

Sample	HPC–R2A (CFU/cm^2)
Influent	1.51 X 10^5
Effluent	1.58 X 10^2
Biofilm	5.13 X 10^5

AR-3

The third AR (AR-3) received water similar to the first two ARs, but with in-line flocculation followed by ozonation and chloramination. It was started on August 6, 1997, and sampled monthly. The FPA results are shown in Tables 3.13 and 3.14.

The sulfurous odor in the influent was probably caused by the onset of anoxic conditions in the lake. GC/MS analysis identified a series of straight-chain nitriles (C9 through C14) at levels of approximately 2–5 mg/L in the effluent water. Although the sensory characteristics of these particular nitriles are not known, lower molecular-weight analogs have been shown in this study to exhibit sweet-bitter characteristics. No nitriles were found in the AR influent water. Nitriles have been reported to form from chlorination or chloramination of aldehydes (White 1992 and LeCloirec and Martin 1985). In this case, the aldehydes may have been produced by biological activity or may have been products of ozonation.

At the time of this sampling, several consumer complaints were received by the water utility for unpleasant tastes and odors. The consumers were presumably receiving a diluted version of water similar to that fed to the AR. These complaints described the water as having a bitter characteristic. Unfortunately, no samples could be collected for further analysis before the problem was alleviated. Because of the organoleptic deterioration of the water and the worsening of the anoxic conditions, the utility decided to switch to an alternate source. The AR-3 experiment was terminated at that time.

Table 3.13

FPA of AR-3 influent (September 24, 1997)

Odor		Flavor	
Descriptor	Intensity	Descriptor	Intensity
Sulfurous	4	Salty	2
Briny	2	Drying	2
Chlorinous	4	Chalky	*

*Other note (the descriptor was reported by less than 50% of the panel)

Table 3.14

FPA of AR-3 effluent (September 24, 1997)

Odor		Flavor	
Descriptor	Intensity	Descriptor	Intensity
Sweet	4	Bitter	4
Sulfurous	2	Drying	2
Chlorinous	2		

*Other note (the descriptor was reported by less than 50% of the panel)

AR Receiving Groundwater

AR- 4

A fourth AR (AR-4) was installed at the effluent of a water utility treating groundwater. It was started in May 1999 and sampled monthly. The water quality parameters of this groundwater were described in the experimental section (Chapter 2) as SoCal. Gr. The water was chlorinated to obtain a free-chlorine residual of approximately 1.5 mg/L as Cl_2. Figure 3.22 shows the chlorine residual profile. The operation of this AR was disturbed a number of times because of required maintenance work at the utility.

Table 3.15 shows a typical sensory profile obtained from AR-4. FPA was also performed on a sample of dechlorinated influent to account for the masking effect of chlorine. It should be noted that because of the high iron concentration of the water, the tubing and walls of AR-4 acquired a deep brown color. These high levels of iron could explain the recurring rusty descriptors in the FPA results for the AR water. A slight yellow color was noted in the effluent water.

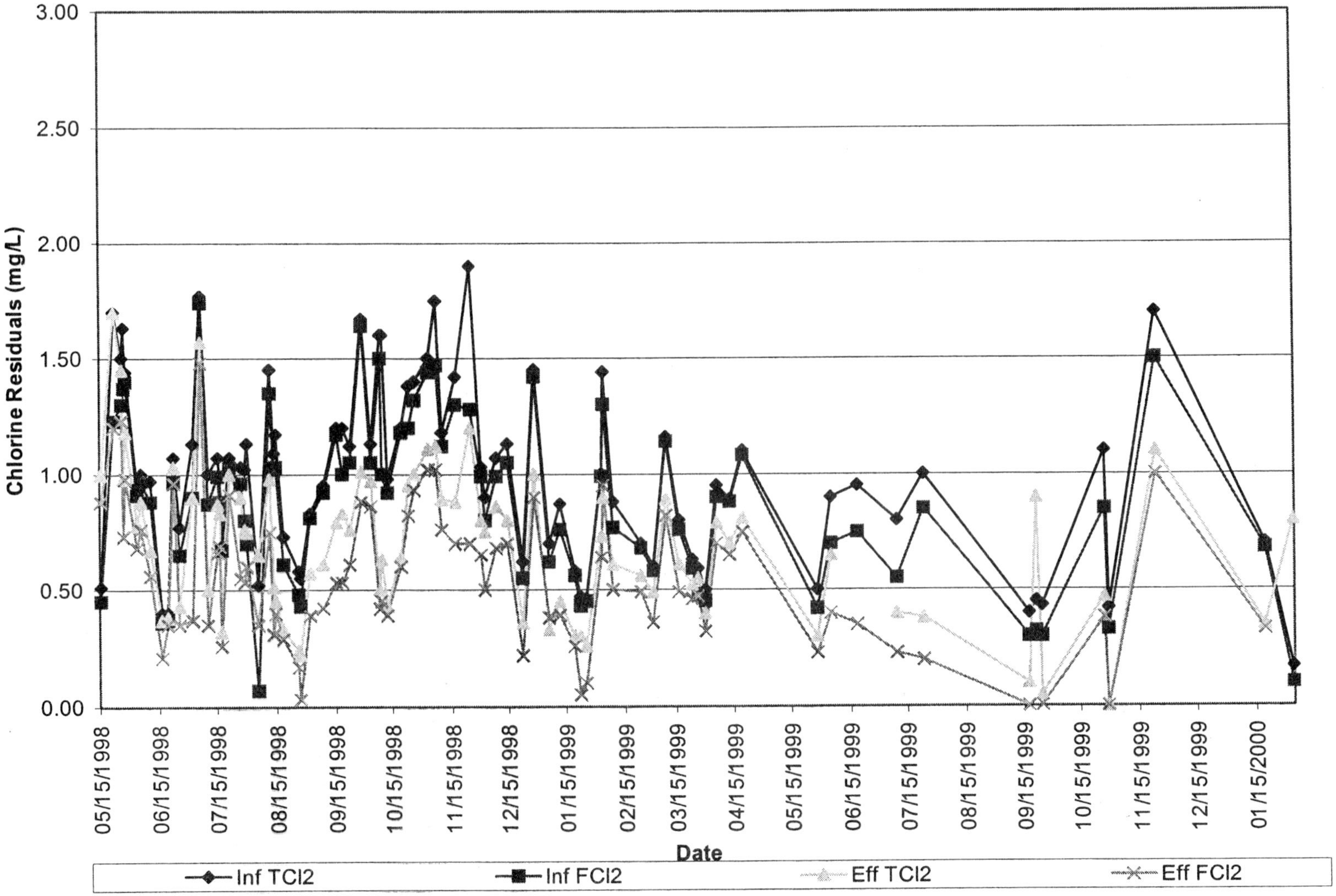

Figure 3.22 Chlorine residual profile for AR-4

In an attempt to simulate stagnant conditions such as those in these dead ends, the rotation of the AR was stopped. Samples were taken two days and three weeks later. Table 3.16 and 3.17 show the FPA results for these two samplings. GC/MS analysis of the November 17, 1999, sample identified the presence of aldehydes, including dodecanal, which is known to have soapy characteristics.

Table 3.15

Typical sensory profile of AR-4 (September 17, 1999)

Influent		Effluent		Dechlorinated Influent	
Descriptor	Intensity	Descriptor	Intensity	Descriptor	Intensity
Chlorine	4	Moldy-earthy	4	Plastic	*
		Rusty	2	Sulfur	*

Table 3.16

Sensory profile of AR-4 (October 29, 1999)

Influent		Effluent	
Descriptor	Intensity	Descriptor	Intensity
Chlorine	4	Plastic	2

Table 3.17

Sensory profile of AR-4 (November 17, 1999)

Influent		Effluent	
Descriptor	Intensity	Descriptor	Intensity
Chlorine	4	Rusty	2
		Soapy	*
		Decaying vegetation	*

Summary and Conclusions

AR-1

- Generally, a chlorine odor was observed at the influent to the AR and no chlorine odor was observed at the effluent. Other odor characteristics were very low in intensity.
- Even during nitrification in the AR, only very little change was observed by FPA in the effluent when the FPA results were compared to periods when nitrification was not occurring. This parallels data in the distribution system for similar circumstances.
- The AR was easier to operate at a higher monochloramine residual and resulted in less nitrification. This also parallels what has been observed in the actual distribution system.

AR-2

- Although the free-chlorine residual at the AR effluent fluctuated between 0.05 mg/L and 2 mg/L as Cl_2, the taste and odor for characteristics other than chlorine remained relatively constant for the entire study.
- The free-chlorine residual at the AR influent fluctuated between 1.7 mg/L and 3.6 mg/L as Cl_2 and was always perceived by FPA to be more intense than the effluent.
- For spiking experiments with trichlorophenol on water dechlorinated to 0.1 mg/L free chlorine at the influent to the AR, no change was observed by FPA at the effluent. However, sensory GC of the same water at the AR effluent detected a moldy odor at a gas-chromatographic retention time of approximately 24 min. This retention time corresponds to the retention time of 2,4,6-trichlorophenol.

AR-3

- GC/MS identified a series of straight chain nitriles (C9 through C14) in the AR effluent. Lower molecular-weight analogs were shown in this study (Chapter 4 and Appendix A) to exhibit sweet and bitter characteristics. Nitriles have been reported by others to form as a result of chlorination or chloramination of aldehydes. In this study with AR-3, the aldehydes may have been produced by biological activity or may have been by-products of ozonation. Consumer complaints describing a bitter taste were received in the distribution system from consumers presumably receiving a diluted version of water similar to that fed to the AR.

AR-4

- FPA results showed a difference in the characteristics of the AR effluent compared to the AR influent.
- GC/MS results for the effluent showed the presence of aldehydes, including decanal, which is known to have soapy characteristics.

CASE STUDIES

A number of case studies were investigated that appear to involve biologically produced taste-and-odor compounds.

Case Study I: Taste-and-Odor Problems Caused by Iodinated THMs (ITHMs)

On November 7, 1996, a series of taste-and-odor complaints were received concerning the off-taste and odor of water in an area of San Joan Despi (Spain), a town with a population of 25,000 near Barcelona, which is mainly supplied with water from the Llobregat River, which is treated at the San Joan Despi Treatment Plant. The affected area corresponded to a very specific part of the town where some stretches of network had standing water that had not moved for a time because of maintenance work being carried out on the distribution system. The tastes and odors were described as medicinal and unpleasant.

Chemical analysis of the various samples taken in the area identified ITHMs that are generally associated with medicinal odors. Chlorobromoiodomethane and dibromoiodomethane were two of the ITHMs identified by chemical analysis. However, their concentrations were low, far below the accepted odor thresholds for iodinated compounds. Upon further investigation, other chemicals were found that could explain the problem: various aldehydes capable of causing tastes and odors in water. Table 3.18 shows the concentration, in µg/L, of the branched aldehydes found in the four samples analyzed. Both problem area A and B samples had the unpleasant sensory characteristics. A third sample (control) was taken from an area of the distribution system where no problem was identified. The fourth sample represented the treatment plant effluent and was similar to the control. Figures 3.23 shows the corresponding chromatograms, except for the fourth sample.

A study by Hrudey and coworkers (1988) reported the identification of 2-methyl propanal and 3-methyl butanal as causes of odor problems in water in Edmonton, Canada. The odor threshold ranges were 0.9–2.3 μg/L for 2-methyl propanal and 0.15–2.0 μg/L for 3-methyl butanal.

In this study, the concentrations of 2-methyl propanal and 3-methyl butanal found in the problem area samples were near or above the reported threshold levels. The cause of the problem was, therefore, attributed to these compounds. It is important to mention that these thresholds refer to individual compounds (Hrudey et al. 1988).

It is important to study the origins of these compounds. They are known to be produced during the oxidation of amino acids and most probably formed in the distribution system, given that the aldehyde concentrations in the water at the treatment plant effluent are very low and below the odor threshold.

Table 3.18

Aldehyde concentrations in water

	Concentration (μg/L)			
Chemical	Problem area A	Problem area B	Control	Plant effluent
2-methyl propanal	1.85	1.35	0.20	0.14
3-methyl butanal	2.05	1.40	0.20	0.17

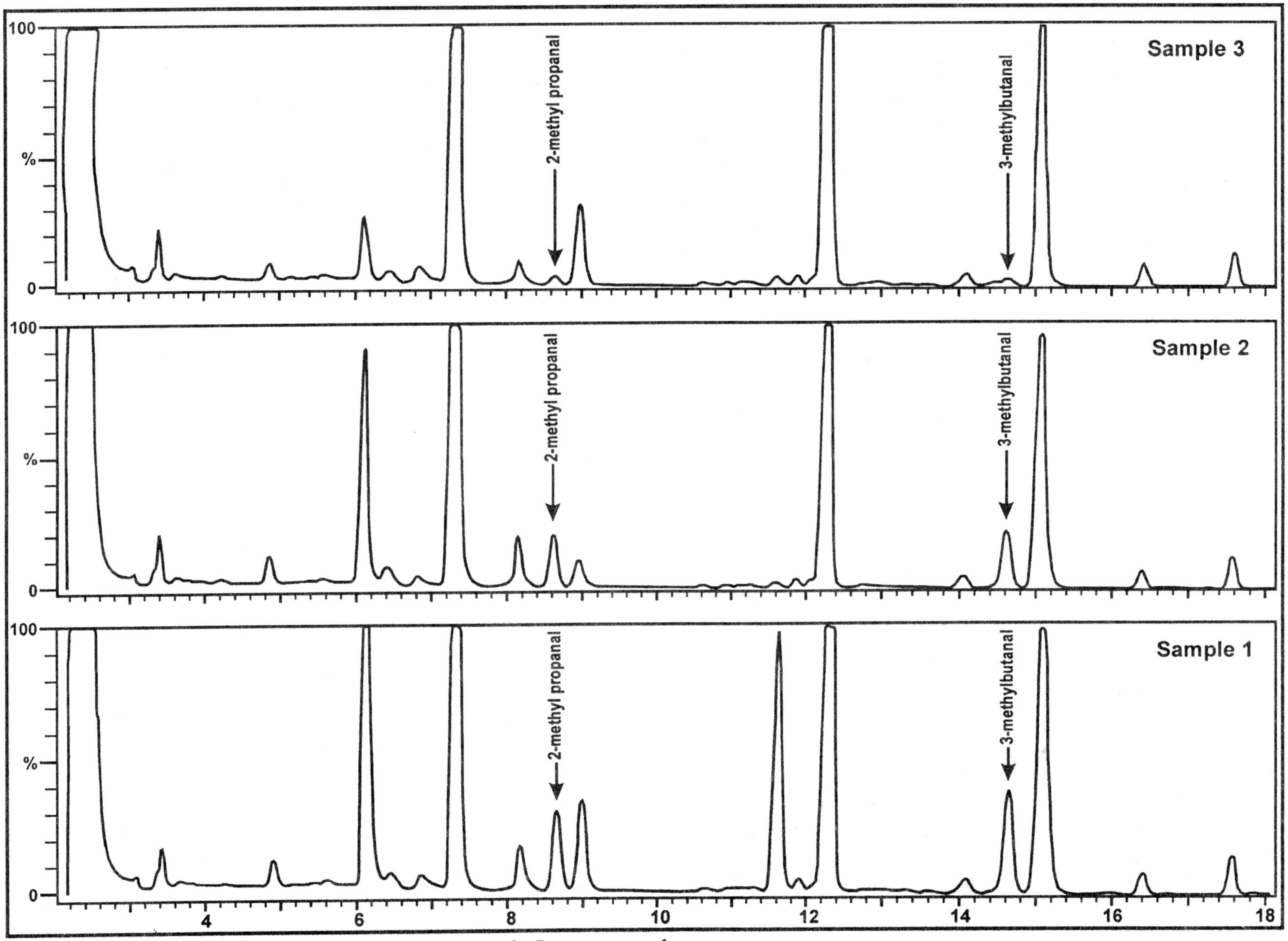

Figure 3.23 Total ion chromatograms for case study I water samples

Case Study II: Degradation of Organoleptic Characteristics After Contact with a Coated Drinking Water Storage Tank

At the end of 1997, the regional center of Lyonnaise des Eaux in western Paris received several complaints of musty flavors. The water originated from a well that was treated by chlorination only and then sent to a storage tank. This tank had been monitored for a number of years, and no problem was reported in previous analyses. Table 3.19 shows the history of the sensory analyses. FTN analysis performed on a sample collected on December 22, 1997, revealed that the chlorinous odor (FTN = 3, which complies with the European guideline) of the treated water had been replaced by intense musty odors (FTN = 5, exceeding the European guideline) after contact with the storage tank. FPA indicated a taste-and-odor intensity of 8 for the musty descriptor. A CLSA–GC/MS analysis did not reveal any compound known to impart a musty flavor. It was suggested in the analytical report that the odor could be caused by compound(s) present at levels below the current 1 ng/L CLSA–GC/MS limit of detection.

Examination of previous FTN data (Table 3.19) indicated that the musty odors had appeared soon after the tank had been newly coated, in October 1992. The first routine analysis after the re-coating was carried out in July 1993 revealed the appearance of the musty odor. No such odor had ever been recorded during the preceding period, between 1980 and 1992.

This coating, called "Barralastic," was prepared on site by mixing two components; the first corresponded to a mortar (sand + cement), while the second component was of unknown composition. However, the mechanical properties of the final product, which had an elasticity similar to that of rubber, indicated that the second component was organic in nature. This material had received approval in 1991 by French health authorities after leaching tests were carried out by the control laboratory of the City of Paris.

Pieces of the coating were sampled on January 5, 1998, and shipped to the laboratory for analysis. The coating (approximately 4 grams) was shipped in a closed 250 mL Erlenmeyer flask. The headspace above the coating presented an intense musty odor upon arrival at the laboratory. Examination of the coating surface with an optical microscope did not reveal any living organism. The whole sample (4 g) was soaked for 96 h 1 L of Evian water.

Table 3.19

Flavor threshold number (FTN) data for chlorinated well water after storage tank

Sampling date	Sensory Profile (FTN)	Descriptor
March 19, 1984	No flavor	
November 8, 1984	No flavor	
March 26, 1985	1	
November 4, 1985	2	
March 19, 1986	2	
October 27, 1986	3	
February 10, 1987	No flavor	
March 17, 1987	No flavor	
October 9, 1987	3	
March 2, 1988	No flavor	
October 18, 1988	No flavor	
February 27, 1989	No flavor	
March 27, 1990	No flavor	
May 6, 992	No flavor	
September 30, 1992	2	
July 5, 1993	3	Musty
November 30, 1993	3	Musty
December 3, 1993	3	Chlorinous
June 16, 1994	5	Chlorinous - Musty
November 9, 1994	1	
June 15, 1995	2	Musty
October 3, 1995	2	Musty
June 26, 1996	5	Musty
July 1, 1996	No flavor	
November 21, 1996	No flavor	
March 20, 1997	5	Chlorinous - Musty
July 2, 1997	1	
December 2, 1997	5	Musty

After soaking, this water showed a musty odor partially masked by another descriptor, which was more chemical in nature but could not be determined precisely. The CLSA extracts obtained were injected in the large-volume injector (LVI)–GC/MS described in Chapter 2. The full-scan chromatograms obtained were searched for the specific ions of target compounds known to impart earthy-musty odors at very low threshold concentrations. The list of target compounds is reported in Table 3.20, together with their specific ions and OTCs.

Figure 3.24 shows selected windows of the chromatograms from the chlorinated water (after the storage tank) acquired under both classical and LVI–GC/MS conditions. The LVI–GC/MS revealed the presence of 2,4,6-tribromoanisole (TBA), which was not visible in the chromatogram originally acquired in December 1997. This compound has a published odor threshold of 30 pg/L at 25°C and a musty descriptor (UKWIR 1996). Its organoleptic characteristics are very similar to that of the chloro-analog, 2,4,6-trichloroanisole (TCA).

Table 3.20

List of target compounds known to impart earthy-musty odors

Target compounds	Specific ions	Odor threshold concentrations (OTCs)
2-Methoxy-3-isopropyl-p yrazine	124, 137, 152	2 ng/L (McGuire et al. 1981)
2-Methoxy-3-isobutyl-pyrazine	94, 124, 151	2 ng/L (McGuire et al. 1981)
2-Methylisoborneol (MIB)	81, 95, 135	9 ng/L (McGuire et al. 1981)
Geosmin	97, 112, 125	4 ng/L (McGuire et al. 1981)
2,4,6-Trichloroanisole (TCA)	167, 195, 210	0.08 to 0.02 ng/L (UKWIR 1996)
2,3,6-Trichloroanisole	167, 195, 210	7 ng/L (McGuire et al. 1981)
2,3,4,6-Tetrachloroanisole	203, 231, 246	4 ng/L (UKWIR 1996)
2,4,6-Tribromoanisole (TBA)	303, 329, 346	0.03 ng/L (UKWIR 1996)

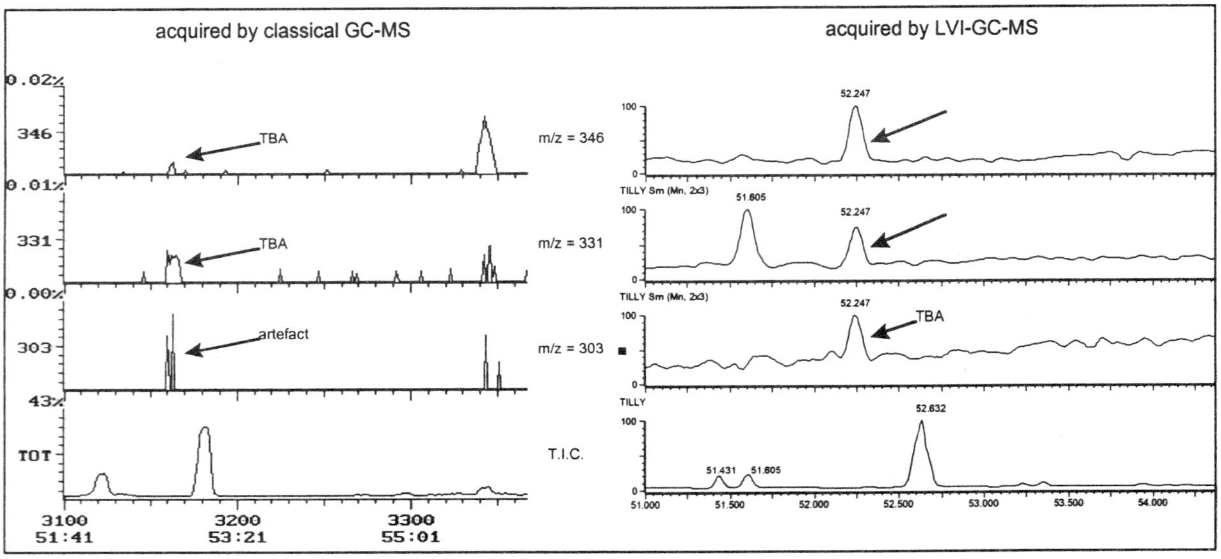

Figure 3.24 Chromatograms from the chlorinated water. TBA = 2,4,6-tribromoanisole

Figure 3.25 shows the same chromatogram windows for the reference water contacted with pieces of the coating. Again, the LVI chromatogram reveals the specific ions of TBA with a high signal-to-noise ratio, yielding a good-quality mass spectrum compared to the mass spectrum of a pure standard (Figure 3.26).

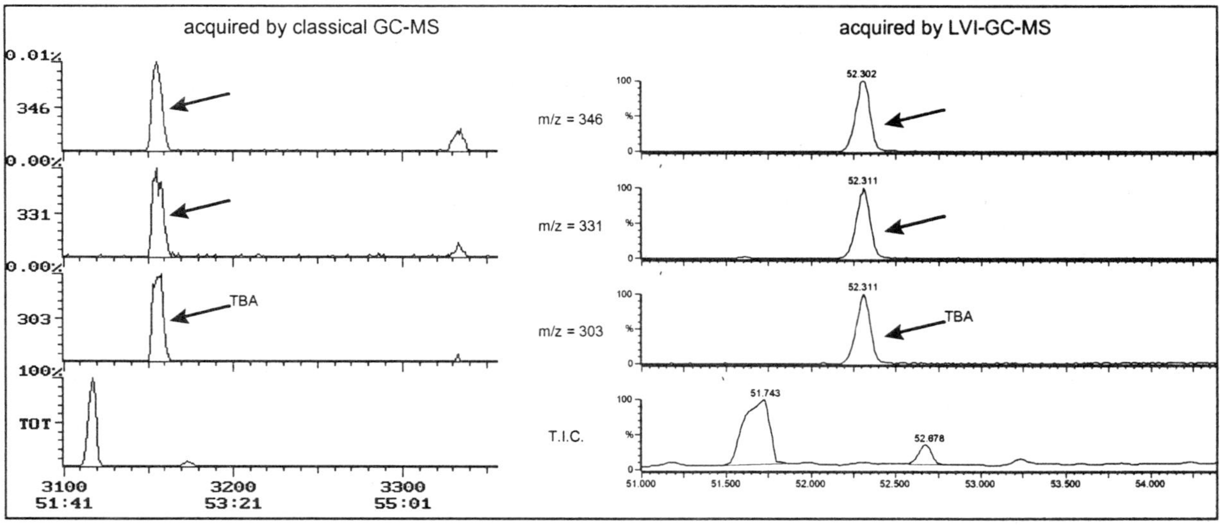

TBA = 2,4,6-tribromoanisole

Figure 3.25 Chromatograms for the reference water contacted with pieces of the coating

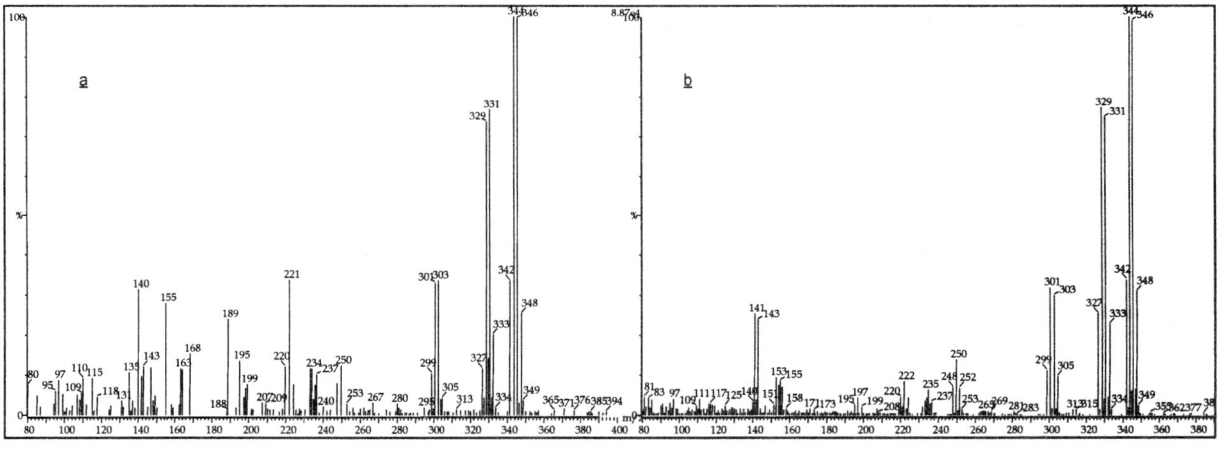

Figure 3.26 Mass spectrum of 2,4,6-tribromoanisole (TBA): (a) for the reference water contacted with pieces of the coating; (b) for a pure standard

It is interesting to note that these characteristic ions were visible in the original chromatogram acquired under standard conditions in January 1998, but they remained unnoticed because no peak was present on the total ion current chromatogram. The concentrations of TBA in the chlorinated and soak water were estimated using an internal standard method (with the nearest chloroalkane as an internal standard) and taking into account the CLSA recovery for this compound. These concentrations were equal to 150 pg/L and 25 ng/L in the chlorinated water and the soak water, respectively. The much higher concentration of TBA in the soak water was a further indication that this compound was released by the coating.

To further confirm this theory, two additional samples that corresponded to the chlorinated water before and after the tank were collected on July 1, 1998. This time, two 1 L aliquots of each sample were extracted by CLSA, and each was eluted with 35 µL of carbon disulfide, thus yielding a total extract volume of 70 µL for each sample. Forty µL of each extract was then injected using LVI–GC/MS, and the total ion chromatograms were searched for the target compounds indicated in Table 3.20. The chromatographic windows encompassing the retention time of TBA are shown in Figure 3.27.

In the chlorinated well water before the tank, the signal-to-noise ratio of the chromatographic peak corresponding to the specific ions of TBA was judged too low to ensure that this sample contained any TBA. Nevertheless, the specific ions were clearly detected in the chlorinated water after contact with the storage tank. The concentration of TBA in water after the tank was estimated at 4.5 ng/L, whereas its level was below the method detection limit of approximately 25 pg/L in the water before the tank.

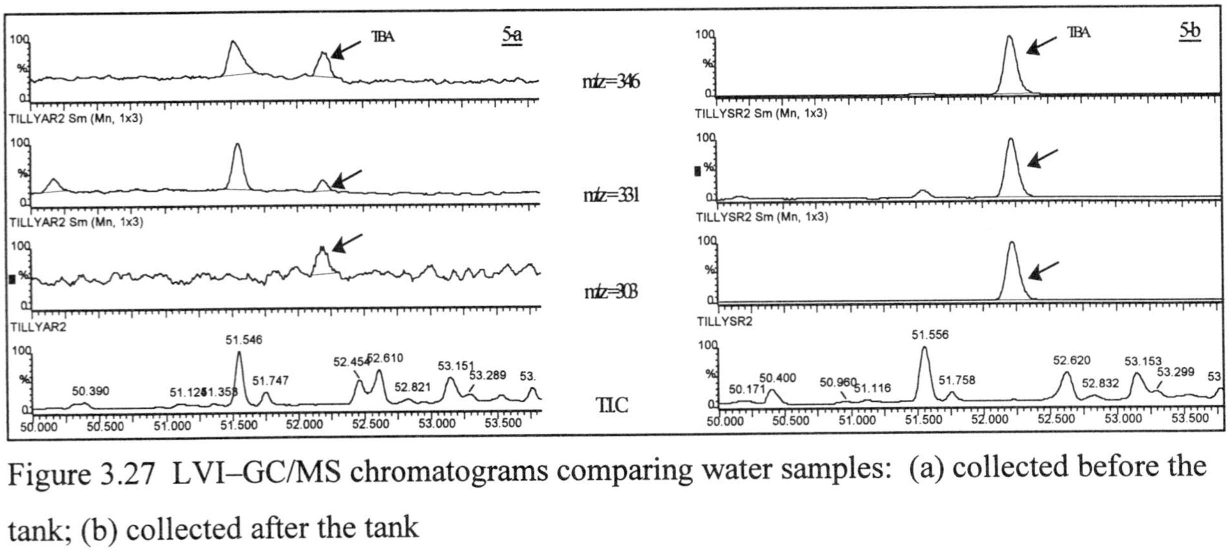

Figure 3.27 LVI–GC/MS chromatograms comparing water samples: (a) collected before the tank; (b) collected after the tank

The water before the tank did not exhibit any flavor, whereas the water after the tank had an FTN equal to 5, with a musty descriptor.

An additional methylene chloride liquid-liquid extraction carried out on these two samples allowed the identification of 2,4,6-tribromophenol in the water after the tank. This compound was not detected in the water before the tank, further confirming that the phenolic derivatives observed were released by the tank coating.

The use of a new analytical approach (LVI–GC/MS) confirmed that TBA was released from the cementicious coating studied at levels exceeding its individual odor threshold. By analogy with the formation of the corresponding trichloroanisole, the best hypothesis is that this material initially contained phenol that reacted with the chlorinated water to yield 2,4,6-tribromophenol at some point. The concentration of tribromophenol probably remained below its odor threshold (approximately 1–6 μg/L at 2025°C, according to the 1996 UKWIR report DW-05/13), as no chlorophenol flavor was ever reported at this site. Then, after a certain contact time no longer than eight months (see Table 3.19), biological activity became sufficient to transform tribromophenol into the corresponding tribromoanisole, which has a much lower odor threshold. In fact, the initial analysis of the soak water, carried out in January 1998, had revealed traces of phenol. Because CLSA is known to yield a very poor recovery for phenol, it can be inferred that the presence of phenol in the soak water was probably significant.

This case study is an example of a taste-and-odor episode that originated from a point source in the distribution system. This case also demonstrates that the official approval of drinking water materials based on a single leaching experiment of a new coating or pipe does not guarantee that the material will remain inert in the long run. The current trend to incorporate more synthetic plastic polymers in cementicious materials makes them more prone to water quality deterioration problems. From an analytical point of view, it is not clear whether the musty odor originated from biological activity or from trace chemicals present in the coating. It does confirm the observation that not all musty odor episodes can be attributed to geosmin and 2-methylisoborneol (MIB) and that new analytical approaches are required to solve such episodes.

As mentioned earlier, TCA has been identified as an important compound produced by biological activity. The descriptor for this compound is earthy-moldy. Figure 3.28 shows the

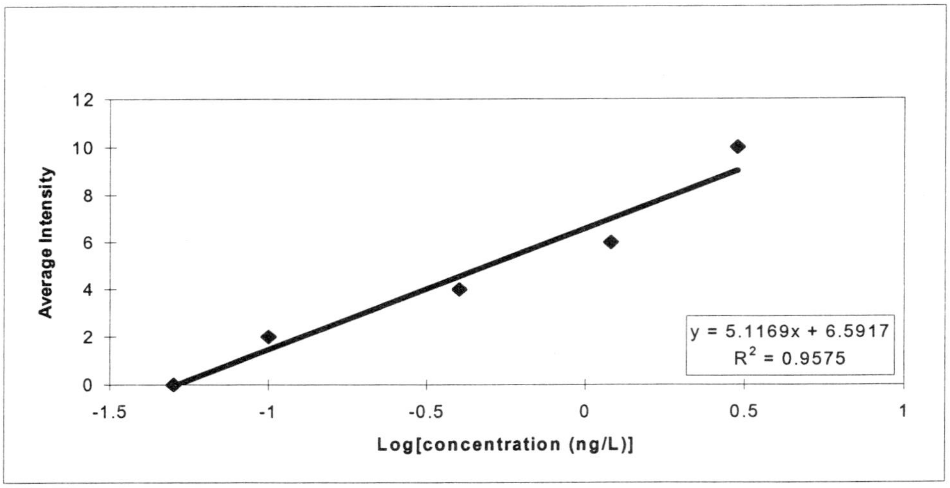

Figure 3.28 Dose-response plot (Weber-Fechner) for 2,4,6-trichloroanisole

dose-response plot using the Weber-Fechner model obtained. The estimated OTC from this plot was 0.08 ng/L. This value agrees with the OTC of 0.02 to 0.08 ng/L reported by UKWIR (1996).

Summary and Conclusions

Case study I

- Branched aldehydes (2-methyl propanal and 3-methyl butanal) at approximately 2 µg/L were found in standing water that resulted from maintenance in the distribution system. These levels were in the range for the OTC determined by Hrudey et al. (1988) and are known to be produced during oxidation of amino acids. Consumers complained of medicinal and unpleasant odors.

Case study II

- A new analytical approach (LVI–GC/MS) confirmed that TBA was released from a cement tank coating at levels exceeding its odor threshold.
- This case study is an example of a taste-and-odor episode that originated from a point source

in the distribution system.

- The case study demonstrates that the official approval of drinking water materials based on a single leaching experiment of a new coating or pipe does not guarantee that the material will remain inert.
- It is not clear whether the musty odor originated from biological activity or from trace chemicals present in the coating.
- This study confirms that not all musty odors can be attributed to geosmin and MIB, and that new analytical approaches are required to solve such episodes.

CONCLUSIONS

The common threads drawn from the pipe loop, annular reactor, and case studies discussed in Chapter 3 are as follows:

- Generation of musty-moldy odors and flavors
- Generation of aldehydes and nitriles
- Transformation of halogenated phenols to halogenated anisoles, which is significant because of the lower OTCs for the halogenated anisoles
- A need for new techniques to measure these lower concentrations of compounds that can cause taste-and-odor changes in the distribution system
- Pipe loops and ARs can produce results similar to those observed in actual distribution systems
- Although the biological causation cannot be positively confirmed, there is a preponderance of evidence pointing to a biological relationship for the work discussed in Chapter 3. More work is necessary to confirm the actual biological relationship to tastes and odors generated in the distribution system.

CHAPTER 4

TASTES AND ODORS PRODUCED BY DISINFECTANTS AND DISINFECTION BY-PRODUCTS

BENCH-SCALE EXPERIMENTS

The objective of these laboratory experiments was to study the role of the composition of the water entering the distribution system on the production of tastes and odors. Various source waters covering a wide range of water quality parameters were treated using various combinations of treatment processes (ozonation prior to and after coagulation), high- and low-dose coagulation, chlorination, and chloramination). The organoleptic characteristics of treated waters were monitored at different detention times (1 and 48 h). Monitoring of these changes was done by flavor-profile analysis (FPA). When organoleptic changes were noted, chemical analysis was performed in an attempt to identify the chemicals produced. See Chapter 2 for more details on these experiments and related analyses.

Effects of Increased Chloramines on Taste-and-Odor Properties

The primary objective of this section was to determine if raising the monochloramine residual in the distribution system from approximately 1.7 up to 2.5 mg/L as Cl_2 would have a negative effect on the taste-and-odor quality of potable water. Four waters were studied: conventionally treated California state project water (SPW), Colorado River water (CRW), a 50/50 blend of CRW/SPW, and a pre-ozonated (ozonated prior to coagulation), conventionally treated, 50/50 blend of CRW/SPW. Chloramination was performed on the four samples to obtain residuals of 1.7 and 2.5 mg/L. FPA was performed on all samples. The experiment was performed in the winter and repeated in the summer to account for seasonal changes.

Conventional Treatment/Post-Chloramination

Tables 4.1 and 4.2 show the FPA results obtained when the individual waters were conventionally treated and then chloraminated at 1.7 and 2.5 mg/L. The seven-point intensity scale, as described in Chapter 2, was used to give the qualitative descriptors (e.g., chalky, chlorine) an intensity rating. Although no results were available for CRW treated at the bench scale, the average FPA for the three-month winter distribution system survey (total chlorine residual, 1.7 mg/L) showed intensity ratings of 4 for the chlorine, chalky, and drying descriptors. Table 4.3 shows the results when these two source waters were blended at 50/50, conventionally treated, and chloraminated to obtain similar residuals. The sign "*" is used for "other note" or for a descriptor that was reported by less than 50% of the panel.

Table 4.1
Sensory characteristics of state project water (100%)

Total Residual	Winter		Summer	
Chlorine (mg/L)	Odor	Flavor	Odor	Flavor
1.7	Sweet* Musty*	Metallic 1.5 Bitter 2 Musty* Chalky* Drying 4	No odor	Unripe fruit* Chlorine* Metallic 3.5 Bitter 2 Drying 7
2.5	Sweet* Citrus* Musty*	Orange/Floral* Sweet* Musty* Chalky* Bitter 1.5 Metallic 1.5 Drying 5	Chlorine 2	Unripe fruit* Chlorine* Chalky* Metallic 4 Bitter 2 Drying 7

*Other note

Table 4.2
Sensory characteristics of Colorado River water (100%)

Total Residual	Winter		Summer	
Chlorine (mg/L)	Odor	Flavor	Odor	Flavor
1.7	Not analyzed	Not analyzed	Chlorine* Floral* Sweet*	Chlorine* Chalky 2 Plastic* Grassy* Drying 2 Slippery (mouthfeel)
2.5	Not analyzed	Not analyzed	Plastic*	Chlorine* Chalky 3 Plastic* Bitter* Drying 3 Tannin (mouthfeel)

*Other note

Table 4.3
Sensory characteristics of 50% blend of CRW and SPW (conventionally treated)

Total Residual Chlorine (mg/L)	Winter		Summer	
	Odor	Flavor	Odor	Flavor
1.7	Sweet* Grassy* Floral*	Chlorine* Chalky 3 Plastic 2 Bitter* Drying 3	Straw* Chalky*	Chlorine* Chalky* Metallic* Bitter* Straw* Musty* Salty* Drying 6
2.5	No odor	Chlorine* Chalky 5 Plastic 1.5 Metallic* Bitter* Drying 3.5	Chalky 2 Plastic* Straw* Musty*	Chlorine* Chalky 2 Bitter* Straw* Plastic* Musty* Drying 6

*Other note

For 100% SPW, although the overall intensity of the descriptors did not change significantly, the number of odors and tastes increased with the increased residuals during winter and summer conditions. Some of these characteristics were perceived by only one panelist. Similar results can be observed for the 100% CRW. However, the plastic odor—not perceived under summer conditions—appeared, while the chlorinous odor, surprisingly, disappeared.

No consensus for odor was reached for the 1.7 mg/L blended sample during winter conditions (Table 4.3). Low levels of sweet, grassy, and floral were observed by individual panelists. No consensus was reached, during summer conditions, for either residual concentration. More descriptors were used by the individual panelists at the higher concentration. Also, the chalky odor was more distinct.

For the flavor analysis, the changes were again very subtle. During winter conditions, the chalky intensity increased with the higher residual, but it was not considered objectionable. A metallic taste appeared as an "other note." Consensus was reached for the drying characteristic for both residual concentrations during summer conditions. At the lower residual concentration, metallic was observed as an "other note," but in the higher residual concentration, metallic was replaced by plastic.

It should be noted that, especially for the summer conditions, characteristics of the individual waters appear in the blend. Additional characteristics (such as straw and musty) that were not perceived in the individual waters also appeared in the blend.

Ozonation/Conventional Treatment/Post-Chloramination

Table 4.4 shows the FPA results of the effects of increased chloramine residual on a 50/50 blend of CRW and SPW that was pre-ozonated and conventionally treated.

There were fewer descriptors used for the ozonated water (Table 4.4) than were used for the water that was conventionally treated (see Table 4.3). No odor was apparent during winter conditions, and no consensus was reached during summer conditions. Only a chalky odor was reported as an "other note."

Consensus was reached during winter conditions for the chalky flavor. Bitter and drying flavors were reported as "other notes." For the summer conditions, no consensus was reached for the lower concentration, but chalky, grassy, and drying flavors were observed as "other notes." For the higher residual, only chalky and drying flavors were observed and appeared to increase at the higher residuals.

Table 4.4
Sensory characteristics of 50% blend of CRW and SPW
(pre-ozonated and conventionally treated)

Total Residual Chlorine (mg/L)	Winter		Summer	
	Odor	Flavor	Odor	Flavor
1.7	No odor	Chalky 2 Drying*	Chalky*	Chalky* Grassy* Drying*
2.5		Chalky 1.5 Bitter* Drying*	Chalky*	Chalky 1.5 Drying 2

*Other note

In summary, there was very little difference in the characteristics of conventionally treated water at the 1.7 and 2.5 mg/L monochloramine residuals. There was also very little difference in odor and flavor for the ozonated water between the two chloramine residuals. In addition, there was very little difference in the results for the simulated summer and winter conditions. Ozonated water appeared to have fewer descriptors and generally lower intensities than conventionally treated water. The chlorine intensities were very similar in SPW, CRW, and the 50/50% blend.

Tastes and Odors Produced by Disinfectants and Disinfection by-Products (DBPs)

Effects of Full Treatment on the Production of Tastes and Odors

A schematic diagram of the treatment process used has been described in Chapter 2 (Figure 2.10). Four waters were studied: Metropolitan Water District of Southern California (MWDSC) SPW, MWDSC CRW, Schuylkill River water (Sch. R.), and a southern California groundwater (SoCal. Gr.). The organoleptic profiles obtained were monitored for various combinations of the following factors:

- Waters (three surface and one groundwater)
- Coagulant (3 and 30 mg/L)
- Ozonation (prior to or after coagulation)
- Disinfection (chlorine and chloramines)
- Detention time (1 and 48 h)

Tables 4.5 through 4.8 show the results obtained and indicate (in the left-hand column) whether ozonation was before or after coagulation.

Table 4.5
Bench-scale experiments for SPW

	Time	Residual Total	Residual Free	FPA Odor
SPW	1 h			Odor-free
	48 h			Hay 2
SPW—O_3	1 h			Caramel 4 Rancid*
	48 h			Caramel 4

		Low Coagulant Dose (3 mg/L)			High Coagulant Dose (30 mg/L)		
		Residual		FPA	Residual		FPA
	Time	Total	Free	Odor	Total	Free	Odor
SPW—O_3—Low coagulant	1 h			Caramel 4 Plastic*			Caramel 2 Rancid*
	48 h			Caramel 2 Hay*			Caramel 2 Rancid*
SPW—O_3—Low coagulant—Chloramines	1 h	1.5	ND	Decaying vegetation 2 Ethanol*	1.6	ND	Cardboard 2 Ethanol*
	48 h	1.0	ND	Caramel 4	1.2	ND	Chlorine 1
SPW—O_3—Low coagulant—Chlorine	1 h	0.5	0.3	Chlorine 4 Plastic*	1.6	ND	Chlorine 4
	48 h	0.5	ND	Caramel 4	1.2	ND	Chlorine 1
SPW—Low coagulant	1 h			Plastic* Straw*			Straw*
	48 h			Cardboard*			Caramel* Plastic* Sour*
SPW—Low coagulant—O_3	1 h			Caramel 2 Sour/vinegar*			Odor-free
	48 h			Caramel 2			Chlorine* Caramel* Rancid 1
SPW—Low coagulant—O_3—Chloramines	1 h	1.6	ND	Decaying vegetation 2 Cardboard 2	1.6	0.1	Decaying vegetation 2
	48 h	1.3	ND	Caramel 4	1.4	ND	Chlorine 1
SPW—Low coagulant—O_3—Chlorine	1 h	0.3	0.2	Chlorine 4	0.8	0.7	Chlorine 2 Decaying vegetation 1
	48 h	0.3	ND	Chlorine 1	0.1	ND	Chlorine 1

ND: Not detected; *Other note

Table 4.6
Bench-scale experiments for CRW

	Time	Chlorine residual Tot.	Chlorine residual Free	FPA Odor
CRW	1 h			Odor-free
	48 h			Odor-free
CRW—O_3	1 h			Decaying vegetation 1
	48 h			(Not analyzed)

		Low Coagulant Dose (3 mg/L)			High Coagulant Dose (30 mg/L)		
		Chlorine residual		FPA	Chlorine residual		FPA
	Time	Tot.	Free	Odor	Tot.	Free	Odor
CRW—O_3—Low coagulant	1 h			Graham cracker 1 Ozone* Citrus*			Caramel 2 Sour*
	48 h			Citrus 1 Graham cracker*			Caramel 2 Rancid*
CRW—O_3—Low coagulant—Chloramines	1 h	1.8	ND	Chlorine 1 Mushroom*	1.8	ND	Odor-free
	48 h	1.4	ND	Rubber 1 Straw* Plastic*	1.4	ND	Chlorine 1
CRW—O_3—Low coagulant—Chlorine	1 h	1.3	1.1	Chlorine 2 Mushroom*	1.7	1.5	Chlorine 1 Mushroom*
	48 h	0.2	ND	Chlorine 2 Cardboard*	0.8	0.7	Chlorine 1
CRW—Low coagulant	1 h			Citrus 2			Straw*
	48 h			Citrus 2 Graham cracker*			Odor-free
CRW—Low coagulant—O_3	1 h			Graham cracker 2 Caramel*			Graham cracker* Mushroom*
	48 h			Citrus 1 Rancid*			Sweet/flowery 2 Rancid 1
CRW—Low coagulant-O_3—Chloramines	1 h	1.8	ND	Graham cracker 2 Caramel*	1.9	ND	Sweet/flowery 2
	48 h	1.4	ND	Caramel 2	1.5	ND	Caramel 2
CRW—Low coagulant—O_3—Chlorine	1 h	1.1	0.9	Chlorine 4	1.7	1.5	Chlorine 2 Rancid*
	48 h	0.1	ND	Chlorine 1 Graham cracker 2	0.8	0.7	Chlorine 1 Caramel 2

ND: Not detected; Other note

Table 4.7
Bench-scale experiments for Schuylkill River water

	Time	Chlorine residual Tot.	Chlorine residual Free	FPA Odor
Sch. R.	1 h			Swampy 2 Musty* Woody*/mint*
	48 h			Not analyzed
Sch. R.—O_3	1 h			Plastic 2 Decaying vegetation* Camphor*
	48 h			Flowery 2 Cardboard*

		Low Coagulant Dose (3 mg/L)			High Coagulant Dose (30 mg/L)		
		Chlorine residual		FPA	Chlorine residual		FPA
	Time	Tot.	Free	Odor	Tot.	Free	Odor
Sch. R.—O_3—Low coagulant	1 h			Plastic 2 Chlorine* Air freshener*			Plastic 2 Decaying vegetation 2 Flowery*
	48 h			Flowery 2 Cardboard*			Cardboard* Rubber*
Sch. R.—O_3—Low coagulant—Chloramines	1 h	1.7	ND	Chlorine 2	1.6	ND	Chlorine 2
	48 h	1.2	ND	Flowery 2	1.3	ND	Flowery 2
Sch. R.—O_3—Low coagulant—Chlorine	1 h	0.9	0.7	Woody* Chlorine* Mint*	1.3	1.5	Chlorine 4
	48 h	0.1	ND	Flowery 2 Cardboard* Chlorine*	0.6	0.5	Flowery 2
Sch. R.—Low coagulant	1 h			Woody* Earthy*			Earthy/musty 2 Mint* Woody*
	48 h			Swampy 2 Fishy 2			Earthy/musty 2 Woody*
Sch. R.—Low coagulant—O_3	1 h			Woody 2 Mint*			Ozone 2 Woody*
	48 h			Flowery* Cardboard* Rubber*			Cardboard* Flowery*
Sch. R.—Low coagulant—O_3—Chloramines	1 h	1.7	ND	Woody 2 Mint*	1.7	ND	Sweet/flowery 2
	48 h	1.3	ND	Medicinal* Chloramines* Flowery*	1.5	ND	Caramel 2
Sch. R.—Low coagulant—O_3—Chlorine	1 h	0.9	0.7	Woody 2	1.2	1.0	Chlorine 2 Decaying vegetation*
	48 h	0.1	ND	Flowery 2 Cardboard* Camphor*	0.6	0.4	Odor-free

ND: Not detected, *Other note

Table 4.8
Bench-scale experiments for a southern California groundwater

	Time	Residual Tot.	Residual Free	FPA Odor
SoCal. Gr.	1 h			Odor-free
	48 h			Odor-free
SoCal. Gr.—O_3	1 h			Odor-free
	48 h			Odor-free

	Time	Low Coagulant Dose (3 mg/L) Chlorine residual Tot.	Low Coagulant Dose (3 mg/L) Chlorine residual Free	Low Coagulant Dose (3 mg/L) FPA Odor	HIgh Coagulant Dose (30 mg/L) Chlorine residual Tot.	HIgh Coagulant Dose (30 mg/L) Chlorine residual Free	HIgh Coagulant Dose (30 mg/L) FPA Odor
SoCal. Gr.—O_3—Low coagulant	1 h			Odor-free			Odor-free
	48 h			Odor-free			Odor-free
SoCal. Gr.—O_3—Low coagulant—Chloramines	1 h	1.5	ND	Ozone* Chlorine*	0.9	0.8	Odor-free
	48 h	1.1	ND	Sweet/flowery*	0.7	0.6	Sweet/flowery*
SoCal. Gr.—O_3—Low coagulant—Chlorine	1 h	0.9	0.7	Chlorine 4 Ozone* Sweet/flowery*	0.9	0.8	Odor-free
	48 h	0.8	ND	Chlorine 2	0.7	0.6	Chlorine 1
SoCal. Gr.—Low coagulant	1 h			Odor-free			Odor-free
	48 h			Rusty 1			Odor-free
SoCal. Gr.—Low coagulant—O_3	1 h			Odor-free			Odor-free
	48 h			Odor-free			Odor-free
SoCal. Gr.—Low coagulant—O_3—Chloramines	1 h	1.5	ND	Odor-free	1.7	ND	Sweet/flowery 2
	48 h	1.6	ND	Sweet/flowery 2	1.6	ND	Sweet/flowery 1
SoCal. Gr.—Low coagulant—O_3—Chlorine	1 h	1.1	1.0	Chlorine 4 Ozone* Sweet*	0.7	0.6	Chlorine 2
	48 h	0.9	0.8	Chlorine 2	0.7	0.6	Chlorine 2

ND: Not detected; *Other note

State Project Water (SPW). This water had moderate levels of total organic carbon (TOC), hardness, and alkalinity. Table 4.5 shows the taste-and-odor characteristics obtained for the various treatment scenarios and controls. A caramel odor that was probably formed during ozonation was the main characteristic of this water. When the water was pre-ozonated, coagulated at a low dosage, and chloraminated, a decaying vegetation odor (apparently the result of the mixture of odors produced by chloramine and caramel odors) was reported after 1 h detention time. Occurrences of a single new odor produced by the mixing of two apparently unrelated odors have been reported in the literature (Khiari et al. 1992). The unaltered caramel odor reappears following partial dissipation of the chloramine residual after 48 h. When the same water was chlorinated, the chlorine at a moderate intensity of 4 seemed to mask the caramel odor at the 1 h detention time. Dissipation of free chlorine after a 48 h detention time allowed the perception of the caramel odor.

When the water was coagulated at a low dosage, post-ozonated, and chloraminated, the same decaying vegetation odor noted earlier appeared. The same behavior of the caramel odor was noted as with the pre-ozonated water. When the same water was chlorinated, the chlorine appeared to destroy the caramel odor, which was not perceived even after the residual had dissipated.

When high doses of coagulant were used, similar trends were observed; however, most of the odors were perceived only as "other notes" (reported by less than 50 % of the panel; in this case, only one panelist). These slight intensities were predictable because of the greater removal of natural organic matter (NOM) resulting from the use of enhanced coagulation.

Colorado River Water (CRW). CRW is a water with moderate TOC and high alkalinity and hardness. A graham cracker odor, apparently related to the caramel odor reported in SPW, was the main characteristic observed (Table 4.6). A recurrent citrus odor was observed in water that was coagulated at a low dose and in pre- or post-ozonated water that was coagulated at a low dose. This odor was not observed when the coagulant was applied at a high dose or when the water was disinfected.

In general, trends similar to the ones observed in SPW, such as the chlorine masking effect, could be seen in CRW.

Schuylkill River Water (Philadelphia). Schuylkill River water is hard with moderate TOC and alkalinity. For this water, the flowery descriptor is the most frequently reported odor (Table 4.7). In addition, a woody characteristic already present in the raw water seems to be enhanced after treatment.

The flowery odor that is usually masked by the presence of disinfectant residuals seems to reappear after long detention times as the residuals dissipate.

Southern California Groundwater. Unlike the three waters described above, which are surface waters, this is a groundwater with moderate TOC and high alkalinity and hardness.

The same descriptions seen in the surface waters were observed (Table 4.8). In this water, the sweet/flowery odors were masked by the disinfectant residuals and reappeared when these residuals decreased.

Summary. In summary, the masking effects of the disinfectant residuals were the single most common finding in this section. In addition, for these particular waters under these specific conditions, longer detention times appeared to improve the organoleptic properties of the water. The odors reported for the four waters described above (flowery, caramel, graham cracker, and citrus) belong to the same category of the flavor wheel (Mallevialle and Suffet 1987) and are usually associated with oxidation processes such as ozonation. Chemical analysis of the samples did not identify any specific chemicals causing the odors. It is not known at this time whether the odors were caused by a single chemical or a combination of chemicals. An additional factor making identification difficult could have been the low odor threshold concentrations (OTCs) of the specific chemicals.

Disinfection of Groundwater

In the previous section, full treatment was performed on all waters, even the groundwater. It was decided to limit this section to a discussion of groundwater disinfection only, because it is common practice for water utilities with groundwater supplies to use minimal treatment (where possible, disinfection alone). For this study, the application of nothing but the disinfectant to the water would make it possible to dissociate the effect of the disinfectant from that of the other treatments.

Three groundwaters were used. These waters were obtained from different locations: from a groundwater system in the northeastern United States (groundwater A) and from wells in southern California (groundwaters B and C).

Groundwater A. Table 4.9 shows the characteristics obtained. After 1 h chlorine detention time, the decaying vegetation descriptor initially present in the raw water decreased, and the swampy odor also present in the raw water totally disappeared. Chlorine has been known to efficiently remove swampy odors that are associated with sulfides and mercaptans. The decaying vegetation descriptor disappeared completely after 48 h detention time, to be replaced by a medicinal descriptor. Closed-loop stripping analysis (CLSA) of the sample, followed by gas chromatographic/mass spectrometric identification, identified the presence of dichloroiodomethane. It is not clear whether this compound was formed later (slow kinetics of formation) or was already present in the sample but was masked by chlorine in the samples with 1 h detention time. Iodinated trihalomethanes (ITHMs) appear to have slow kinetics of

Table 4.9
Bench-scale experiments for groundwater A

		Chlorine residual		FPA
	Time	Tot.	Free	Odor
Groundwater A—Raw	1 h			Decaying vegetation 4 Swampy 2
	48 h			Decaying vegetation 4
Groundwater A—Chlorine	1 h	1.6	1.6	Chlorine 4 Decaying vegetation 2
	48 h	1.2	1.1	Chlorine 2 Medicinal*
Groundwater A—Chloramines	1 h	1.7	ND	Swimming pool 4
	48 h	1.3	ND	Swimming pool 2

ND: Not detected

*Other note

formation and may have been produced later. They have been shown to form during chlorination or chloramination of NOM in the presence of iodide. Although no iodide analysis was performed on the raw water, many groundwaters have been reported to contain substantial concentrations of iodides.

The corresponding chloraminated sample exhibited a swimming pool descriptor that is consistent with chloramines, or more specifically, the dichloramine descriptor. This swimming pool descriptor decreased after 48 h detention time, along with the decrease of the chloramine residual.

Groundwater B. Table 4.10 shows the characteristics obtained. The sulfur-burnt match descriptor found in groundwater B was more resistant to removal by disinfection than the swampy odor perceived in groundwater A. Although no particular compound with a sulfur-burnt match odor could be found in the literature, it is assumed that this descriptor is caused by oxidized sulfur species. The chemical(s) causing this type of odor may be at such an oxidation state that they cannot undergo further oxidation, and thus persist even at longer detention times. Similar trends were observed during both chlorination and chloramination.

Table 4.10

Bench-scale experiments for groundwater B

		Chlorine residual		FPA
	Time	Tot.	Free	Odor
Groundwater B—Raw	1 h			Sulfur-burnt match 4
	48 h			Decaying vegetation 4
Groundwater B—Chlorine	1 h	1.1	1.1	Sulfur-burnt match 2 Chlorine*
	48 h	0.4	0.4	Sulfur-burnt match 2
Groundwater B—Chloramines	1 h	1.4	ND	Sulfur-burnt match 2 Chlorine*
	48 h	0.8	ND	Sulfur-burnt match 2

ND: Not detected

*Other note

Chemical analysis did not allow for a positive identification of the specific compounds. However, colloidal sulfur was found in the CLSA extract of the chlorinated sample at 48 h detention time. No sulfur was found in the other samples. The origin of the sulfur could not be clearly attributed to the sample or to the thermal degradation of the carbon disulfide that is used as a solvent during CLSA. In addition, it was not possible to confirm that the sulfur compounds found in the raw water were the same as the one formed during disinfection. Although the odors were identical, the chemical composition could have been different.

It is reported in the literature (White 1992) that the low chlorine dosages and high pH conditions of water containing hydrogen sulfide favor the formation of elemental sulfur.

Groundwater C. Table 4.11 shows the characteristics obtained. The sulfur descriptor identified in groundwater B was also found in groundwater C. Surprisingly, a medicinal odor that was present in the raw water of groundwater C at short detention times disappeared after 48 h. At this time, no explanation can be given, and the compound causing the medicinal odor cannot be identified.

Table 4.11
Bench-scale experiments for groundwater C

		Chlorine residual		FPA
	Time	Tot.	Free	Odor
Groundwater C—Raw	1 h			Sulfur 2 Medicinal 2 Decaying vegetation*
	48 h			Sulfur 2
Groundwater C—Chlorine	1 h	1.4	ND	Chlorine 2 Medicinal 2 Sulfur*
	48 h	0.8	ND	Sulfur 2 Chlorine 2
Groundwater C—Chloramines	1 h	1.6	ND	Sulfur 2 Medicinal* Rubber*
	48 h	1.4	ND	Sulfur 2

ND: Not detected

*Other note

Considering that the reported OTC of monochloramine is about 0.65 mg/L, the concentration of chloramines at both detention times in the chloraminated samples (1.6 and 1.4 mg/L) would have been expected to produce a bleach odor. It is likely that they were masked by the sulfur odors. The reported OTCs were done in odor-free water with no consideration of a possible matrix effect. As was found with groundwater B, CLSA of the 48 h sample showed the presence of colloidal sulfur. Again, no sulfur was found in the other CLSA extracts.

Summary. In summary, with the exception of the appearance of the medicinal odor in groundwater A, the decay of the disinfectant residual after longer detention times appeared to improve the organoleptic properties of the three groundwaters studied as part of the work on disinfection only. However, it should be remembered that the sensory analysis performed by trained panelists does not always reflect the consumer's acceptance of the water. Trained panelists perform these tests to identify and rate the organoleptic characteristics objectively so as to provide clues to the cause of any problems. More work should be performed to identify the sulfur odors, their mechanisms and kinetics of formation, and possibly their stability in treated waters.

Kinetics of ITHM Formation

The objective of this section was to study the effects of treatment, chlorine-to-nitrogen (Cl_2:N) ratio, disinfectant residual, and detention time on the formation of ITHMs.

The parameters studied were as follows:

Water matrix: CRW and SPW

Treatment: conventional and pre-ozonated/conventional

Disinfectant: chloramines

Cl_2:N ratio; 3:1 and 5:1

Chloramine residual: 0.5 and 4 mg/L

Detention time: 1, 48, and 192 h

In an earlier work (Khiari, Ventura, and Barrett 1999), it was reported that ozonation produces only negligible amounts of ITHMs. Although ozonation experiments were performed in this study, no quantifiable amounts of ITHMs were found, and only the results of conventional treatment will be reported. Because of the low amounts of iodide in CRW, the ITHMs produced in this water were too low to be reliable. Similarly, the amounts of ITHMs in pre-ozonated waters were very low.

Tables 4.12 and 4.13 show the results obtained for conventionally treated SPW. The results presented below are only semi-quantitative and were obtained prior to the full development of the method. However, despite these restrictions, the trends are real and can be used in the scope of these experiments.

Table 4.12
Formation of iodinated trihalomethanes (ITHMs) at Cl_2:N ratio of 3:1

		ITHM Concentration (μg/L)					
Residual (mg/L)	Time (h)	$CHCl_2I$	CHBrClI	$CHBr_2I$	$CHClI_2$	$CHBrI_2$	CHI_3
0.5	1	0.1	0.05	--	0.03	--	--
	48	0.1	0.1	--	0.1	--	--
	192	0.15	0.1	--	0.1	--	--
4	1	0.8	0.5	--	0.3	--	--
	48	1.4	0.8	--	0.6	--	--
	192	1.8	0.7	--	0.8	0.1	0.1

-- Not quantitated

Table 4.13
Formation of ITHMs at Cl_2:N ratio of 5:1

		ITHM Concentration (μg/L)					
Residual (mg/L)	Time (h)	$CHCl_2I$	CHBrClI	$CHBr_2I$	$CHClI_2$	$CHBrI_2$	CHI_3
0.5	1	0.1	0.1	--	0.05	--	--
	48	0.15	0.1	--	0.5	--	--
	192	0.15	0.15	--	0.1	--	--
4	1	0.7	0.5	--	0.1	--	--
	48	1.2		--	0.5	0.1	--
	192	1.3	0.6	--	0.6	0.2	0.1

-- Not quantitated

Figures 4.1 and 4.2 summarize the results obtained for dichloroiodomethane and bromochloroiodomethane under the conditions studied. Chlorodiiodomethane followed a similar trend, and no additional work was deemed necessary.

It can be observed that the Cl_2:N ratio is an important factor in the formation of ITHMs. Higher concentrations of the compounds are generally formed when the 5:1 ratio is used at the low residual. Higher concentrations are generally formed when the 3:1 ratio is used at the high residuals. An increase in dichloroiodomethane and bromochloroiodomethane concentrations was seen with longer detention times. Although the concentrations found in these waters were too low to create taste-and-odor problems because of the initial iodide content (16 μg/L), a clear trend can be observed. The organoleptic properties of all ITHMs are described later in this chapter.

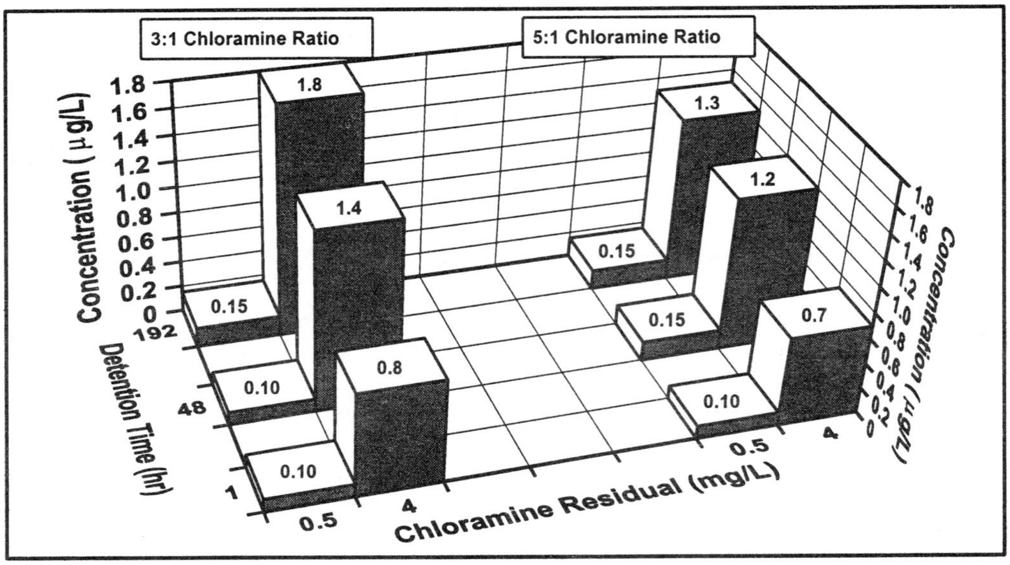

Figure 4.1 Dichloroiodomethane evolution with detention time and Cl_2:N ratio

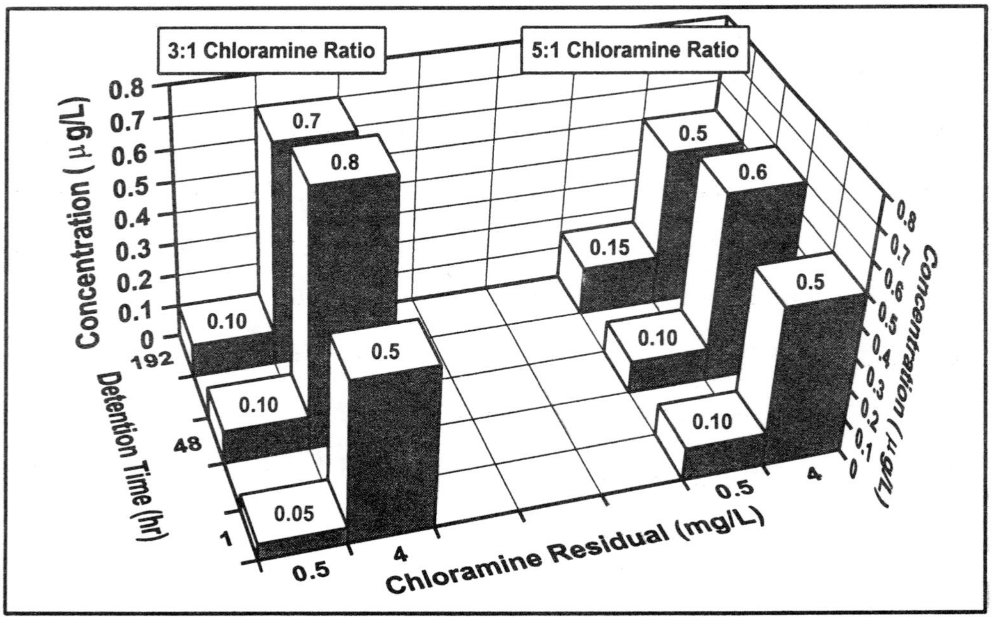

Figure 4.2 Bromochloroiodomethane evolution with detention time and Cl_2:N ratio

CASE STUDIES OF TASTE-AND-ODOR PROBLEMS ASSOCIATED WITH DISINFECTANTS AND DBPS

The following case studies illustrate taste-and-odor problems that occurred in distribution systems.

Case Study I

Several complaints were received by a water utility located in the northeast region of France (Alsace). The complaints were characterized as intense medicinal odors. Two compounds generally associated with medicinal odors were identified in the complaint samples: tribromophenol and bromoform. Both of these compounds are known to form during chlorination of phenol in the presence of bromide.

The system consisted of a variable blend of five different source waters. The blend is stored in a reservoir prior to distribution. In an effort to identify the source and the conditions of the problem, bench-scale chlorination experiments were performed on all five source waters (S1 through S5) feeding the system and the distribution reservoir. The parameters studied were chlorine dose (0.1 and 0.3 mg/L) and detention time (2 and 14 h). The same conditions were used again with chlorine dioxide as the disinfectant.

Table 4.14 and Figure 4.3 show the taste-and-odor characteristics obtained. Source waters S1 and S4 appeared to contribute more than the other waters. Although the percentage of S1 and S4 in the blend is not known, it may be inferred from the high intensity of the medicinal odor in the distribution reservoir that it was a high percentage. It can be seen that longer detention times favored the production of medicinal types of odors in the water. This can be explained not only by the slow kinetics of formation of the particular brominated compounds, but also by the dissipation of the chlorine residual, which may mask any medicinal odor present at shorter detention times.

Another characteristic reported in most of the samples after longer detention times was the chloramine descriptor. In this instance, "chloramine-swimming pool" describes the odor but not the chemical and may be associated with a mixture of chlorine and medicinal odors.

Table 4.14

Effects of disinfectant dose and detention time on taste-and-odor characteristics (chlorine and chlorine dioxide)

	S1		S2		S3		S4		S5		Distribution Reservoir	
Chlorine dose: (mg/L)	0.10	0.30	0.10	0.30	0.10	0.30	0.10	0.30	0.10	0.30	0.10	0.30
Residual at 2 h												
Free	0.00	0.5	0.00	0.15	0.00	0.06	0.00	0.22	0.00	0.15	0.04	0.22
Total	0.06	0.25	0.08	0.20	0.70	0.15	0.07	0.25	0.04	0.18	0.07	0.25
Taste at 2 h	Taste-free	Taste-free	Chlorine	Chlorine	Chlorine	Chlorine	Chlorine	Taste-free	Taste-free	Taste-free	Taste-free	Taste-free
Intensity			++	+	+	++	+					
Taste at 14 h	Medicinal	Chloramine (swim.pool)	Chloramine (swim.pool)	Chloramine	Chlorine	Chloramine (swim.pool)	Medicinal	Medicinal	Taste-free	Chloramine (Swim.pool)	Medicinal	Medicinal
Intensity at14 h	++++	+++	+++	+++	++	++++	++++	++++		++++	++	++
ClO_2 dose (mg/L)	0.10	0.30	0.10	0.30	0.10	0.30	0.10	0.30	0.10	0.30	0.10	0.30
Residual at 2 h: (mg/L)	0.04	0.17	0.04	0.15	0.04	0.17	0.02	0.15	0.00	0.15	0.08	0.18
Taste at 2 h	Taste-free	ClO_2^{+}	Taste-free	ClO_2^{+}	Taste-free	ClO_2^{+}	Taste-free	ClO_2^{+}	Taste-free	ClO_2^{+}	Taste-free	ClO_2^{++}
Taste at 24 h	Taste-free	ClO_2^{+}	Taste-free	ClO_2^{+}	Taste-free	ClO_2^{+}	Taste-free	ClO_2^{+}	Taste-free	ClO_2^{+}	Taste-free	ClO_2^{++}

++++ Strong

+++ Moderate to strong

++ Weak to moderate

\+ Weak

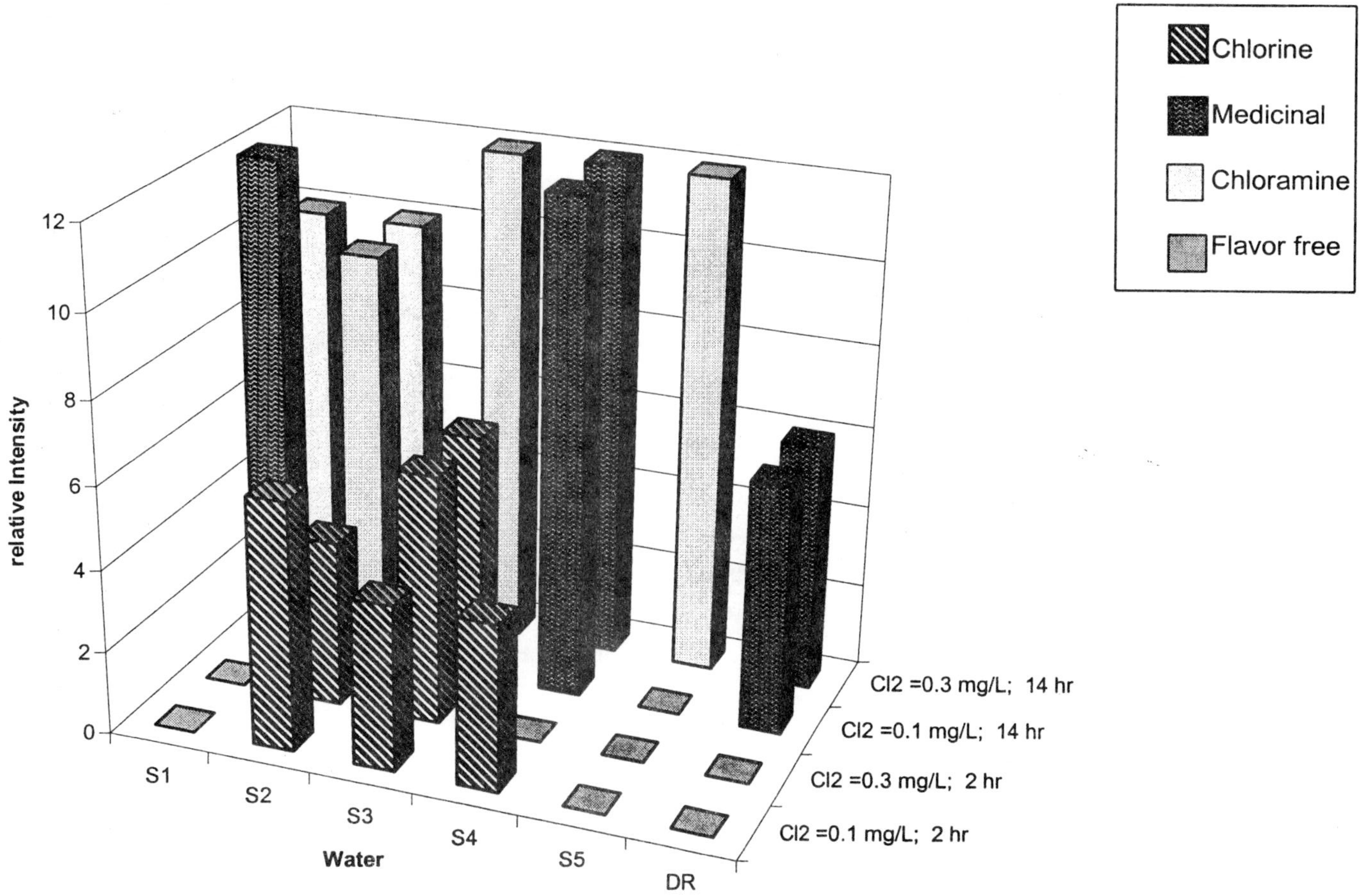

Figure 4.3 Effects of disinfectant dose and detention time on taste-and-odor characteristic

Bench-scale treatment with chlorine dioxide prepared from hydrochloric acid was performed on the same sources used for chlorination. The taste-and-odor characteristics were monitored. Table 4.14 shows that no off flavor was produced except for a slight chlorine dioxide flavor at higher dosages. Chlorine dioxide was proven to be the best alternative to avoid taste-and-odor production for this water.

Case Study II

Consumers in the southeastern region of France complained regularly of "hard to describe" tastes and odors that persisted in coffee or tea. On site, it could be easily realized that the chlorophenolic taste was more or less intense, depending on the pH in the distributed water. As in the previous case, chlorine dioxide prepared from hydrochloric acid was the best alternative to prevent the formation of objectionable taste-and-odor compounds.

TASTE-AND-ODOR CHARACTERISTICS OF AMMONIA

The sensory characteristics of disinfectants such as chlorine and chloramine species have been investigated and reported numerous times in the literature. However, the taste-and-odor characteristics of ammonia have been overlooked. When chloramines are used for the disinfectant residual in a distribution system, a free ammonia residual is often carried. For one reason, the feed of ammonia to form chloramines may not be finely tuned. For another reason, the production of dichloramines is kept to a minimum. Free ammonia residuals can approach 0.5 mg/L and have been associated with biological nitrification. Suffet et al. (1995) reported an OTC of 0.037 mg/L for ammonia. In this study, the OTC (at 25°C) was estimated at 0.9 mg/L. The descriptors for odor were ammonia, rubbery, plastic, cardboard, and reedy, while the taste descriptors were bitter and tingling sensation.

Figure 4.4 shows the dose-response plot based on the Weber-Fechner model for ammonia.

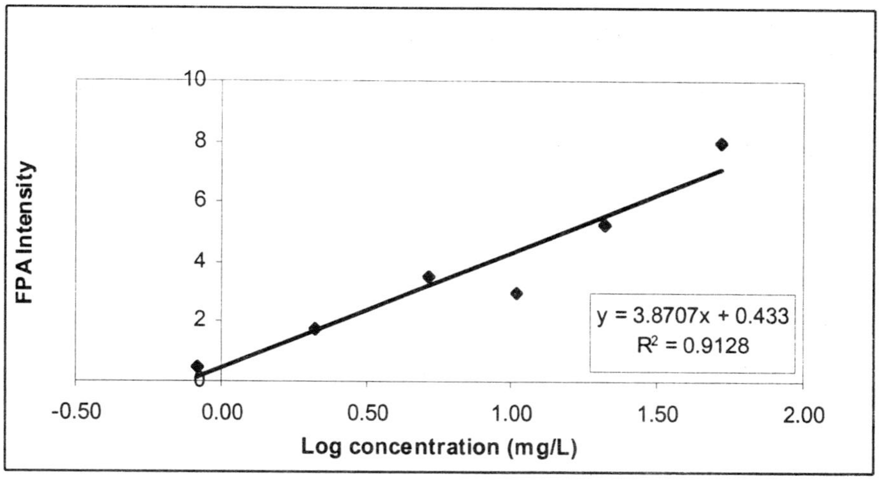

Figure 4.4 Dose-response plot for ammonia

TASTE-AND-ODOR CHARACTERISTICS OF DBPS

The taste-and-odor characteristics of numerous DBPs have been investigated. Table 4.15 summarizes the OTCs (at 45°C) and descriptors obtained. More detailed descriptions and dose-response plots can be found in Appendix A. Different classes of DBPs have been investigated:

- Aldehydes
- Nitriles
- Chloroketones
- Halonitriles
- Iodinated halomethanes
- Trihalomethanes
- Nitromethanes

Aldehydes

Although aldehydes are commonly known as ozonation DBPs, they have also been reported to form from chlorination of amino acids. Aldehydes can also have biological origins;

this study shows that they can be leached from materials used in distribution systems. Low-molecular-weight aldehydes generally show rancid/unpleasant odors, whereas aldehydes of higher molecular weight have more pleasant/fruity odors. Their OTCs range from 0.03 μg/L for undecanal to 47 μg/L for benzaldehyde.

Chloroketones

The OTCs determined for chloroketones were rather high (80 to 500 μg/L) and should not represent a major taste-and-odor problem.

Nitriles

As reported in Chapter 3, numerous nitriles have been identified in chloraminated drinking water. Nitriles can be formed upon further chlorination of aldehydes. Except for propionitrile and valeronitrile, nitriles appear to have some fruity-sweet charcteristics. Their OTCs range from 0.13 μg/L for octanenitrile to 64 μg/L for isobutyronitrile.

Halonitriles

Halonitriles are known DBPs of chlorination and chloramination. As is the case with chloroketones, halonitriles exhibit rather high OTCs (50 to 400 μg/L) and should not be of concern in terms of sensory properties. Surprisingly, no chlorinous odors were perceived for the haloacetonitriles. Instead, all of them had a fruity, piney type of odor, and no chlorinous characteristic could be found.

Iodinated Halomethanes

Both di- and tri-iodinated halomethanes are known to be produced by chlorination and chloramination of drinking water in the presence of iodide. For ITHMs, the medicinal characteristics appeared to increase with the number of iodine atoms in the compound. Iodoform was the most medicinal-smelling of them all. Their OTCs were at low μg/L levels. Iodoform exhibited the lowest OTC, at only 0.032 μg/L.

Iodinated dihalomethanes did not present any medicinal characteristics. Their OTCs were higher than those of the ITHMs.

Trihalomethanes

Prior to this study, it was believed that the unidentified chlorinous odors perceived in some distribution networks in Europe, when no (or very low) chlorine residuals were present, could be caused by halogenated DBPs. A systematic investigation of the sensory characteristics of DBPs was undertaken, and the results are shown in Appendix A. In light of the results, it appears that the unidentified chlorinous odors perceived in some distribution networks in Europe (with no or very low chlorine residuals) are not caused by halogenated DBPs. The OTCs of some regulated trihalomethanes were found to be high (for example, chloroform has an OTC of 90 μg/L). Dibromochloromethane and bromodichloromethane have OTCs of 50 and 40 μg/L, respectively. Medicinal characteristics were identified with both chemicals.

Table 4.15
Sensory characteristics of DBPs

COMPOUNDS	Exp. OTC (μg/L)	DESCRIPTORS
		ALCOHOLS
n-Butanol		Cheese (Roquefort), Putrid, Sweet.
		ALDEHYDES
Propanal	12	Below 200 μg/L: Lemon, Citrus, Rancid, Milky. Above 200 μg/L: Solvent, Gum, Painting, Glue, Fruity.
Butanal	2	Sweet, Marking pen, Solvent, Fruity, Dehydrated milk.
Pentanal	5	Below 32 μg/L: Sweet, Green. Above 32 μg/L: Fruity, Sour, Rancid, Sweet.
Hexanal	0.32	Below 80 μg/L: Green, Grassy, Earthy. Above 80 μg/L: Green apple, Fruity.
Heptanal	0.25	Rancid oil, Fishy, Fruity /Citrus.
Octanal	0.32	Rancid, Fruity, Citrus, Rancid oil, Sweet.
Nonanal	0.32	Rancid, Fruity.
Decanal	0.08	Citrus (lemon), Rancid.
Undecanal	0.03	Insecticide, Citrus (lemon), Irritating, Penetrant.
Dodecanal	0.13	Lemon-flavored tablet, Insecticide.
Benzaldehyde	46.6	Sweet almonds, Rancid, Strawberry candy.
3-Methyl propanal	0.32	Below 32 μg/L: Decaying vegetation, Marshy, Fruity.
2-Methyl butanal	0.8	Above 32 μg/L: Valerian, Bitter. Sweet, Sweet almonds, Rancid.
Phenylacetaldehyde	0.3	Below 29.6 μg/L: Honey, Sweet, Pollen. Above 29.6 μg/L: Flowery, Honey, Green, Sweet.
		CHLOROKETONES
1,1-Dichloropropanone	80	Below 500 μg/L: Woody, Fodder, Boiled vegetables. Above 500 μg/L: Detergent, Disinfectant.
1,1,1-Trichloropropanone	500	Below 3200 μg/L: Pleasant, Sweet, Fresh, Citrus. Above 3200 μg/L: Solvent, Insecticide, Rancid, Rotten.
1,1,3-Trichloropropanone	240	Lemon, Orange, Apple, Soapy.
		NITRILES
Propionitrile	12	Decaying vegetation, Unpleasant. Above 3200 μg/L: Paint thinner, Decaying vegetation.
Butyronitrile	32	Below 1200 μg/L: Fruity, Sweet, Citrus. Above 1200 μg/L: Sweet organic solvent.
Isobutyronitrile	64	Citrus, Sweet.
		Above 1000 μg/L: Rubbery, Hydrocarbon, Paint, Turpentine, Acetone.
Hexanenitrile	3.2	Geranium, Plastic, Bitter.
Octanenitrile	0.13	Geranium, Plastic, Pungent, Rancid, Sweet, Penetrant. Above 30 mg/L: Geranium, Plastic, Coconut, Earthy.

(continued)

Table 4.15 (cont'd)
Sensory characteristics of DBPs

NITRILES		
Dodecanenitrile	0.32	Citrus, Tangerine, Sweet.
Valeronitrile	13	At low concentrations: Putrid, Unpleasant, Decaying vegetation, Rotten. At higher concentrations: Sweet, Fruity. Above 1500 μg/L: Sweet almonds, Green, Sweet.
Isovaleronitrile	3.2	Oniony, Solvent. Above 200 μg/L: Solvent, Gum, Painting, Glue, Fruity.
HALONITRILES		
Chloroacetonitrile	350	Fruity.
Dichloroacetonitrile	50	Resin, Pine, Perfume.
Trichloroacetonitrile	300	Pine oil, Fruity.
Bromochloroacetonitrile	150	Fruity.
Dibromoacetonitrile	250	Fruity, Sweet.
Chloroacetonitrile	400	Sweet, Rancid, Solvent, Toasty, Ginseng, Valerian.
IODINATED HALOMETHANES		
Dichloroiodomethane	5.76	Below 790 μg/L: Sweet, Syrup. Above 790 μg/L: Solvent, Glue, Paint, Varnish.
Dibromoiodomethane	2.9	Bitumen, Solvent, Sweet, Fruity, Drying, Rancid. Above 1000 μg/L: Different but no descriptor could be found.
Bromochloroiodomethane	5.12	Below 220 μg/L: Fresh grass, Sweet, Alcohol. Above 220 μg/L: Solvent, Acetone, Alcohol.
Chloroiodomethane	13	Solvent, Grassy, Alcohol.
Chlorodiiodomethane	9.42	Sweet, Rancid, Medicinal, Solvent.
Bromodiiodomethane	0.13	Sweet, Medicinal product, Solvent
Diiodomethane	30	Alcohol, Solvent, Rubbery.
Iodoform	0.032	Medicinal (dentist's office, aspirin), Gum, Sweet.
TRIHALOMETHANES		
Dibromochloromethane	50	Medicinal, Solvent, Alcohol.
Bromodichloromethane	40	Chemical, Medicinal, Flowery.
NITROMETHANES		
Chloropicrin	100	Fruity, Camphor, Grassy.

CHAPTER 5
BLENDING

In this chapter, two types of blending experiments are presented. The first part deals with blending two chlorinated waters of different origins and different water quality. The second part deals with blending of a chlorinated groundwater with chloraminated surface water.

EFFECTS OF BLENDING DIFFERENT TREATED WATERS ON THE ORGANOLEPTIC PROFILE OF WATER

The purpose of these experiments was to determine the organoleptic profiles obtained when blending two waters with different characteristics at various ratios, and to evaluate the contribution of specific descriptors from the individual waters to the final blend. The final objective of these experiments was to determine the optimum blend that contains the highest percentage of Llobregat water while maintaining acceptable organoleptic properties. Llobregat water presents sensory characteristics such as decaying vegetation that are rather objectionable, while Ter water is considered more acceptable. This work lays out (1) the benefits of performing sensory studies when blending is routinely performed and (2) a procedure for making such studies.

As mentioned earlier in the experimental section, although the blends and blending experiments were performed at bench scale in the laboratory, the two waters (Ter and Llobregat) were actual treated samples obtained from the Barcelona (Spain) distribution system.

Comparison of Individual Waters with Blends

Comparison of Ter and Llobregat Waters with Corresponding 70/30 Blends

A preliminary experiment was performed to establish the composition of the blending and to test the general procedure. The tests were performed at bench scale during the month of October 1997. Both the organoleptic and the chromatographic profiles were done in duplicate,

48 h after sampling and blending the samples. The following four blends (A, B, C, and D) were studied:

A. 100% Ter water

B. 70% Ter water and 30% Llobregat water

C. 30% Ter water and 70% Llobregat water

D. 100% Llobregat water

The organoleptic profiles were evaluated at the time of blending (t = 0) and after 48 h detention time.

Results. Figures 5.1 through 5.6 show the bar diagrams summarizing the flavor-profile analysis (FPA) results for the individual samples and blends. All results reported by the panelists were evaluated statistically using the analysis of variance (ANOVA) and separation of averages by the Student-Newman-Keuls (SNK) method.

Discussion. The basic organoleptic profiles of the two individual waters were well known already, as every week a control was analyzed by the FPA method. The characteristic descriptions of the Ter distribution water were fresh grass odor and flavor and metallic taste. As for the Llobregat water, the characteristic descriptors were decaying vegetation odor and flavor and salty taste. Statistical treatment of the data (Tables 5.1 and 5.2) shows that there was no significant difference between blends C (30% Ter and 70% Llobregat) and D (100% Llobregat). It can also be noted that blend B (70% Ter and 30% Llobregat) presented characteristics of the two individual waters.

Table 5.1 shows the FPA results for the four blends at 0 and 48 h. Table 5.2 shows the change in intensity from time t = 0 to 48 h detention time for each descriptor. Only "descriptors" (tastes or odors reported by 50% or more of the panel) are reported.

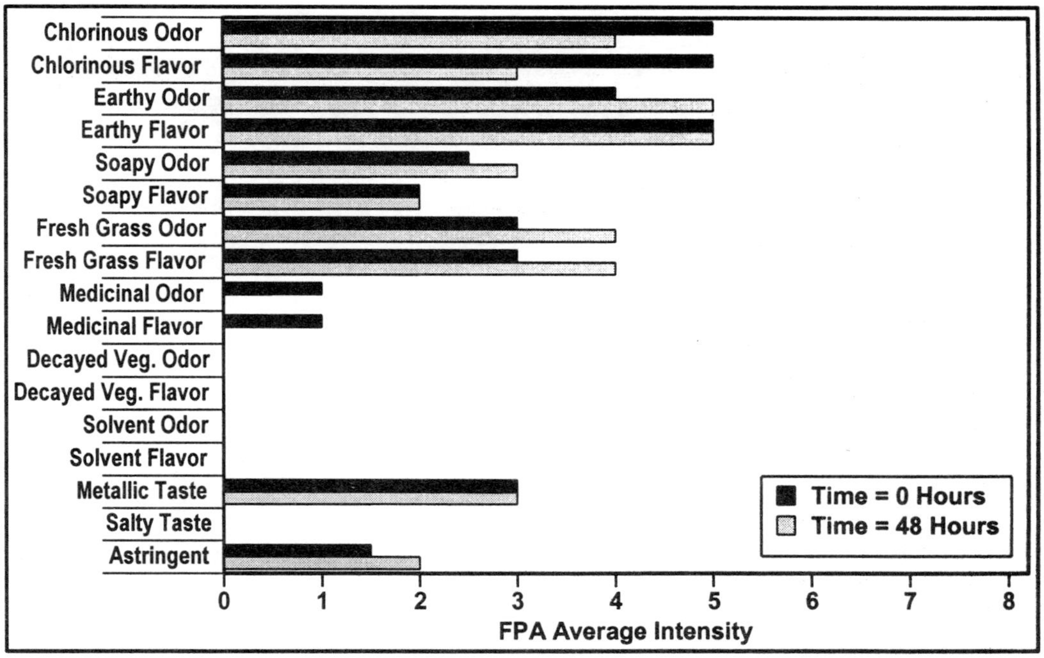

Figure 5.1 FPA for 100% Ter water

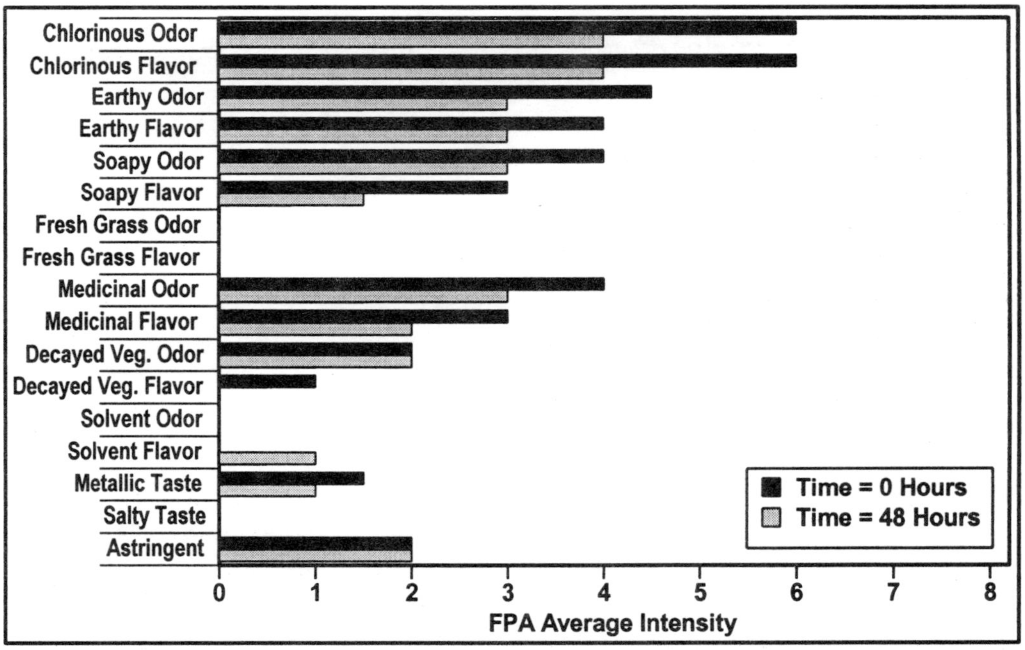

Figure 5.2 FPA for the 70% Ter water + 30% Llobregat blend

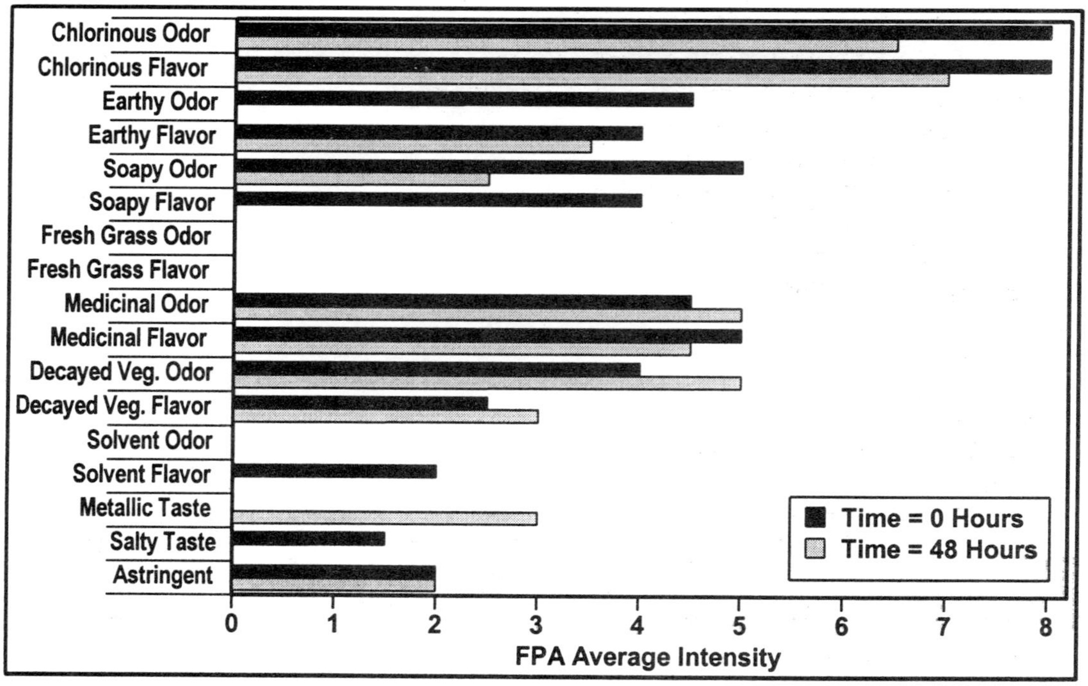

Figure 5.3 FPA for the 30% Ter water + 70% Llobregat blend

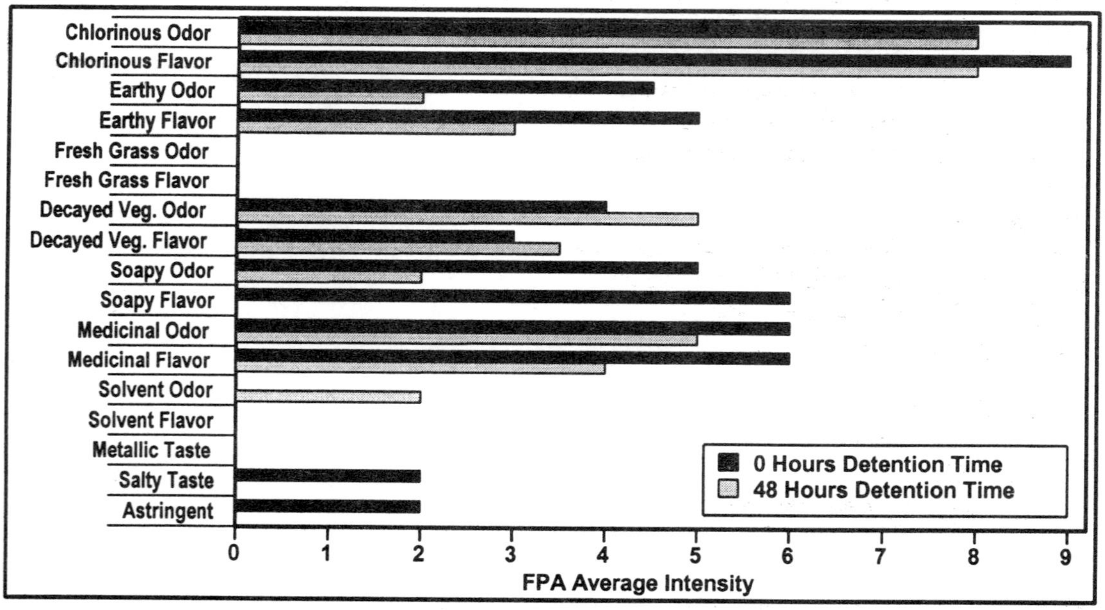

Figure 5.4 FPA for 100% Llobregat water

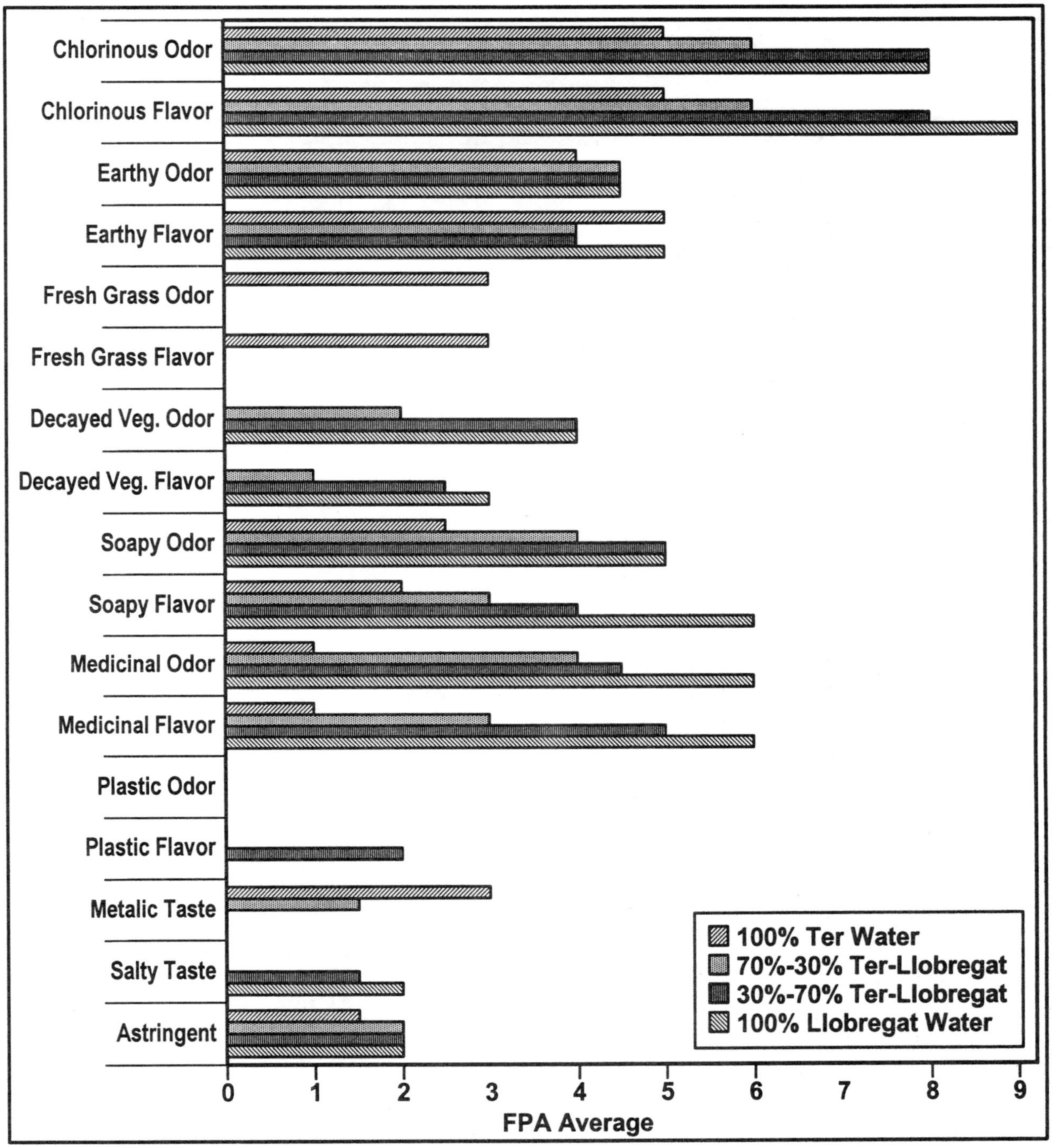

Figure 5.5 Sensory characteristics for blends at 0 h detention time

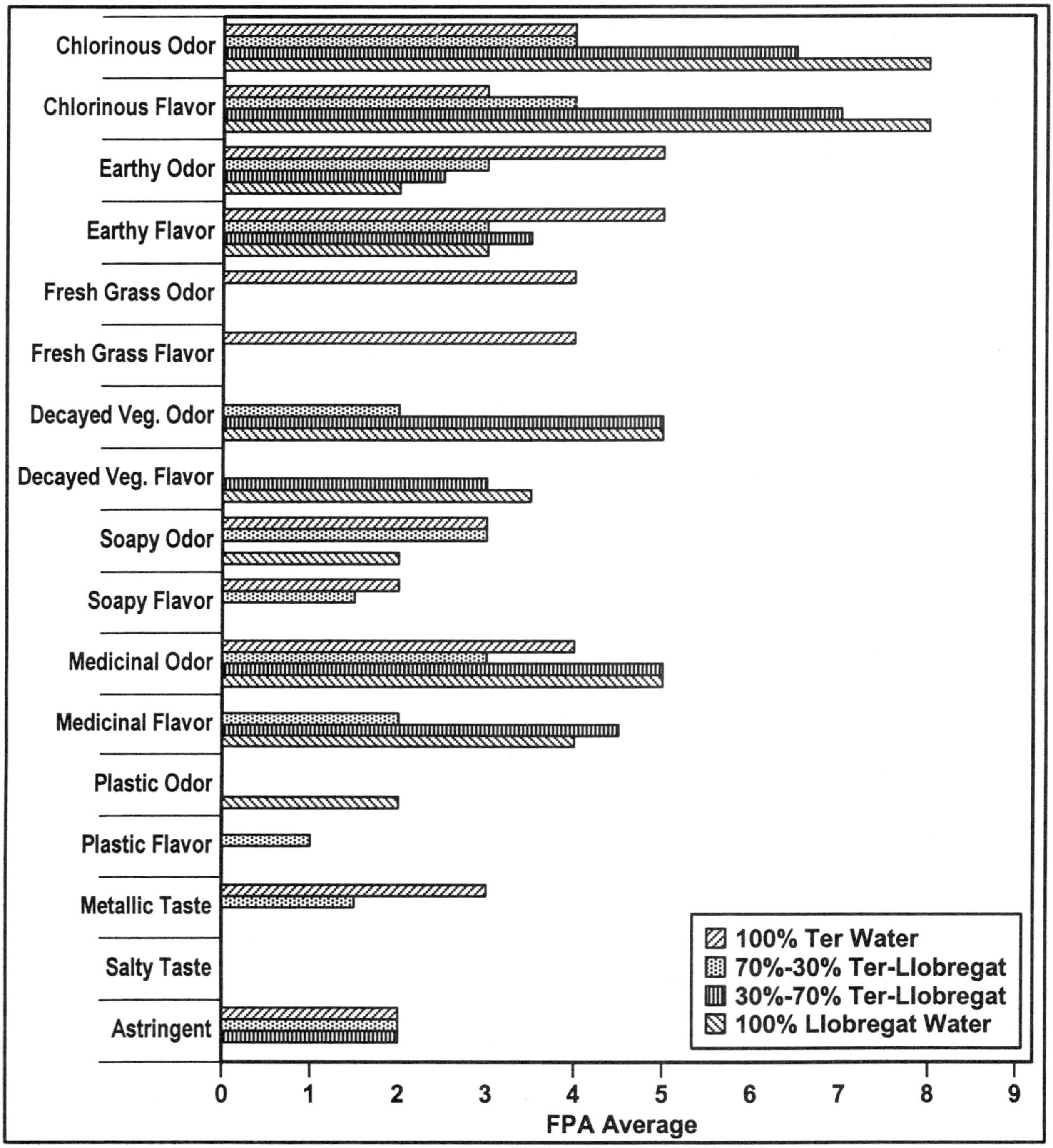

Figure 5.6 Sensory characteristics for blends at 48 h detention time

Table 5.1
Flavor-profile analysis (FPA) for blending experiments

Detention time: 0 h

	Descriptor	A	B	C	D
ODOR	Chlorinous	5	6	8	8
	Earthy	4	4.5	4.5	4.5
	Soapy	2.5	4	5	5
	Fresh grass	3			
	Medicinal	1	4	4.5	6
	Decaying vegetation		2	4	4
	Solvent				
FLAVOR	Chlorinous	5	6	8	9
	Earthy	5	4	4	5
	Soapy	2	3	4	6
	Fresh grass	3			
	Medicinal	1	3	5	6
	Decaying vegetation		1	2.5	3
	Solvent			2	
TASTE	Metallic	3	1.5		
	Salty			1.5	2
	Astringent mouthfeel	1.5	2	2	2
	Odor SNK factor	6	4	2.8	2.4
	Flavor SNK factor	6	4.6	2.8	2

Detention time: 48 h

	Descriptor	A	B	C	D
ODOR	Chlorinous	4	4	6.5	8
	Earthy	5	3	2.5	2
	Soapy	3	3		2
	Fresh grass	4			
	Medicinal	4	3	5	5
	Decaying vegetation		2	5	5
	Solvent				2
FLAVOR	Chlorinous	3	4	7	8
	Earthy	5	3	3.5	3
	Soapy	2	1.5		
	Fresh grass	4			
	Medicinal		2	4.5	4
	Decaying vegetation			3	3.5
	Solvent		1		
TASTE	Metallic	3	1.5		
	Salty				
	Astringent mouthfeel	2	2	2	
	Odor SNK factor	5.8	4.6	3.3	2.2
	Flavor SNK factor	6.7	5.8	2.8	2.1

Table 5.2
Change of intensity by descriptors for the four samples

SAMPLE A	Time (h)			SAMPLE B	Time (h)	
Descriptor	0	48		Descriptor	0	48
Chlorinous	5	4	ODOR	Chlorinous	6	4
Earthy	4	5		Earthy	4.5	3
Soapy	2.5	3		Soapy	4	3
Fresh grass	3	4		Fresh grass		
Medicinal	1			Medicinal	4	3
Decaying vegetation				Decaying vegetation	2	2
Solvent				Solvent		
Chlorinous	5	3	FLAVOR	Chlorinous	6	4
Earthy	5	5		Earthy	4	3
Soapy	2	2		Soapy	3	1.5
Fresh grass	3	4		Fresh grass		
Medicinal	1			Medicinal	3	2
Decaying vegetation				Decaying vegetation	1	
Solvent				Solvent		1
Metallic	3	3	TASTE	Metallic	1.5	1
Salty				Salty		
Astringentl	1.5	2		Astringent	2	2
Odor SNK factor	6	5.8		Odor SNK factor	4	4.6
Flavor SNK factor	6	6.7		Flavor SNK factor	4.6	5.8

SAMPLE C	Time (h)			SAMPLE D	Time (h)	
Descriptor	0	48		Descriptor	0	48
Chlorinous	8	6.5	ODOR	Chlorinous	8	8
Earthy	4.5			Earthy	4.5	2
Soapy	5	2.5		Soapy	5	2
Fresh grass				Fresh grass		
Medicinal	4.5	5		Medicinal	6	5
Decaying vegetation	4	5		Decaying vegetation	4	5
Solvent				Solvent		2
Chlorinous	8	7	FLAVOR	Chlorinous	9	8
Earthy	4	3.5		Earthy	5	3
Soapy	4			Soapy	6	
Fresh grass				Fresh grass		
Medicinal	5	4.5		Medicinal	6	4
Decaying vegetation	2.5	3		Decaying vegetation	3	3.5
Solvent	2			Solvent		
Metallic		3	TASTE	Metallic		
Salty	1.5			Salty	2	
Astringent	2	2		Astringent	2	
Odor SNK factor	2.8	3.3		Odor SNK factor	2.4	2.2
Flavor SNK factor	2.8	2.8		Flavor SNK factor	2	2.1

Figures 5.5 and 5.6 show the differences for FPA results at 0 and 48 h detention time. The medicinal odor reported in 100% Ter water increased after 48 h contact time but decreased for 70% Ter water. For the soapy descriptor, the relative intensity increased as the percentage of the Ter water decreased at zero contact time. However, at 48 h contact time, the intensity decreased as the percentage of Ter water decreased. The solvent descriptor which was not perceived in 100% Llobregat water at zero contact time appeared after 48 h contact time. These results demonstrated that changes in taste-and-odor quality were occurring in both blended and nonblended water in the distribution system.

Table 5.3 shows the percent contribution of the distinctive descriptors (fresh grass odor and flavor and metallic taste) of Ter water to the two blends. It is interesting to note that none of the contributions of the specific descriptors equaled 70% in blend B when 70% of the water was from the Ter.

Table 5.4 shows the contribution of the Llobregat descriptors to the blends. Descriptors such as "medicinal" that appeared also in Ter water were not taken into consideration.

Summary. The FPA made it possible to study the effects on the organoleptic properties of blending waters from the two main origins of the Barcelona water supply and to monitor changes resulting from a 48-h detention time. By comparing the FPA of the blended waters and that of the original samples, it can be seen that at these particular blend ratios, the water from the Llobregat River had a higher influence than what could be expected from its proportion of the mixture.

Table 5.3
Contributions of distinctive descriptors of Ter distribution water to various blends

	Time (h)	A (100% T)	B (70%T:30%Llo)	C (30%T:70%Llo)	D (100% Llo)
Fresh grass Odor	0	100	23	0	0
	48	100	15	0	0
Fresh grass Flavor	0	100	0	0	0
	48	100	22	0	0
Metallic Taste	0	100	45	0	0
	48	100	58	19	0

Table 5.4
Contributions of distinctive descriptors of Llobregat distribution water to various blends

	Time (h)	A (100% T)	B (70%T:30%Llo)	C (30%T:70%Llo)	D (100% Llo)
Dec. veg. Odor	0	0	46	103	100
	48	0	46	115	100
Fresh grass Flavor	0	0	46	90	100
	48	0	0	94	100

Comparison of Ter and Llobregat Waters with Corresponding 10/90 and 20/80 Blends

The second experiment for the determination of organoleptic profiles of blends used the same waters but at different percentages (10/90 and 20/80) and a longer detention time. In order to take into account the effect of chlorine decay, a detention time of seven days was used. The samples prepared were as follows:

A. 100% Ter water

B. 90% Ter water and 10% Llobregat water

C. 80% Ter water and 20% Llobregat water

D. 100% Llobregat water

Initial FPA testing was performed during the month of January, and the FPA was repeated 7 days after sampling and blending of samples. Table 5.5 shows the free and total chlorine residuals and the conductivity of the samples.

Results. Table 5.6 and Figures 5.7 through 5.12 show the organoleptic profiles obtained. As mentioned in the previous section, the distinctive descriptors of the Ter distribution water were fresh grass odor and flavor and metallic taste. For the Llobregat water, the distinctive descriptors were medicinal odor and flavor, decaying vegetation odor and flavor, and salty taste.

Table 5.5
Chlorine residuals and conductivity

	Conductivity (μS/cm)	Free Chlorine (mg Cl_2/L)		Total Chlorine (mg Cl_2/L)	
		Day 0	Day 7	Day 0	Day 7
A (100% T)	491	0.71	0	0.97	0.21
B (90%T:10%Llo)	652	0.54	0	0.88	0.12
C (80%T:20%Llo)	810	0.51	0	0.81	0.1
D (100% Llo)	2030	0.78	0.42	0.99	0.54

Discussion. It can be observed that the blends with 10% and 20% Llobregat water show a contribution from the Llobregat water. For example, this can be seen in the appearance of medicinal odor, decaying vegetation odor, and medicinal flavor in these blends.

Similarly, the characteristics of Ter water could be found also in the blend with 90% Ter water. However, in the water with 80% Ter water, only the metallic taste could be found.

The organoleptic profiles did not change significantly after a week's detention time. It is interesting to note that although the free and total chlorine residuals dropped to zero, or nearly zero, after a week, the chlorinous odor and flavor remained as strong as on the first day (intensity 6).

From Table 5.5, the difference in measured total and free chlorine can be calculated. For day 0, the difference range was 0.3 for sample A and 0.2 for sample D. For day 7, the differences ranged from 0.2 for sample A to 0.1 for sample D. The compounds causing these differences between free and total chlorine may be organic chloramines or other compounds. Because of the FPA results, the compounds causing the false free-chlorine residual measurement of 0.42 for sample D at day 7 were probably causing the chlorinous odor. The free and total chlorine residuals for the blends were less than for the individual waters, possibly because of volatilization during the laboratory preparation of the blends.

In March of the same year, the experiment was repeated. However, the results were not as straightforward as the ones presented above because nontypical descriptors for the Ter water, such as medicinal and decaying vegetation odors, were found in the Ter distribution water. The typical descriptors of the Ter water, fresh grass odor and flavor, were observed again in the subsequent samples that had contributions of 80% and of 90% Ter water.

Summary. Bench-scale results for blended waters, with 10% and 20% Llobregat water and 90% and 80% Ter water, maintained distinctive organoleptic profiles of the Llobregat water. The flavor and odor intensities decreased or disappeared after blending and over time.

Table 5.6
FPA results for blended waters

	Day 0				Day 7			
	Ter	10% Llo	20% Llo	Llo	Ter	10% Llo	20% Llo	Llo
ODOR								
Chlorinous	5	6	6	6	4.5	4	5	6
Earthy	3	4	3	2.5	4	4.5	4	2
Fresh grass	2	0	0	0	2.5	1	0	0
Decaying vegetation	0	1	1.5	3	0	0	0	4.5
Soapy	2.5	3	3	1.5	1	2.5	2.5	2
Medicinal	0	3	3.5	3	0	1	4	3
Plastic	0	0	0	1.5	0	0	0	0
FLAVOR								
Chlorinous	4.5	5	5	7	5	4	5.5	7
Earthy	4	3	3	2	4	3.5	4	0
Fresh grass	2	2	0	0	2	0	0	0
Decaying vegetation	0	0	0	2	0	0	0	4
Soapy	1.5	3.5	3	3	2	1	2	2
Medicinal	0	3.5	4	6	0	1	3	3
Plastic	0	0	0	0	0	1	1	3
TASTE								
Metallic taste	2	0	1	0	1.5	1	1	0
Salty taste	0	0	0	1	0	0	0	0
Astringent mouthfeel	2.5	1	3	0	1	1	0	2

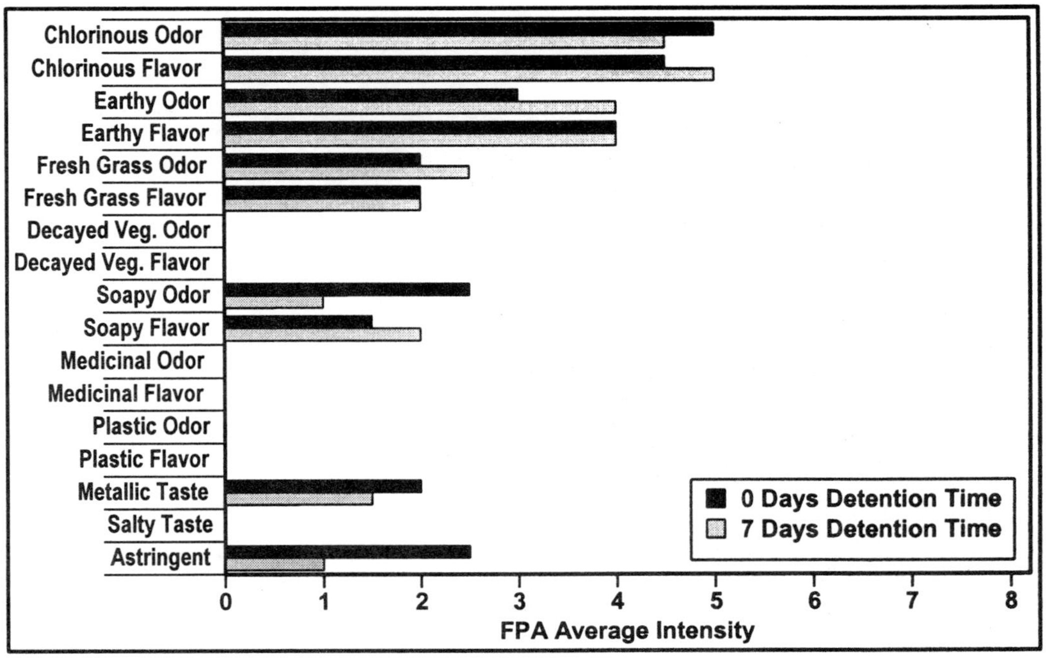

Figure 5.7 FPA for 100% Ter water at 0 and 7-day detention times

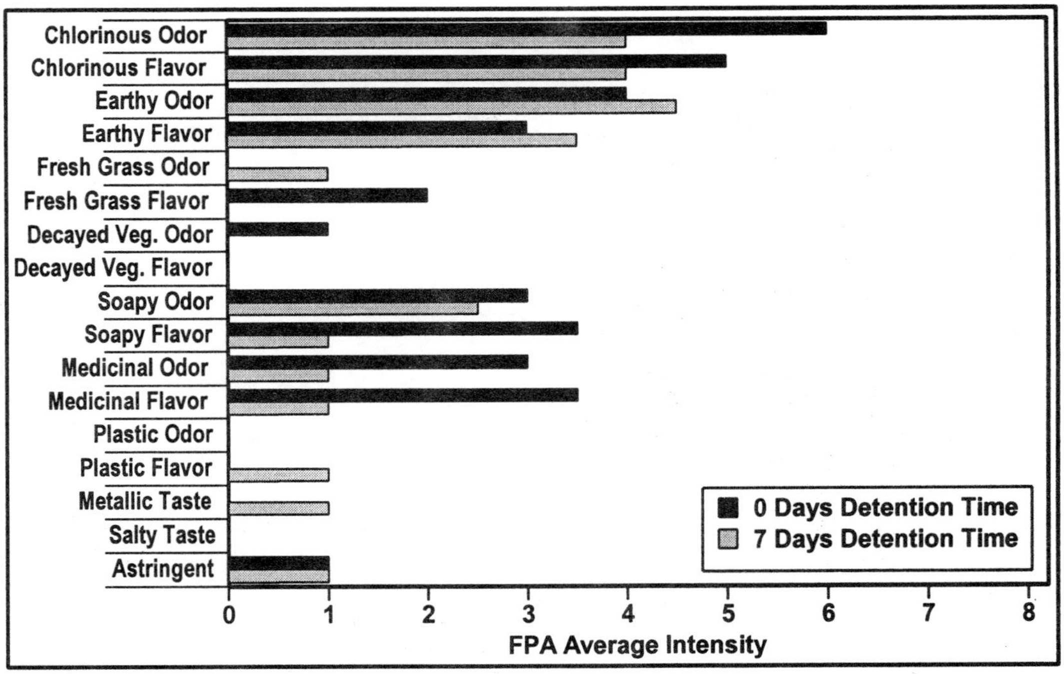

Figure 5.8 FPA for 90% Ter + 10% Llobregat blend

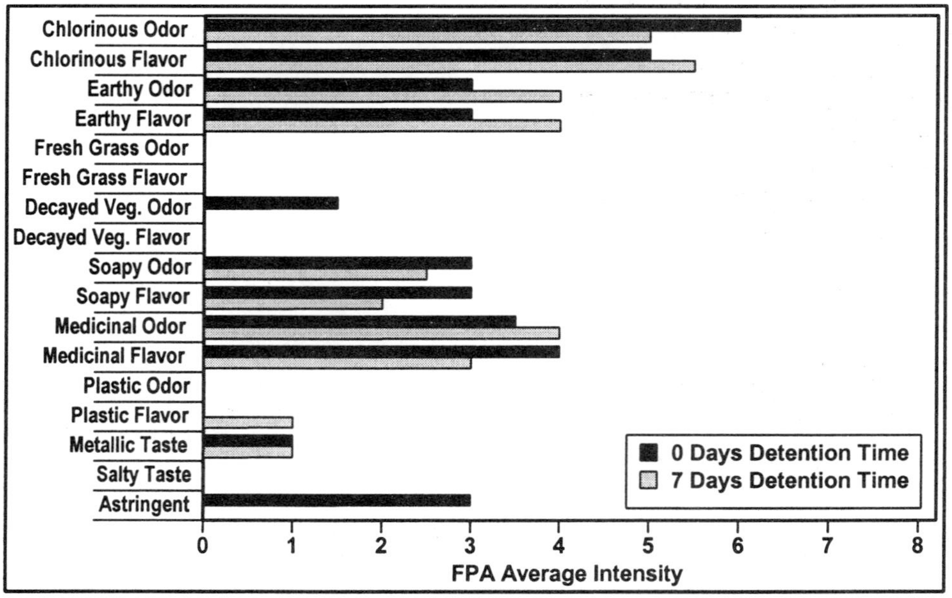

Figure 5.9 FPA for 80% Ter + 20% Llobregat blend

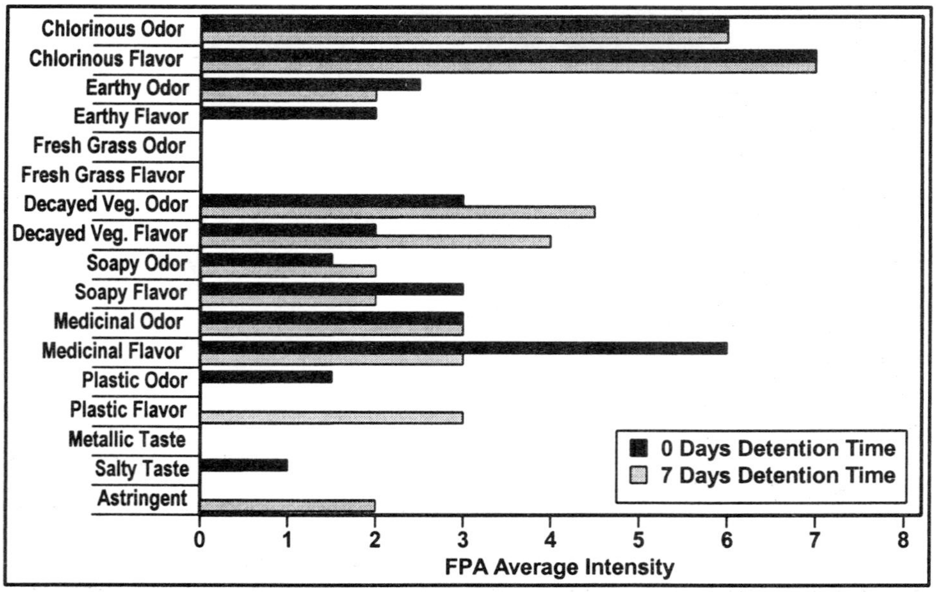

Figure 5.10 FPA for 100% Llobregat water

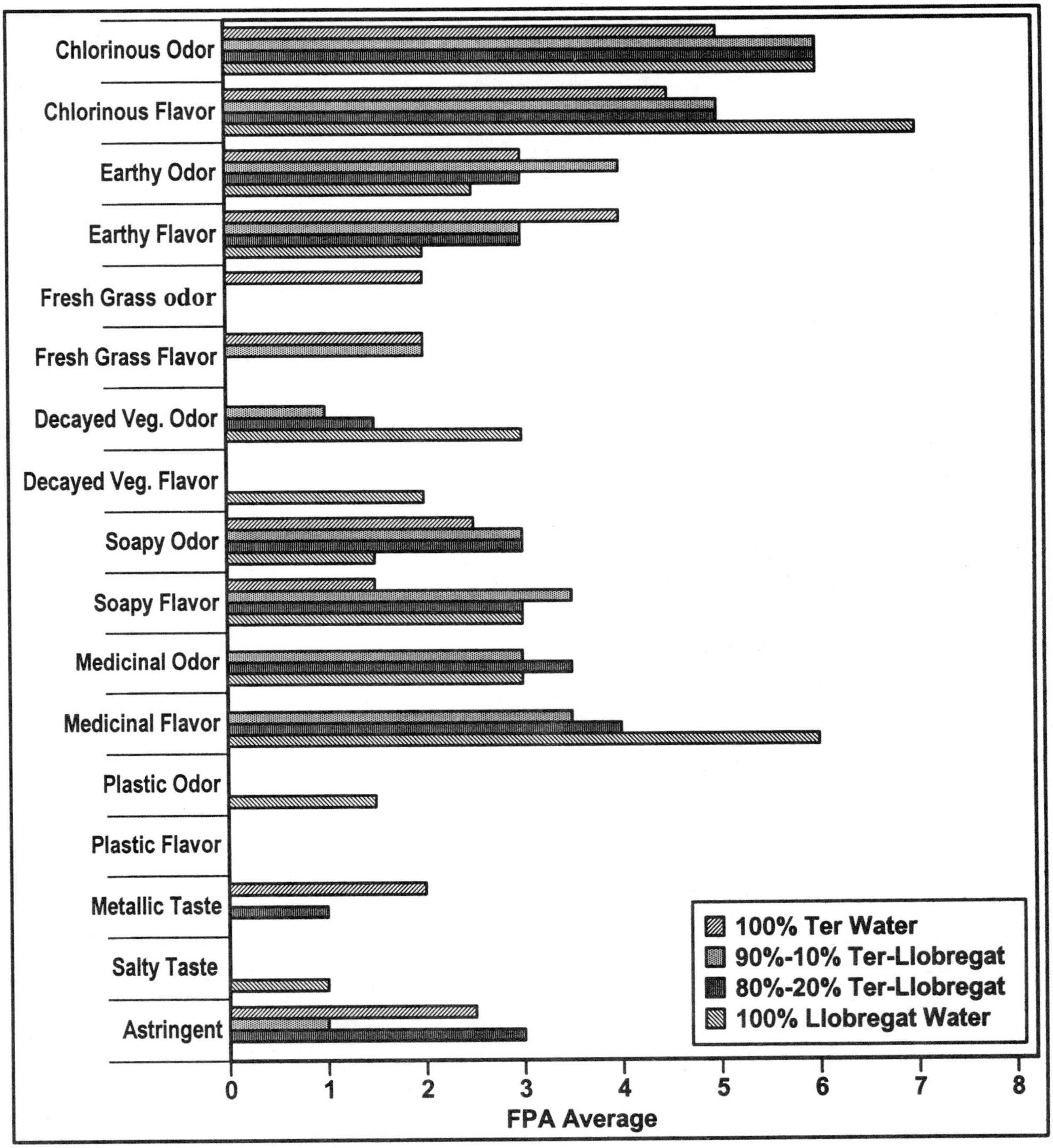

Figure 5.11 FPA for the various blends at day 0

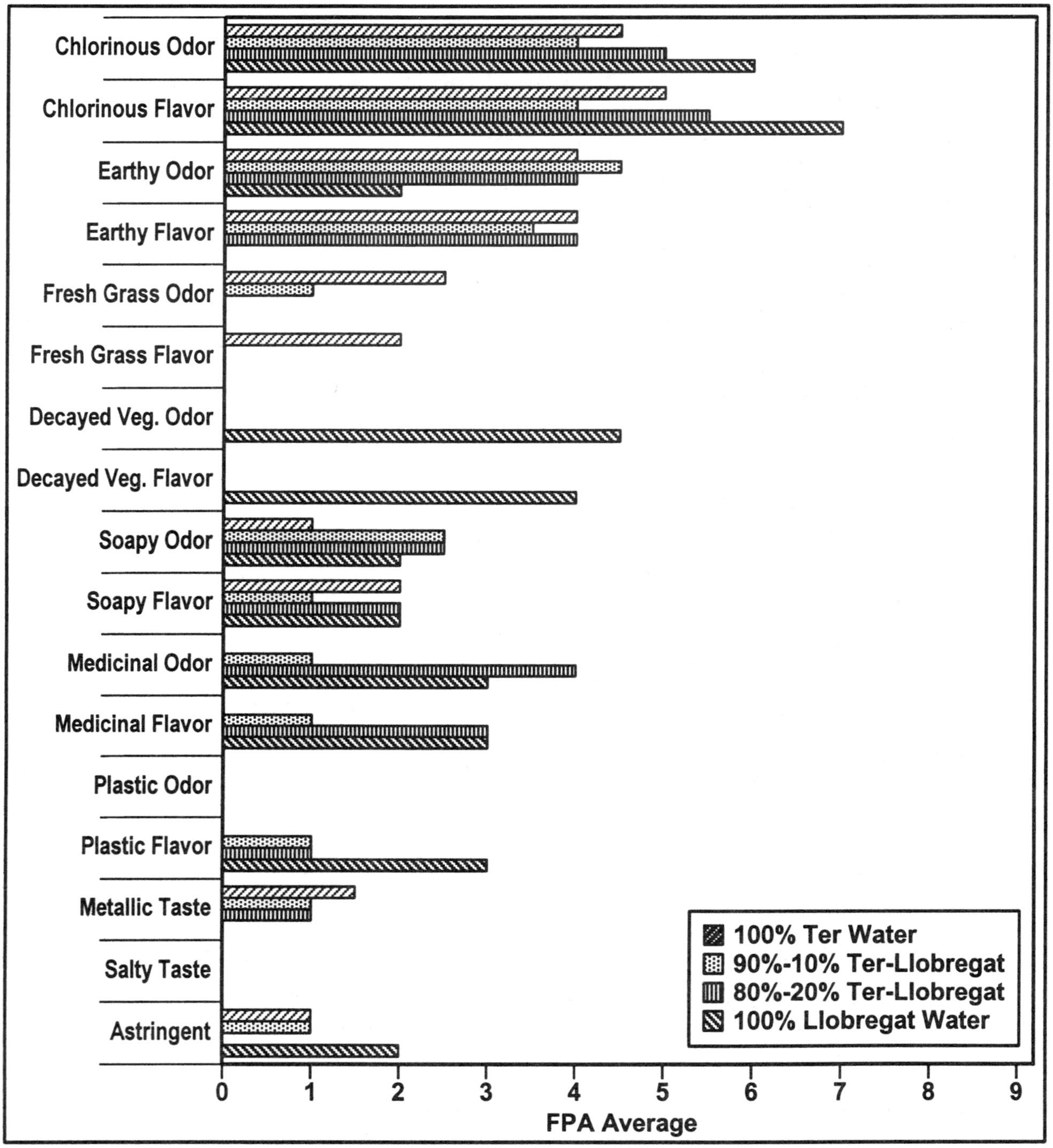

Figure 5.12 FPA for the various blends at day 7

Determining the Percentage of Llobregat Water in the Blend (Ter + Llobregat) That Differs from the Reference

Initially, the following blends were studied:

70% Ter and 30% Llobregat

30% Ter and 70% Llobregat

80% Ter and 20% Llobregat

90% Ter and 10% Llobregat

When comparing the FPA results of the blended waters with those of the original waters, it could be observed that the blended waters presented characteristics of both original waters. However, it was not readily apparent whether an overall sensory difference existed between these blended waters and their corresponding original waters.

Subsequently, an overall difference test called the "A"/"Not A" test (ASTM 1996) was carried out to determine the percentage of Llobregat water in a blend that differed organoleptically from the Ter water and to make a similar determination when the Llobregat water was taken as the reference water. Several different conditions were tested (Table 5.7).

In the "A"/"Not A" method, panelists first get familiarized with sample "A" until he or she believes the control can be identified consistently. Other samples may be tested so that the

Table 5.7
Samples and conditions used for "A"/"Not A" tests

			ODOR	FLAVOR
Test	Reference	Percent of Llobregat in Blend	Temperature	Temperature
1	Ter	0,10, 20, 40, 60	Room	Room
2	Ter (Cl_2=0)	0, 10, 20, 40, 60	Room	Room
3	Ter	0, 20, 40, 60, 100	45°C	25°C
4	Llobregat	100, 75, 50, 25, 10, 5, 0	45°C	25°C

panelist can understand what differences may be present. Then, some samples are presented to each panelist, some of which are product "A" while others are "Not A." The odor is determined randomly and the number of "A's" and "Not A's" is the same. Panelists have to determine which ones are "A" and which are "Not A." The notation "A" was used for the reference water: Ter or Llobregat treated waters The notation "Not A" was used for the blended waters to be analyzed: blended waters (Ter + Llobregat) at different percentages.

Table 5.8 shows the results for the first two tests, which are expressed in terms of percentage of Llobregat in the blend (Ter + Llobregat), that differed from the reference (treated Ter water). The results are reported when two or more panelists (six performed the test) detected differences between a sample and the reference. These first two tests were very similar. The only difference was that test 1 was carried out with residual chlorine in the samples, while in test 2 samples were dechlorinated with sodium thiosulfate. The results of the tests performed with the dechlorinated waters were much more difficult to interpret.

Table 5.8 shows that, in the odor test, in order to detect a change with the dechlorinated Ter water, it was necessary to have a water with a much higher percentage of Llobregat water (60%) compared to the chlorinated waters.

In test 3, the reference water was Ter treated water and in test 4, the reference was Llobregat treated water. Table 5.9 and 5.10 show the results of these tests. In Table 5.9, the results are expressed in percentages of the Llobregat water, and in Table 5.10 they are in percentages of the Ter water.

Table 5.8
Chlorinated versus dechlorinated water results (ambient temperature)

			Minimum Percent Contribution of Llobregat to the Blend to Detect a Change	
Test	Reference	Percent of Llobregat in Blend	Odor	Flavor
1	Ter	0, 10, 20, 40, 60	40	40
2	Ter (Cl_2=0)	0, 10, 20, 40, 60	60	40

Table 5.9
Llobregat versus Ter results

			Minimum Percent Contribution of Llobregat to the Blend to Detect a Change	
Test	Reference	Percent of Llobregat in Blend	Odor (45°C)	Flavor (25°C)
3	Ter	0, 20, 40, 60, 100	20	40
4	Llobregat	100, 75, 50, 25, 10, 5, 0	50	25

Table 5.10
Ter versus Llobregat results

			Minimum Percent Contribution of Ter to the Blend to Detect a Change	
Test	Reference	Percent of Ter in Blend	Odor (45°C)	Flavor (25°C)
3	Ter	100, 75, 50, 25, 10, 5, 0	80	60
4	Llobregat	0, 25, 50, 75, 90, 95, 100	50	75

Summary

From the above results, it is apparent that at least for a blend of Ter water with Llobregat water, 20% Llobregat water must be present before a change in odor is noticed when using Ter water as the reference; and 40% Llobregat water must be present before a change in flavor is noticed. However, it is apparent that 50% Ter water is necessary before a change in odor is noticed in the blended water when Llobregat water is used as a reference; and 75% Ter water is necessary before a change in flavor is noticed.

From the perspective of the consumer, these results mean that the consumer would notice a change in the odor with the addition of a very small percentage of Llobregat water if they were used to drinking Ter water. To the contrary, a much larger amount of Ter water would have to be added to the blend of water consumed by a customer accustomed to Llogbregat water before a change in odor would be noticed.

From the results, it is apparent that a larger amount of water needs to be blended with the reference water (Ter or Llobregat) before a change is perceived in the flavor of the blended water compared to the amount needed to perceive a change in odor.

As a general summary, it can be observed that the differences between odors appear before the differences between flavors.

Four main conclusions can be drawn from the previous tests:

- Differences in odors with respect to the chlorinated waters appear before differences in the dechlorinated waters. That is, dechlorination of the blended samples did not help in the detection of odor quality.
- Differences with respect to the Ter water appear before differences with respect to the Llobregat water.
- Differences between odors appear before differences between flavors.
- Differences between odors from Ter appear earlier at higher temperatures.
- When Ter water was blended with Llobregat water, the use of 45°C for odor testing enhanced the detection of the changes made.

DETERMINATION OF THE BEST BLEND FROM THE LLOBREGAT AND TER TREATED WATERS

The purpose of these experiments was to determine the best blend of the Ter and Llobregat treated waters in terms of odors and flavors using FPA. As mentioned earlier, the optimal blend would contain the greatest percentage of Llobregat water, such that the intensity of each of the descriptors would remain at an acceptable level. Llobregat water presents sensory characteristics such as decaying vegetation that are rather objectionable, whereas Ter water was generally considered more acceptable. However, Llobregat water was desirable for reasons other than water quality.

Methodology

In the first experiment, various dilutions of Llobregat treated water with bottled water were prepared, and the FPA of the Llobregat water and each dilution were determined. The bottled water used was found to be free of taste and odor and did not impart additional descriptors to the blend.

A second experiment was set up to study various blends of Llobregat and Ter water. All samples were tested by FPA at the time of preparation and retested on different days.

The best blend was determined based on the characteristics obtained and their overall acceptability. Certain characteristics are less acceptable than others (e.g., decaying vegetation compared to fresh grass). The followed descriptors were monitored:

- Chlorinous and earthy odor and flavor for both waters
- Medicinal and decaying vegetation odor and flavor and salty taste for the Llobregat water
- Fresh grass odor and flavor and metallic taste for the Ter water

Results and Discussion

Study of Llobregat Water Blended with Bottled Water

Table 5.11 shows the results of FPA obtained for each of the various dilutions of Llobregat water with bottled water. A zero value indicates that this characteristic was not found at all or was found by less than half the panelists (other notes).

Figure 5.13 presents the intensity/index diagram for the chlorinous, medicinal, and decaying vegetation odors and flavors. It can be observed that the chlorinous descriptor does not change significantly with dilution when analyzed by the flavor method, and the odor disappears only at a 12.5% dilution. The fact that this descriptor appeared in water in the absence of free chlorine, as has been mentioned earlier, may result from the presence of organic chloramines or other compounds and not chlorine itself. The medicinal odor and flavor disappeared at 17.5% and 25% dilution, respectively. The decaying vegetation odor disappeared at 17.5% dilution and the flavor at 25% dilution. The salty taste disappeared at 35% dilution.

Table 5.11
FPA results for the various dilutions of Llobregat water

Dilution Index (DI)	DI–1	DI–2	DI–3	DI–4	DI–6	DI–8
% Llobregat in each dilution	100	50	35	25	17.5	12.5
Free chlorine (mg Cl_2/L)	0.40	0.15	0.1	0.1	--	--
			FPA Intensity			
ODOR						
Chlorinous	6.5	5	4	4	2	ND
Decaying vegetation	4	2	3	1	ND	2
Medicinal	5	3	2	2	ND	ND
Plastic	1	ND	1.5	1.5	ND	ND
FLAVOR						
Chlorinous	7	5.5	5	4.5	3	3
Decaying vegetation	4	ND	1.5	ND	ND	ND
Soapy	2	3	ND	2	ND	2
Medicinal	5	3	ND	2	ND	2
Plastic	1.5	ND	ND	ND	ND	ND
TASTE						
Salty	2	2	ND	ND	ND	ND
Astringent mouthfeel	ND	ND	1.5	1.5	ND	ND
Spicy mouthfeel	2	1	ND	ND	ND	ND

ND: Non-detected

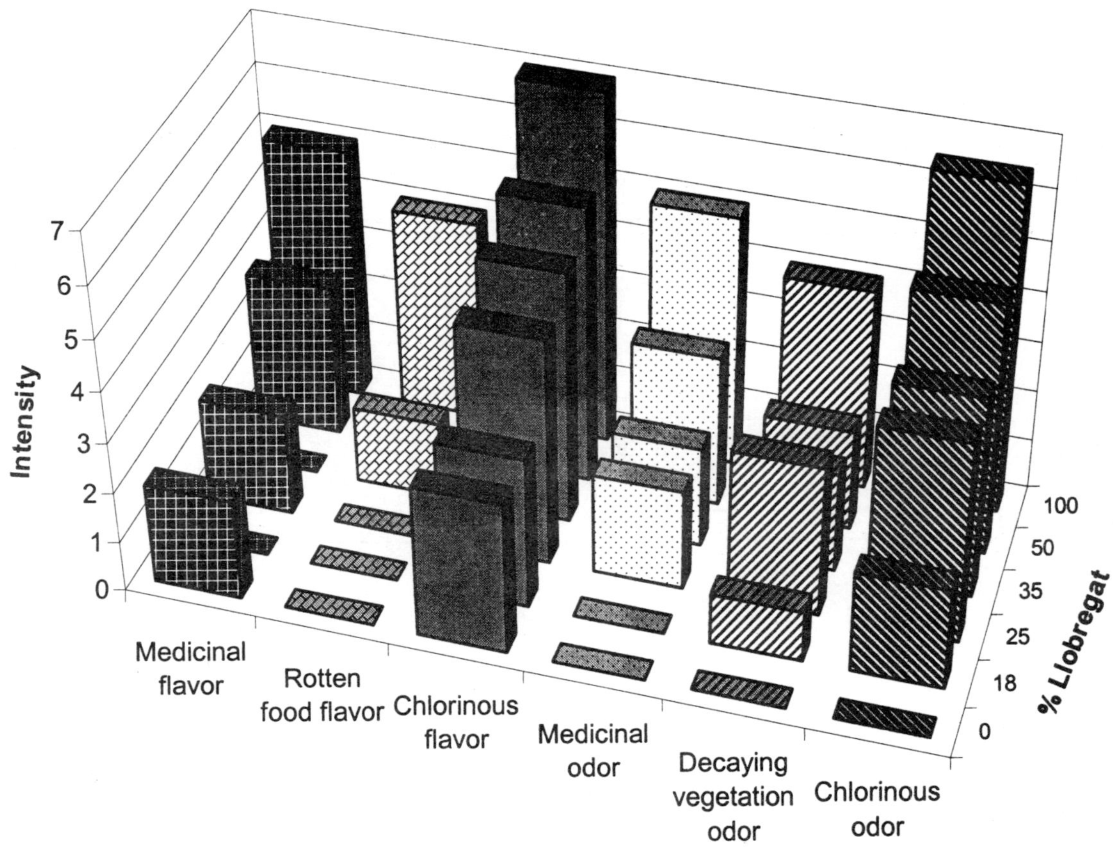

Figure 5.13 FPA for various dilutions of Llobregat water with bottled water

Study on Blending Llobregat and Ter Treated Waters

Table 5.12 shows the averages of the results from the FPA of the samples analyzed from the Llobregat and Ter waters and from the 70%, 30%, 20%, and 10% blends analyzed on four different dates.

The intensity/dilution diagram was plotted for these attributes and is shown in Figure 5.14. Using this diagram, it was possible to determine the best blend for Llobregat and Ter treated waters to minimize the intensity of the descriptors of each individual water. Although the best blend appears to be the one with 10% Llobregat water, the range between 20% and 30% was considered acceptable.

Table 5.12
Average FPA results for Ter, Llobregat, and blended water samples

% Llobregat in each blend	0%	10%	20%	30%	70%	100%
			FPA Intensity			
ODOR						
Chlorinous	5.0	5.5	5.5	5.5	7	7
Earthy	4	3.5	3.5	3.5	3	2.5
Medicinal	0.5	2.5	3	3	4	4
Decaying vegetation	ND	0.5	2	2	4.5	3.5
Fresh grass	2	1	1	1	ND	ND
FLAVOR						
Chlorinous	5	5	5	5.5	7	7.5
Earthy	4	3	3.5	3.5	4	3
Medicinal	ND	1.5	2.5	2.5	5	5
Decaying vegetation	ND	ND	ND	0.5	2.5	1.5
Fresh grass	2.5	2	1	1.5	ND	ND
TASTE						
Metallic	2	1	1	2	ND	ND

ND: Non-detected

In summary, except for the chlorinous odor, the flavor descriptors disappeared before odor descriptors when water from the Llobregat was diluted with bottled water.

The disappearance of the medicinal flavor after the change in dilution from 35% to 25% was particularly noteworthy, as was the reduction of the chlorinous odor after the change in dilution from 25% to 17.5%. Except for the chlorinous descriptor, the intensities of these attributes were already low (≤3) at 50% dilution and very low (≤2) at 25% dilution. Three features are highlighted:

1. The persistent presence of the chlorinous descriptor
2. The greater reduction in a particular flavor descriptor compared to its corresponding odor
3. The dropping of odor descriptors (except for the chlorinous odor) to low intensities at 50% and 25% dilutions

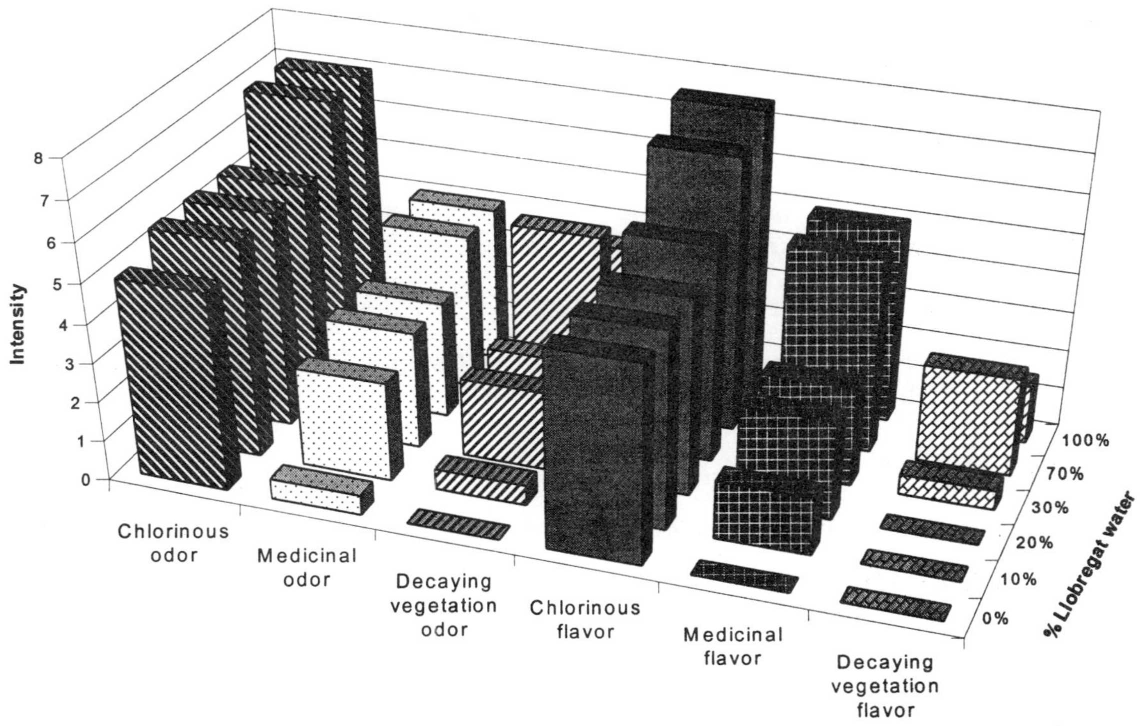

Figure 5.14 Intensity of descriptors for different blends of Llobregat + Ter waters

The chlorinous attribute was stronger in the Llobregat than in the Ter water. An intensity of 7 (that of the Llobregat) was not reduced until a 30% Llobregat/70% Ter blend was obtained.

Minimal changes were observed for both earthy odor and flavor for the various blends because the descriptor was already present in both sources of water at similar intensities. For most characteristics found in each water source, the intensities of those in Llobregat water declined as the percentage of Ter water increased, while most of the Ter characteristics increased, as expected.

Nevertheless, the Llobregat characteristics predominated and remained in the blends, although at weak intensities, even when the blend contained only 10% Llobregat water. In contrast, characteristics of the Ter were not noticeable even when the blend contained 30% Ter water. Very weak values (0–2.5) for all descriptors except chlorinous and earthy were found in the 10% Llobregat dilution, with weak intensities still being reported in the 30% dilution. Therefore, a 30% Llobregat blend was the ideal blend in order to obtain water of a good organoleptic quality.

In summary, the following observations regarding the study of the blending of Llobregat and Ter waters should be noted:

- Chlorinous and earthy descriptors were found in both sources of water and remained in the blended water.
- Llobregat attributes were more persistent than Ter attributes upon blending.
- Based on the FPA intensity/dilution diagram, a 30% Llobregat/70% Ter blend was the optimum percentage of Llobregat water that can be used to obtain the best water.

STUDY OF BLENDED SAMPLES OF TER AND LLOBREGAT TREATED WATERS FROM THE DISTRIBUTION SYSTEM

The purpose of this study was to compare the organoleptic properties of blends of the Llobregat and Ter water samples taken from the supply system to the organoleptic properties obtained from blends prepared in the laboratory.

Methodology

The FPA was used to profile five blends of distribution-system samples collected on different days from the supply system with different percentages of Llobregat and Ter treated waters. The results from these samples were compared with the results from the FPA of Llobregat and Ter treated waters samples analyzed during the last six months and blends of 30% Llobregat prepared in the laboratory.

Table 5.13 shows the minimum, maximum, and median values for conductivity and total residual chlorine in the 19 samples of Ter water and in the 20 samples of Llobregat water analyzed during the last six months of 1998.

Table 5.14 shows the general water quality parameters for the waters studied. The percentages in the blends were calculated using the conductivity values of the Ter and Llobregat distribution samples taken the same day that the sample was run. The residual chlorine values were sometimes higher in the blended samples than in the original water sources. This was because the water had been boost-chlorinated at other points in the distribution system.

Table 5.13
Conductivity and total chlorine residual of Ter and Llobregat samples*

	Conductivity (μS/cm)			Total Chlorine (mg Cl_2/L)		
	Min.	Med.	Max.	Min.	Med.	Max.
Ter supply	489	555	602	0.42	0.75	1.02
Llobregat supply	1611	1683	1968	0.5	0.98	1.38

*Second half of 1998

Table 5.14
Characteristics of the water samples studied

% Llobregat	35%	35%	40%	55%	70%
Date	06/30/98	12/14/98	07/03/98	12/10/98	12/14/98
Conductivity (μS/cm)	884	1031	1084	1101	1444
Free chlorine (mg Cl_2/L)	0.63	0.25	1.01	0.62	0.23
Total chlorine (mg Cl_2/L)	1.06	0.39	1.18	0.87	0.46
Turbidity (NTU)	0.36	0.22	0.30	0.69	0.32
Temperature (°C)	19	14.1	21	12.2	14.4

Results

Table 5.15 shows the FPA results for the five samples analyzed. The differences in FPA intensities for the various blends are very subtle.

Comparison with Prepared Blended Waters

Table 5.16 shows the FPA results obtained from the actual distribution-system samples and for the blends prepared in the laboratory. The actual distribution-system samples containing

Table 5.15
FPA results for actual distribution-system samples

Percent of Llobregat Water in Blend	35%	35%	40%	55%	70%
Date	06/30/98	12/14/98	07/03/98	12/10/98	12/14/98
	FPA Intensity				
ODOR					
Chlorinous	5.5	6.5	7	6	6
Earthy	4	4	4	3	4.5
Fresh grass	1	1.5	2	ND	3.5
Decaying vegetation	2	4	3	3	2
Soapy	3	3	2	2	ND
Medicinal	2	4	4	3	2
FLAVOR					
Chlorinous	5	6	6	6	5
Earthy	3.5	4	5	3	4
Fresh grass	ND	2	2	2	3
Decaying vegetation	2	2	2	2	2
Soapy	2	2	3	2	ND
Medicinal	2	3	4	2	3
TASTE					
Metallic	2	ND	ND	ND	1
Salty	ND	ND	ND	ND	ND
Astringent mouthfeel	1	ND	4	2	2

ND: Non-detected

35% Llobregat and 65% Ter water were compared with two samples prepared in the laboratory containing 30% Llobregat and 70% Ter water.

It can be observed that, in terms of FPA characteristics, there were no significant differences between the 30% Llobregat blend taken from the system and the 30% samples prepared in the laboratory. Meng et al. (1992) discuss the consistency error (using replicates) of the major descriptors for a specific set of data as ranging from 10 to 44 percent, and the overall average to be 42 percent. There may have been differences in the results taken from the system compared to those in the laboratory, but the differences were less than the error of the method.

Comparison with Ter and Llobregat Waters

Figure 5.15 and 5.16 show the average organoleptic profiles obtained for the 20 samples analyzed (one for Llobregat distribution water and 19 for Ter water) during the second half of 1998. The largest differences were as follows:

- In the Ter water, the fresh grass descriptor and the metallic taste were much more important than in the Llobregat water.
- In the Llobregat water, the attributes that differentiated this water from the Ter water were the medicinal, decaying vegetation, and salty descriptors.

Fresh grass, decaying vegetation, and medicinal odors and flavors were used to compare the profiles of blends from the Llobregat and Ter waters.

It can be observed that:

- There were no new characteristics in the samples obtained from the supply system and those prepared in the laboratory compared to those in the water from the Ter and Llobregat rivers.
- Similar to the results obtained in the laboratory, all the Llobregat descriptors appeared clearly, despite the low percentage of water from the Llobregat river.

Table 5.16
FPA results for "real" and prepared water samples

	"Real" Samples 35% of Llobregat		Prepared Samples 30% of Llobregat	
DATE	06/30/98	12/14/98	07/16/97	10/13/97
	FPA Intensity			
ODOR				
Chlorinous	5.5	6.5	5	6
Earthy	4	4	3	4.5
Fresh grass	1	1.5	2	ND
Decaying vegetation	2	4	1.5	2
Soapy	3	3	2	4
Medicinal	2	4	2	4
FLAVOR				
Chlorinous	5	6	5	6
Earthy	3.5	4	3.5	6
Fresh grass	ND	2	2	nd
Decaying vegetation	2	2	2	1
Soapy	2	2	2	3
Medicinal	2	3	2	3
TASTE				
Metallic	2	ND	ND	1.5
Salty	ND	ND	ND	ND
Astringent mouthfeel	1	ND	2	ND

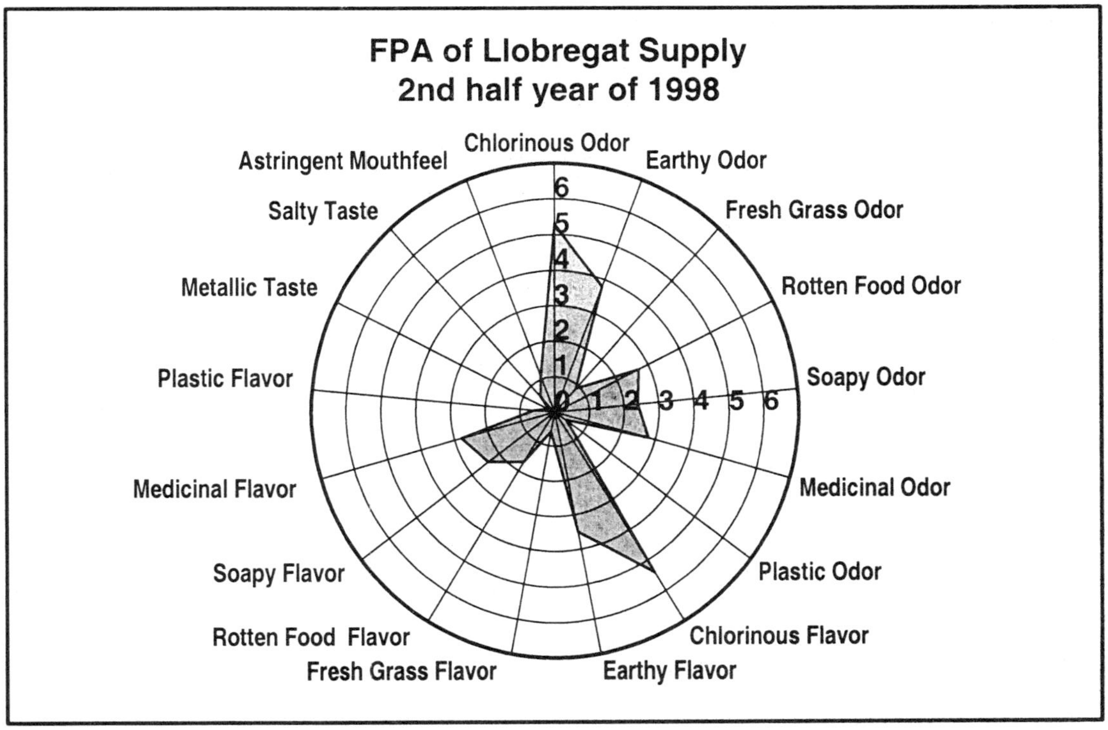

Figure 5.15 Average FPA from Llobregat water samples

Summary

- The blended samples collected from the actual distribution system had no characteristics other than those found in the water from the Llobregat and Ter sources.
- The Llobregat characteristics had a strong influence on the organoleptic profile of the resulting blends, despite only a low percentage of Llobregat water compared to Ter water.
- The Ter characteristics were also very significant in the blended "real" waters.
- There were no significant differences between blends prepared in the laboratory and tested at zero time and blends collected from the actual distribution system at unknown times of detention.

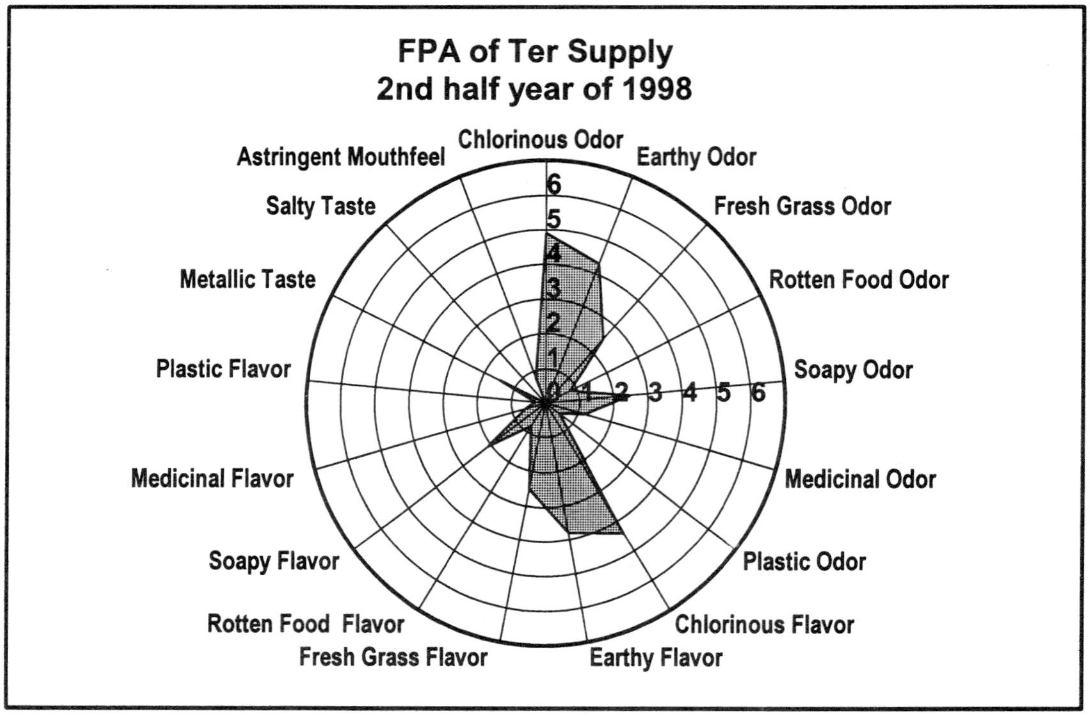

Figure 5.16 Average FPA from Ter water samples

EFFECTS OF INCREASED CHLORAMINE RESIDUAL ON TASTES AND ODORS IN BLENDS OF CHLORINATED AND CHLORAMINATED WATERS

The objective of this section was to evaluate the effects of blending chlorinated water with chloraminated water on the organoleptic characteristics of the blends. The parameters studied were the chlorine-to-nitrogen ratio of the chloraminated water, the disinfectant residuals in both waters, and the detention time of the blend. The relationships between the taste-and-odor characteristics and the concentrations of the four chlorine species (free, mono-, di-, and tri-chloramine) in each blend were examined. Because dichloramine and trichloramine have been reported to cause a more unpleasant odor than free chlorine or monochloramine, it was therefore desirable to study their formation and their effect on the organoleptic characteristics on the blends.

Previous work by Barrett et al. (1985) has shown that, for blended chlorinated and chloraminated water, three of the four species may be present in the same solution at short contact times (e.g., 15 min). Trichloramine was found in water that was chlorinated past breakpoint (15:1 ratio), but not in blended water that had an equivalent 15:1 ratio of chlorine to ammonia-nitrogen. Dichloramine has been shown to be present at 7:1 to 15:1 ratios in both blended and directly chlorinated or chloraminated water. Monochloramine has been shown to be present at all three ratios in the blended water and at the 5:1 and 7:1 ratios of chlorinated water. Positive values for the determination of monochloramine or dichloramine at the 15:1 ratio have been attributed to organic chloramines not destroyed by free chlorine (Jolley et al. 1983). The absence of trichloramine in the blending experiments may have resulted from the fact that the estimated total ammonia-nitrogen concentration was quite low to the right of the breakpoint curve.

Chlorine Residual Measurements

Table 5.17 summarizes the chlorine residual measurements for free chlorine (free), monochloramine (NH_2Cl), dichloramine ($NHCl_2$), estimated trichloramine (NCl_3) (each determined individually), and the sum of the four species measured at one time (total) for the blended samples. The corresponding FPA results are presented in Table 5.18. These experiments were designed (see Chapter 2) to observe the effects of increasing ratios of chlorine to ammonia-nitrogen on the taste-and-odor characteristics of the blended water. The ratio of chlorine to ammonia-nitrogen and the initial chloramine and/or chlorine residuals are shown in the left three columns of Table 5.18 for SPW and well water, respectively.

No dichloramine concentration was greater than or equal to 0.1 mg/L as Cl_2 (the detection limit) in the prepared blends. This fact suggested that the presence of dichloramine would not cause taste-and-odor complaints at less than 0.1 mg/L as Cl_2. Similarly, only one of the blended samples had an estimated trichloramine concentration greater than 0.2 mg/L, the detection limit. The trichloramine level of 0.5 mg/L is suspected to be an erroneous value because trichloramine is not expected to be present in a sample when the ratio of chlorine to ammonia-nitrogen is 5:1. (Palin 1975). Moreover, a more intense chlorine, geranium-like, or rotting grapefruit odor would be expected (Williams 1966) if trichloramine is actually present. The FPA results are discussed below.

Table 5.17

Results of *N,N*-diethyl-*p*-phenylenediamine–ferrous ammonium sulfate (DPD-FAS) titrimetric method for blending experiments (mg/L as Cl_2)

	15 min					24 h					120 h				
Test #	Free	NH_2Cl	$NHCl_2$	NCl_3	Total	Free	NH_2Cl	$NHCl_2$	NCl_3	Total	Free	NH_2Cl	$NHCl_2$	NCl_3	Total
A-1	<0.05	0.50	<0.1	<0.2	0.54	<0.05	0.38	<0.1	<0.2	0.40	<0.05	0.23	<0.1	<0.2	0.26
A-2	<0.05	0.48	<0.1	<0.2	0.53	<0.05	0.45	<0.1	<0.2	0.50	<0.05	0.28	<0.1	<0.2	0.32
A-3	<0.05	0.80	<0.1	<0.2	1.00	<0.05	0.83	<0.1	<0.2	0.86	<0.05	0.70	<0.1	<0.2	0.74
A-4	<0.05	1.54	<0.1	0.50	2.10	<0.05	1.60	<0.1	<0.2	1.64	<0.05	1.43	<0.1	<0.2	1.48
B-1	<0.05	0.35	<0.1	<0.2	0.37	<0.05	0.30	<0.1	<0.2	0.30	<0.05	0.17	<0.1	<0.2	0.19
B-2	<0.05	0.43	<0.1	<0.2	0.50	<0.05	0.30	<0.1	<0.2	0.35	<0.05	0.19	<0.1	<0.2	0.21
B-3	<0.05	0.62	<0.1	<0.2	0.72	<0.05	0.46	<0.1	<0.2	0.50	<0.05	0.40	<0.1	<0.2	0.51
B-4	<0.05	1.39	<0.1	<0.2	1.46	<0.05	0.95	<0.1	<0.2	1.00	<0.05	0.73	<0.1	<0.2	0.83
B-5	<0.05	0.43	<0.1	<0.2	0.46	<0.05	0.34	<0.1	<0.2	0.36	<0.05	0.16	<0.1	<0.2	0.22
B-6	<0.05	0.58	<0.1	<0.2	0.61	<0.05	0.47	<0.1	<0.2	0.52	<0.05	0.27	<0.1	<0.2	0.31
B-7	<0.05	1.17	<0.1	<0.2	1.20	<0.05	0.91	<0.1	<0.2	0.93	<0.05	0.76	<0.1	<0.2	0.90
B-8	<0.05	2.04	<0.1	<0.2	2.08	<0.05	1.53	<0.1	<0.2	1.57	<0.05	1.38	<0.1	<0.2	1.54
C-1	<0.05	0.35	<0.1	<0.2	0.39	<0.05	0.25	<0.1	<0.2	0.29	<0.05	0.18	<0.1	<0.2	0.24
C-2	<0.05	0.55	<0.1	<0.2	0.68	<0.05	0.42	<0.1	<0.2	0.52	<0.05	0.29	<0.1	<0.2	0.32
C-3	0.12	0.21	<0.1	<0.2	0.35	<0.05	<0.1	<0.1	<0.2	0.06	<0.05	<0.1	<0.1	<0.2	<0.05
C-4	0.25	0.38	<0.1	<0.2	0.73	<0.05	<0.1	<0.1	<0.2	0.08	<0.05	<0.1	<0.1	<0.2	<0.05
C-5	<0.05	0.37	<0.1	<0.2	0.40	<0.05	0.28	<0.1	<0.2	0.31	<0.05	0.22	<0.1	<0.2	0.22
C-6	0.60	<0.1`	<0.1	<0.2	0.66	0.27	0.23	<0.1	<0.2	0.54	<0.05	0.38	<0.1	<0.2	0.38
C-7	0.12	0.26	<0.1	<0.2	0.42	<0.05	<0.1	<0.1	<0.2	<0.05	<0.05	<0.1	<0.1	<0.2	<0.05
C-8	0.35	0.28	<0.1	<0.2	0.69	<0.05	<0.1	<0.1	<0.2	0.10	<0.05	<0.1	<0.1	<0.2	<0.05
Control	<0.05	0.49	<0.1	<0.2	0.55	<0.05	0.39	<0.1	<0.2	0.46	<0.05	0.25	<0.1	<0.2	0.34
Control	1.70	1.63	<0.1	<0.2	3.42	1.50	1.50	<0.1	<0.2	3.06	1.44	1.34	<0.1	<0.2	2.87
Control	<0.05	0.43	<0.1	<0.2	0.54	<0.05	0.43	<0.1	<0.2	0.48	<0.05	0.21	<0.1	<0.2	0.26
Control	1.18	2.04	<0.1	<0.2	3.28	1.77	1.31	<0.1	<0.2	3.14	0.75	1.88	<0.1	<0.2	2.70
Control	0.35	0.1	<0.1	<0.2	0.40	0.29	<0.!	<0.1	<0.2	0.38	0.24	<0.1	<0.1	<0.2	0.30
Control	0.76	0.06	<0.1	<0.2	0.84	0.77	<0.1	<0.1	<0.2	0.93	0.67	<0.1	<0.1	<0.2	0.73

Sample codes (A, B, and C) and the conditions used are defined in Table 2.12.

Table 5.18
FPA results for blending experiments

Blend Cl_2:N ratio, 5:1

Sample				Contact Time		
SPW*		Well				
Cl_2:N	Res.	Res.		15 min	24 h	120 h
A-1			ODOR	Straw/hay 2	OF**	Swimming pool 2 Sweet*** Straw***
3:1	0.5	0.5	FLAVOR	Metallic 2 Drying 2	Chalky 2 Bitter 2 Drying 2	Drying 2
A-2			ODOR	Straw/hay 2 Briny***	Straw/hay 2	Sweet/plastic 2 Rosebush***
3:1	0.5	1.0	FLAVOR	Metallic 2 Drying 2	Bitter 2 Drying 2	Drying 2
A-3			ODOR	Chlorine 2 Hay/straw***	Straw/hay*** Solvent***	OF**
3:1	4.0	0.5	FLAVOR	Chalky 2 Bitter 2 Drying 2	Drying 2	Drying 2
A-4			ODOR	Chlorine 2	OF**	OF**
3:1	4.0	1.0	FLAVOR	Bitter 2 Drying 2	Bitter 2 Drying 2	Drying 2

*SPW: State project water
**OF: Odor-free
***: Other note

(continued)

Table 5.18 (continued)

Blend Cl_2:N ratio, 7:1

Sample				Contact Time		
SPW*		Well				
Cl_2:N	Res.	Res.		15 min	24 h	120 h
3:1	B-1 0.5	0.5	ODOR	OF**	Swimming pool 2 Straw/hay 2	Sweet 2 Plastic 2
			FLAVOR	Drying 2 Bitter 2	Metallic 2 Drying 2	Drying 2
3:1	B-2 0.5	1.0	ODOR	OF**	Swimming pool 2 Straw/hay 2	Plastic 2
			FLAVOR	Metallic 2 Drying 2	Drying 2	Drying 2
3:1	B-3 4.0	0.5	ODOR	Chlorine 2 Flowery***	Chlorine 2 Straw/hay 2	Swimming pool 2 Sweet 2
			FLAVOR	Bitter 2 Drying 2	Drying 2	Drying 2
3:1	B-4 4.0	1.0	ODOR	Chlorine 2	OF**	Sweet*** Hay/straw***
			FLAVOR	Drying 2	Drying 2	Drying 2
5:1	B-5 0.5	0.5	ODOR	Hay/straw*** Swimming pool***	Swimming pool*** Rosebush***	Swimming pool*** Hay/straw***
			FLAVOR	Bitter 2 Drying 2	Chalky 2 Bitter 2 Drying 2	Drying 2
5:1	B-6 0.5	1.0	ODOR	OF**	Swimming pool*** Rosebush***	Swimming pool*** Citrus***
			FLAVOR	Drying 2 Bitter 2	Drying 2	Drying 2
5:1	B-7 4.0	0.5	ODOR	Chlorine 2 Hay/straw 2	OF**	Sweet*** Briny***
			FLAVOR	Drying 2	Bitter 2 Drying 2	Drying 2
5:1	B-8 4.0	1.0	ODOR	Chlorine 2	OF**	Sweet*** Briny*** Swimming pool***
			FLAVOR	Bitter 2 Drying 2	Bitter 2 Drying 2	Chalky 2 Bitter 2 Drying 2

*SPW: State project water
**OF: Odor-free
***: Other note

(continued)

Table 5.18 (continued)

Blend Cl_2:N ratio, 15:1

Sample SPW* Cl_2:N	SPW* Res.	Well Res.		Contact Time 15 min	24 h	120 h
3:1	C-1 0.5	0.5	ODOR	Sweet/vanilla*** Chlorine 2	Sweet/citrus***	OF**
			FLAVOR	Drying 2	Bitter 2 Drying 2	Bitter 2 Drying 2
3:1	C-2 0.5	1.0	ODOR	Sweet/vanilla*** Chlorine 2	Swimming pool***	Swimming pool*** Hydrocarbon***
			FLAVOR	Chalky 2 Drying 2	Chalky 2 Drying 2	Drying 2
3:1	C-3 4.0	0.5	ODOR	Chlorine 2 Woody***	Solvent***	Chlorine 2 Flowery***
			FLAVOR	Chalky 2 Drying 2	Drying 2	Drying 2
3:1	C-4 4.0	1.0	ODOR	Chlorine 2	OF**	Chlorine 2
			FLAVOR	Drying 2	Drying 2	Drying 2
5:1	C-5 0.5	0.5	ODOR	Flowery/ vanilla***	Swimming pool 2	Swimming pool 2
			FLAVOR	Bitter 2 Drying 2	Drying 2	Drying 2 Bitter 2
5:1	C-6 0.5	1.0	ODOR	Chlorine 2 Hydrocarbon***	OF**	Not done
			FLAVOR	Drying 2	Chalky 2 Drying 2	Not done
5:1	C-7 4.0	0.5	ODOR	Straw 2 Chlorine 2	Swimming pool 2	OF**
			FLAVOR	Drying 2	Bitter 2 Drying 2	Bitter 2 Drying 2
5:1	C-8 4.0	1.0	ODOR	Straw/hay 2 Chlorine 2	Hay/straw 2 Swimming pool***	Sweet 2
			FLAVOR	Drying 2	Drying 2	Bitter 2 Drying 2

*SPW: State project water
**OF: Odor-free
***: Other note

(continued)

Table 5.18 (continued)

Controls

Sample		Contact Time 15 min	24 h	120 h
Control 1				
SPW*	ODOR	OF**	OF**	Not done
Cl_2:N, 3:1	FLAVOR	Drying 2	Drying 2	
Initial Res. 0.5 mg/L				
Control 2		Chlorine 2	Chlorine 2	
SPW*	ODOR		Sweet***	
Cl_2:N, 3:1	FLAVOR	Drying 2	Drying 2	
Initial Res. 4.0 mg/L				
Control 3				
SPW*	ODOR	OF**	OF**	
Cl_2:N, 5:1	FLAVOR	Drying 2	Drying 2	
Initial Res. 0.5 mg/L			Bitter 2	
			Chalky 2	
Control 4			Hydrocarbon 2	
SPW*	ODOR	Chlorine 2	Plastic 2	
Cl_2:N, 5:1	FLAVOR	Drying 2	Drying 2	
Initial Res. 4.0 mg/L				
Control 5				
Well	ODOR	Chlorine 2	Chlorine 2	
Cl_2:N, 5:1	FLAVOR	Drying 2	Drying 2	
Initial Cl_2 Res. 0.5 mg/L		Bitter 2		
Control 6				
Well	ODOR	Chlorine 2	Chlorine 2	
Cl_2:N, 5:1	FLAVOR	Drying 2	Drying 2	
Initial Cl_2 Res, 1.0 mg/L				

*SPW: State project water
**OF: Odor-free
***: Other note
Res.: residual

As noted in *Standard Methods* (APHA 1998) regarding *N,N*-diethyl-*p*-phenylenediamine–ferrous ammonium sulfate (DPD-FAS) titration, the need for thioacetamide for high levels of combined chlorine was observed for samples with chloramine concentrations of approximately 2.5 mg/L as Cl_2 in a separate experiment. During these blending experiments, some control samples contained high values of monochloramine. Because no thioacetamide was used in the DPD-FAS titration, the monochloramine might have interfered with the free chlorine readings; therefore, for discussion purposes, it was considered appropriate to add the free chlorine to the monochloramine value.

FPA Odor Results

Blend at 5:1 Ratio of Chlorine to Ammonia-Nitrogen

For the 5:1 samples (A-series), the odor of chlorine at an intensity of 2 was perceived in two out of 12 samples at a 15-min contact time, and a swimming pool odor at intensity 2 was perceived for one sample at a 120-h contact time. The chlorine odor was present at the 15-min contact time for the samples where the initial monochloramine residual for chloraminated California state project water (SPW) was 4 mg/L. The actual monochloramine residuals were 0.80 and 1.54 mg/L for the blended samples. The swimming pool and sweet plastic odors at intensity 2 and "other notes" of sweet, straw, and rosebud were usually present where the initial monochloramine residual for SPW was 0.5 mg/L and where the monochloramine residual ranged from 0.23 to 0.50 mg/L.

For sample A-3 at a 15-min contact time, where the odor of chlorine was observed, the monochloramine residual was 0.8 mg/L and the trichloramine reading was less than 0.2 mg/L. For the sample at a 24-h contact time, the monochloramine residual was still 0.8 mg/L, but the odor of chlorine was no longer present. For samples A-2 and A-3 at 15-min contact times, the dichloramine reading was less than 0.1 mg/L, but only A-3 had the chlorine odor. For sample A -4 at 15-min and 24-h contact times, the monochloramine reading was 1.54 mg/L or greater, but only the A-4 sample at the 15-min contact time had the chlorine odor. This sample also had a trichloramine reading of 0.5 mg/L. As stated above, the odor of trichloramine reminds most people of geraniums or rotting grapefruit. Williams (1966) states that under certain conditions other nitrogen-chlorine compounds may be present with the trichloramine. It is suspected that compounds other than trichloramine are responsible for the trichloramine reading for the A-4 sample because the chlorine odor was normal for this sample.

Blend at 7:1 Ratio of Chlorine to Ammonia-Nitrogen

Performing a similar evaluation of the results for the 7:1 samples (B-series), no values for dichloramine or trichloramine were above the detection limits. A method with lower detection limits that is specific for dichloramine and trichloramine is needed. Swimming pool odors have been associated with the presence of dichloramine, but because of the high detection limit for dichloramine with the DPD-FAS method, no reading for dichloramine was obtained. However, because the swimming pool odor characteristic was observed in a number of samples at the 7:1 chlorine-to-nitrogen ratio, dichloramine is probably present.

For the 7:1 samples (B-series), the odor of chlorine at intensity 2 was observed in five of the 24 samples. All but one was at a 15-min contact time. A swimming pool odor at intensity 2 was observed in three samples, and an "other note" was reported in six samples. Only one of these samples was at a 15-min contact time. The other samples were spread between the 24- and 120-h contact times, suggesting that the swimming pool characteristic took longer than 15 min to form and was slightly more prevalent at a 24-h contact time. A chlorine odor at intensity 2 was only present for blended SPW samples with an initial monochloramine residual of 4 mg/L in SPW. The actual monochloramine residuals in the blended waters ranged from 0.46 to 2.04 mg/L. An odor characteristic of sweet (also flower, rosebush, and citrus) was obtained for nine samples out of 24, but only two were at an intensity of 2. The others were identified as "other notes." The sweet characteristic was observed in all but two samples at the 120-h contact time, and was observed in one sample at 15 min and two samples at 24-hr contact times. A sweet characteristic became more apparent after longer detention times. The straw/hay odor characteristic was observed in seven of 24 samples. Four were at an intensity of 2; the others were "other notes." One of the intensity 2 observations was at 15 min, and the others were at 24 h.

For the blended samples with 7:1 ratios of chlorine to ammonia-nitrogen, no obvious trend was observed where the SPW's initial ratio of chlorine to ammonia-nitrogen had been 3:1 or 5:1 rather than 7:1. Also, the initial concentration of free chlorine in well water did not seem to make an observable difference on the odor of the blend.

Blend at 15:1 Ratio of Chlorine to Ammonia-Nitrogen

For the C-series samples, the chlorine odor in most samples was probably caused by the free chlorine. For the C-4 sample, it is unknown what caused the chlorine odor. Again, the dichloramine and trichloramine readings were below the detection limits, yet swimming pool odors were present in almost a third of the samples indicating the probable presence of dichloramine. Perhaps most of these samples were near the breakpoint, rather than actually to the right of it.

For the 15:1 blends, the chlorine odor at intensity 2 was present in nine samples. Seven of these nine samples were at the 15-min contact time, and two were at the 120-h contact time. As was the case with the 5:1 and 7:1 blends, the 15:1 blend had a swimming pool odor that appeared at 24- and/or 120-h contact times. Again, the initial ratio of chlorine to ammonia-nitrogen and the free-chlorine residual of the SPW and well water, respectively, did not appear to affect the results. The sweet (vanilla, citrus, flowery) characteristic was observed in six out of 24 samples; one of these, at a 120-h contact time, was at an intensity of 2. This characteristic was observed four times in samples where SPW had an initial 3:1 ratio of chlorine to ammonia-nitrogen.

The straw (hay, woody) odor characteristic was observed four times, three of which were at intensity 2. No hay or straw characteristics were observed at the 120-h contact time.

Controls

For the controls where the initial SPW monochloramine residual was 0.5 mg/L, the water was odor-free and the monochloramine residuals after the 24-h and 15-min contact times ranged from 0.39 to 0.49 mg/L. Where the initial monochloramine residual was 4 mg/L, three out of four control samples had a characteristic chlorine odor at an intensity of 2, and the total chlorine after the 24-h and 15-min contact times ranged from 0.48 to 3.28 mg/L. For all the controls at the 24-h contact time, other odors that were observed at an intensity of 2 or less were hydrocarbon, plastic, and sweet. The well water controls consistently had a chlorine odor of intensity 2.

The odor intensity values for the control samples with a free-chlorine residual (controls 5 and 6) and with a monochloramine residual (controls 1, 2, 3, and 4) are similar to those reported by Krasner and Barrett (1984). However, the intensity values reported by Krasner and Barrett are on a seven-point scale, and those reported in Table 5.18 are on a 12-point scale. Therefore, an intensity of 2 in Table 5.18 would be similar to an intensity of 0.5 on the 7-point scale. In addition, the work reported by Krasner and Barrett was conducted with Sparkletts® crystal-fresh™ drinking water (Los Angeles, California) at pH 5 for hypochlorous acid and pH 8–9 for monochloramine. Actual treated drinking water was used for the work reported in Table 5.18, and the pH was between 7 and 8.5.

FPA Flavor and Taste Results

The most frequently observed taste characteristic (other than drying) in the blended samples was bitter with an intensity of 2. Bitter was also observed in two of 12 control samples. For the 15:1 blend samples, bitter was observed at 15-min, 24-h, and 120-h contact times. For the 7:1 blend samples, bitter was more prevalent in the samples that initially had SPW at a 5:1 ratio of chlorine to ammonia-nitrogen, and this characteristic was present at each contact time for at least one set of samples. For the samples (7:1 blend samples) that had SPW at an initial 3:1 ratio of chlorine to ammonia-nitrogen, bitter was only observed at the 15-min contact time in two samples. For the 15:1 blend samples, bitter was not observed at the 15-min contact time where the initial SPW chlorine-to-nitrogen ratio was 3:1 and was most prevalent at the 120-h contact time where the initial SPW chlorine to-nitrogen ratio was 5:1. For the 5:1 blend samples, bitter was observed at 15-min and 24-h contact times. Bitter at intensity 2 was observed in two of the 12 control samples.

For the 24 blended samples, chalky was observed eight times at intensity 2; only one of these samples was at the 120-h contact time. Chalky at intensity 2 was observed in one control sample. Metallic at intensity 2 was only observed four times, three times at 15-min contact time, once at 24-h contact time and none at 120-h contact time. Metallic was not observed in the control samples.

Discussion

The FPA results of this study generally confirmed what was expected from an interpretation of the corresponding chlorine residual results. The swimming pool odor observed for various samples was not expected from the determined dichloramine values. This suggested that a species other than dichloramine was responsible for the odor or that a lower detection limit is needed. During this project, sensory gas chromatography has shown that a strong chlorine odor was present where none of the traditional chlorine species were present as measured with the DPD-FAS method. Additional work needed to identify the compounds responsible for the chlorine odor. From the work during the study, it appears to be a compound with two chlorine atoms. Researchers in France (Welté and Montiel 1999) have reported similar findings, although trichloramine was suspected to be causing the odor there.

The samples at the 7:1 and 15:1 ratios appeared to have flowery and sweet descriptors observed more frequently than at the 5:1 ratio. These descriptors have been associated with aldehydes. Le Cloirec and Martin (1985) discuss a direct relationship between the molar ratio of chlorine to amino acid and the formation of aldehydes and nitriles. Because the ratio of chlorine to amino acid is higher as the chlorine-to-nitrogen weight ratio increases from 5:1 to 15:1, more aldehydes would be expected at the 7:1 and 15:1 ratios compared to the 5:1 ratio. The controls did not have the flowery and sweet descriptors. These descriptors were observed more often at the 24-h and 120-h contact times than at the 15-min contact time. This may have been because of masking by the stronger chlorine descriptors at short contact times or because it took longer for the flowery and sweet odor-causing compounds to form.

SUMMARY

- Trends for the chlorine residuals generally followed the trends observed by Barrett et al. (1985).
- The changes in the organoleptic properties of the blended water were very subtle.

- The organoleptic characteristics of the water changed with the various contact times, with the swimming pool and sweet characteristics becoming more prevalent at 24 and 120 h and the straw odor characteristics decreasing at longer detention times.
- The most frequently observed taste characteristic was bitter, which seemed to be more prevalent for the blends that had SPW at an initial 5:1 ratio of chlorine-to-ammonia-nitrogen.

CHAPTER 6

THE EFFECTS OF PIPE MATERIALS ON TASTE AND ODOR PROBLEMS IN DISTRIBUTION SYSTEMS

INTRODUCTION

The component parts of the water distribution systems are (a) pipes and their linings, (b) pipe-joint gaskets for metal pipes, (c) pipe-joint lubricants for metal pipes, and (d) primers and cements to join plastic pipes. It has been difficult to pinpoint the problem of tastes and odors generated by these materials because of the localized and transient nature of the taste-and-odor problems. Taste-and-odor episodes usually occur for a brief period in areas disturbed by maintenance projects or new construction. By the time the complaints are logged and investigated, the causative compounds are not easily detected.

The most apparent taste-and-odor sources are the materials of which the piping network is constructed. Piping material can effect the quality of drinking water in a direct and indirect manner. Constant contact of water with residual chlorine or chloramine on piping materials can have the direct effect of leaching of taste-and-odor compounds into the drinking water. Permeation of compounds through the piping material can have an indirect effect on drinking water quality. In fact, one of the first indications that something may be wrong with the distribution pipes is the presence of a taste or odor in the water from the pipe. The two most common materials used in distribution systems are (a) polyvinyl chloride (PVC) pipe and its primers and cements and (b) ductile-iron pipe with cement–mortar lining and its rubber gaskets and joint lubricants. The use of PVC pipes and cast-iron pipes (both ductile iron pipe with cement–mortar lining and gray cast-iron pipe) has become more commonplace since the 1980s.

Few formal studies have investigated the contribution of distribution-system components to tastes and odors in drinking water supplies. Fadel et al. (1993) found that an oleate-based lubricant, TYTON® joint lubricant (TJL) (Seacord Corp., Orangeburg, S.C.), introduced wet paper, chalky, or wood putty odors from water in new water mains. Mixed straight-chained aldehydes (C-6 to C-9) were extracted from the water and identified by gas chromatography/ mass spectrometry (GC/MS). However, the odors of the aldehydes are different from the odors identified by the flavor-profile analysis (FPA) panels. The odors identified by the same FPA panel for the individual aldehydes were hexanal (lettuce hearts, pumpkin), heptanal (rancid walnut oil) and nonanal (hay-like). Although the odors were not exactly the same, the authors during this study carefully evaluated the aldehydes' contribution to the final odor. TJL was found to release high-molecular-weight organic compounds, increase the pH and turbidity, and react readily with chlorine, and increase the odor (Fadel et al. 1993). Burlingame (1993) conducted a laboratory-scale comparison of three lubricants: TJL, Blue Lube™ lubricant (BLL) (J.C. Whitlam Mfg. Co., Wadsworth, Ohio), and Aquaflow VA-SP 10 (AVA or AVA 10) (Astra Industrial Services, Newbury Park, Calif.). In 1998, Philadelphia chose BLL as the lubricant that performed the best. BLL was described as free of "odor-causing" compounds, and it produced the least odors compared to its competitors when the manufacturer's applicator was clean and when a discrete amount of lubricant was applied.

Anselme et al. (1985b) reported the appearance of aldehydes, ketones, and phenolic compounds in water passed through high-density polyethylene pipe. The presence of these compounds was associated by the authors with a plastic taste. High-density polyethylene pipe is not usually used in the distribution system in the United States and was not included in this study.

Guo (1997) and Guo, Toomuluri, and Eckert (1998) reported a significant amount of barium, cadmium, and chromium leached from the cement–mortar linings in two of the three test pipes during the first 14 days of water stagnation. Berend and Trouwborst (1999) found high concentrations of aluminum (>300 μg/L) in drinking water from cement–mortar lining of the cast-iron pipes. Aluminum concentrations of <100–200 μg/L in drinking water do not appear to be an important health risk for the normal population. However, a high concentration (>200 μg/L) of aluminum in drinking water used for kidney dialysis causes neurological disorder in hemodialysis patients (Alfrey 1993).

OBJECTIVES

The objective of this study is to help define the contribution of distribution-system components to tastes and odors in drinking water. The leaching of the most commonly used pipes (made from PVC and iron) and the associated materials that enable connection of the pipes —e.g., pipe-joint lubricants, gaskets, and primer and cement—are studied.

This study had three specific objectives:

1. To design a testing protocol for off-odors, which could be used by water utilities in determining the suitability of materials for use in distribution systems.
2. To characterize odors produced by each material, as perceived by an odor panel, under the testing protocol.
3. To identify some of the chemicals leached from these materials and verify, if possible, if the odor or taste perceived was caused by the chemical(s) identified.

MATERIALS

The materials selected for this study included three lubricants, a rubber gasket, and four types of pipe.

Lubricants and Rubber Gasket

TJL (Seacord Corp., Orangeburg, S.C.) is the most commonly used lubricant in the United States. It is an oleate-based brown paste of dense consistency, which the manufacturer claims does not impart odors, taste, or cloudiness to water at a concentration of 10 mg/L (Fadel et al. 1993). The manufacturer's label states that TJL is a nontoxic, vegetable soap that contains potassium oleate and propylene glycol.

BLL is a smooth, grit-free paste, light blue in color, and described by the manufacturer as odorless. AVA10 is a smooth, grit-free white paste. The TYTON® rubber gasket is a black rubber ring of 4- or 10-in. inner diameter.

Pipes

The four types of pipes selected for this study (Hirsch Pipe Co., Santa Monica, Calif.) are the most widely used in the United States:

- PVC white

- PVC gray
- Ductile iron pipe (DIP) with cement–mortar lining
- Gray cast-iron pipe (CIP)

For the plastic pipe, a pipe primer and cement are needed.

- PVC pipe primer—WELD•ON™ P-68 Purple Primer (IPS Corp., Compton, Calif.)
- PVC pipe cement—Red Hot Blue Glue™ (T. Cristy Enterprises, Orange, Calif.)

The PVC pipes were studied alone as well as with (1) PVC pipe primer, (2) PVC pipe cement, and (3) both PVC pipe primer and cement.

White and Gray PVC Pipes

Plastic pipes were first introduced in the United States around 1940. White and gray PVC pipes are widely used for potable water and drainage systems. PVC pipe is a rigid pipe manufactured by an extrusion process. Fittings are made by a molding process. It is available in diameters up to 36 in. and lengths of 20 to 40 ft, with various types, grades, and pressure ratings. PVC pipes are connected by a solvent-cement joint or by a bell-and-spigot push-on joint. A solvent-cement joint is made by spreading a combination of primer and cement on the surfaces to be joined. Solvents react chemically by dissolving the surface of the pipe and fitting. The two components are put together while wet. After drying, the two components are fused into a homogeneous mass, producing a leak-proof joint. Sources of off-tastes and odors may result from the aqueous leaching and permeation of the chemicals used to join the pipes. The organoleptic properties associated with the connection of two pipes are solvent, medicinal, gasoline, and plastic. Therefore, the PVC primer joint solvent and cement were tested as a contributing factor to taste-and-odor problems in distribution systems.

Plastic pipes themselves can contribute to taste-and-odor problems primarily by dissolution of polymer additives within the pipe; oxidation of the internal surface of pipe during extrusion and dissolution of the resulting polar compounds during contact with water; and as permeation, or the migration of external contamination through the pipe wall (Mallevialle and Suffet 1987). The first two mechanisms generally involve simple dissolution of the chemicals, which are in contact with the water. The third way, permeation, involves several steps of solution diffusion, where the contaminants on the outside surface will dissolve in the polymer, diffuse through the plastic pipe wall, and desorb from the inner surface (Mallevialle and Suffet 1987). The rate of penetration of organic chemicals through PVC pipes can be determined using equations developed by Berens (1985) in his study of solvent transport. Berens (1985) found that based on Fick's law, the rate of penetration was predictable when known diffusivities and solubilities of the solvents were known. The Flory-Huggins equation describes the relationship between solubility and activity (*P/Po*):

$$\ln(P/Po) = \ln(V_1) + V_2 + x(V_2)^2 \tag{1}$$

where V_1 = the volume fraction solvent in the swollen polymer

V_2 = the polymer volume fraction

x = the polymer-solvent interaction parameter, a measure of the swelling or solvent power of a liquid

If x is less than 0.5, the liquid will completely dissolve the polymer, and an x greater than 0.5 connotes decreasing swelling power. Organic chemicals commonly found in the ambient environment generally have activity below 0.5; thus, Fick's law of diffusion applies. An activity of 0.25 would be considered an extremely high level of environmental pollution in the range in which permeation through PVC pipe walls is very slow (about 100,000 years). Activity levels above 0.5 can be detected in the immediate vicinity of gross spills or leaks of pure solvents. Another model, by Desmond (1985), described the same type of processes for tetrachloroethylene leaching from vinyl-lined asbestos cement pipe. Therefore, permeation can cause offensive tastes and odors in drinking water when soil around a pipe has been saturated with gasoline or other solvents.

Cast-Iron Pipes (CIPs)

CIPs have been the standard for water distribution systems worldwide for many years. There are more miles of this pipe in use today than any other pipe (Nayyar 1992). Two types of CIPs have been used in distribution systems: DIP and gray CIP. DIP has diameters of 4 in. and larger and lengths of 18 and 20 ft. DIP is lined with a thin coating of cement–mortar lining. Cement–mortar lining minimizes tuberculation and maintains the carrying capacity of the pipe. Aggressive waters can cause the CIP to lose carrying capacity through corrosion and tuberculation. The cement–mortar lining is approximately 1/8 in. thick and adheres to the pipe. Various thermoplastic and other epoxy-lined pipes are also available and used to handle fluids considerably more corrosive than water.

Lining the pipe interior with cement–mortar protects the metal from corrosion by providing a coating of electrochemically inert material between the metal and water. Although cement–mortar lining successfully protects against electrochemical corrosion, water can leach cementitious materials from this lining for as long as four years. Extended contact time under low-flow conditions can increase pH, alkalinity, and concentrations of leached components such as calcium, barium, cadmium, chromium, and aluminum. Water that is of low pH and alkalinity and that contains little calcium appears to be the most corrosive to the cement–mortar lining (Berend and Trouwborst 1999). Thus, taste-and-odor problems can originate from emission of organic compounds and metals from the internal surface of pipes and their lining and coating materials.

Leaching of organic compounds and metals from the lining of the CIPs can be described by mathematical models (Guo 1997). The model takes into consideration the following parameters: the initial wetting process, dissolution, diffusion and advection of metals in the lining, accumulation of metals in static pipe water, and dilution of metals by the flowing water. Diffusion and dissolution processes are considered simultaneously, and a gradual leaching front is assumed.

The transport mechanism in a cement–mortar lining was described by the water seepage velocity as a function of time and the distance, which the wetting front has moved. The metal concentration in the pore water of the cement–mortar lining is time-dependent (with initial concentrations $t = 0$) and can be described using an advection-diffusion equation. Equations were also developed to predict the metal concentrations in static and flowing pipe water. In flowing pipe water, the time variation of the metal concentration is affected by the rate at which metal leaches out from the lining, advection (dilution) by the incoming fresh water, and diffusion-dispersion in the water,Guo (1997).

Based on the modeled results of Guo (1997), solubility is the most sensitive factor leading to a potential increase in metal concentrations in the use of cement–mortar-lined pipe. Overall, diffusion is the primary mechanism by which permeation and leaching of organic compounds and metals occur from materials in the drinking water distribution system. Thus, minimizing trace metal impurities in the pipe and coating material minimizes the leaching of hazardous chemicals and taste-and-odor problems in the drinking water.

EXPERIMENTAL DESIGN

Original Protocol: Leaching of Lubricants and Rubber Gasket

A method modified from Fadel et al. (1993) was used for the leaching protocol. Autoclaved tapwater was used as the medium. All glassware used in the experiment was cleaned by acid-washing for 24 h in an acid bath (1:1 HNO_3:HCl), rinsed with Milli-Q water (Millipore Corp., Bedford, Mass.), and oven-dried. In order to minimize biological activity, the water was autoclaved in 2 and 4 L jars with Teflon inserts in the covers. Two-inch sections of the gasket or the lubricant in 10-mL sterile glass vials were submerged in 4 L of the autoclaved water. The transfer of each material to the 4 L autoclaved jars of water was done across an open flame to maintain sterile conditions. Also, the lids of the jars were flamed and were only opened long enough to drop in the sterilized vials containing lubricant, or the gasket sections. The jars were filled to the rim excluding any headspace.

The increase in pH caused by autoclaving and the addition of materials to the water was corrected by the addition of dilute sulfuric acid to adjust the pH to between 7 and 8. Each material was prepared in three different mediums: autoclaved water with (1) 2 mg/L chlorine, (2) 3 mg/L monochloramine, or (3) no disinfectant. The jars were wrapped in aluminum foil and shaken in shaker water baths (Vega Bath model 224, Fisher Scientific, Pittsburgh, Pa.) at room temperature, away from the sunlight, for a period of 10 days.

The initial amounts of the TJL, BLL, and AVA 10 that were dissolved in the water were 10, 20, and 10 mg/L, respectively.

Samples from each jar were analyzed on the third, fifth or sixth, eighth, and tenth days. Each sample was analyzed for water quality changes as discussed in Chapter 2: pH, conductivity, oxygen content, disinfection residual, total organic carbon (TOC), and ultraviolet absorbance at 254 nm (UVA_{254}). FPA was conducted as discussed in Chapter 2, except that free or combined chlorine residual was removed by the addition of a stoichiometric amount of sodium sulfite before any FPA. Closed-loop stripping analysis (CLSA), followed by GC/MS and sensory GC, as discussed in Chapter 2, was used to analyze extracts of the water samples to determine ng/L levels of trace organic chemicals dissolved in the water. The compounds in the water samples were extracted primarily by CLSA (Mallevialle and Suffet 1987, Young and Suffet 1999). In this study, CLSA concentrated compounds 50,000 times (1 L of water to 20 μL of methylene chloride). Simultaneous distillation-extraction (SDE) was used on selected samples (Mallevialle and Suffet 1987, Young and Suffet 1999).

For the GC/MS analysis, for the results discussed in this chapter, a tentative identification of a compound is considered to occur when the data system indicates a percent identification of above 70%. Any identification less than 70% is considered to be a less reliable identification. A column bleed identification is indicated for all compounds containing silyl groups or is described as a silane.

For the sensory GC analysis, quality assurance was performed once a week for calibration of the olfactory port. Intensities of the odors eluting from the gas chromatograph were rated on a five-point scale: 1—threshold; 2—very weak; 3—weak to moderate; 4—moderate to strong; and 5—strong. This scale was a modification of the odor intensity scale of the FPA of 1—threshold; 2—very weak; 4—weak; 6—weak to moderate; 8—moderate; 10—moderate to strong; and 12—strong (APHA 1995).

SDE was selected to complement the CLSA technique by extracting organoleptic compounds of greater polarity and higher molecular weight than those extracted by CLSA. In this study, the SDE concentrated compounds 40,000 times: 4 L water to 0.1 mL dichloromethane (DCM). SDE was used only in the metal pipe studies.

Confirmation Test Protocol: Leaching of Lubricants and Rubber Gasket

A confirmation test protocol was used that differed from the original protocol in the following ways: Milli-Q water adjusted with 0.1 M orthophosphate buffer to pH 7.3 was chlorinated with 2.8 mg/L chlorine and shaken for 10 days. The Milli-Q water was not autoclaved or dropped into an open glass jar. The lubricants were made up to a 4 L solution by adding water to them after dissolving and premixing in smaller amounts of water. Then CLSA and GC/MS; sensory GC; and FPA before and after CLSA analyses were performed. Thus, autoclaving and tapwater were not used. The concentrations of the TJL and BLL used were 13 mg/L.

Because the TJL had the highest odor potential, a third run of TJL was completed on a more qualitative basis to look at the consistency of the compounds produced in the second run. A portion of TJL was added to 50 mL of Milli-Q water and mixed ultrasonically. The mixture was then diluted to about 2 L with Milli-Q water. The final concentration of the TJL solution was between the concentration of the first and second runs, that is, 10–13 mg/L.

The GC columns were changed for the second and third runs, as two matched capillary DB-5 columns of 60 m length, instead of 48 m length-columns, were used. This increased the retention time by about 5 min and was compensated for by adding 5 min to all retention times to maintain consistency of the data for the whole study.

Leaching of PVC Pipes

The four different pipes were analyzed for production of tastes and odors in drinking water under the worst water quality conditions that could cause the most leaching. These parameters were chosen on the bases of practical worst case. The parameters selected were

- pH = 6.5
- Mineral concentration (NaCl) = 500 mg/L
- Oxygen concentration = 8 mg/L
- Cl_2 concentration = 2.7–3.0 mg/L or
- NH_2Cl concentration = 3.0 mg/L

- Leaching time = 2 and 7 days

These are typical concentrations of total dissolved solids, oxygen, and monochloramine that could occur in drinking water; however, the Cl_2 concentration is higher than usually seen, and the pH is lower than usually observed. The most acidic pH found in the drinking water was used, as this is the most corrosive water quality condition.

PVC pipes were cut with a hacksaw into 2.5" X 3" X 0.5" pieces. The PVC pipes were then disinfected with a chlorine concentration of 20 mg/L for 24 h in pH 7.0 Milli-Q water. The pipes were subsequently rinsed three times with Milli-Q water before beginning the pipe leaching procedure. After the disinfection procedure, each pipe was submerged in Milli-Q water in a dark glass container and leached with 2.7 mg/L of Cl_2 for 2 and 7 days. All glassware used in the experiment was cleaned by acid-washing for 24 h in an acid bath (1:1 HNO_3:HCl), rinsed with Milli-Q water, and oven-dried.

White and gray PVC pipes were leached with or without the cement (Cristy's Red Hot Blue Glue™) and/or the primer (IPS' WELD•ON™ plastic pipe primer). Seven experimental set-ups were employed for each pipe. After 2 and 7 days of leaching, most experiments were analyzed as discussed in the above section except for CLSA and sensory GC, in which cases the analyses were performed only after 7 days of leaching.

Leaching of CIPs

DIP with cement–mortar lining and gray CIP were cut into 2-ft pieces. The diameters of the DIP and gray CIP were 4 and 3 in., respectively. The pipes were disinfected as described above for 24 h before use. Three experimental set-ups were employed for the two pipes. After the disinfection procedure, each pipe was filled with 2 L of the Milli-Q water. The pipes were closed by placing a silicone stopper on each end of the pipe. Each silicone stopper was wrapped in aluminum foil and covered with Teflon tape to stop any leaching from the silicon stopper. Each pipe was shaken for 7 days (Orbit Shaker, Lab-Line, Melrose Park, Ill.).

After 7 days of leaching each experimental set-up was analyzed as was done for the plastic pipes, with the addition of inductively coupled plasma (ICP) analysis for metals.

Weber-Fechner Curves of Iron (II) Sulfate Heptahydrate in Milli-Q Water

Milli-Q water was used for the FPA analysis. FPA samples were prepared in brown glass bottles. A 100-mg portion of iron (II) sulfate heptahydrate ($FeSO_4{\cdot}H_2O$) was dissolved in 1 L of Milli-Q water. The concentration of iron (II) solution (100 mg/L) was further diluted with Milli-Q water to the desired concentrations: 2.2, 6.0, 10.0, and 24.0 mg/L. Diluted standards were analyzed by FPA at a pH range of 5.5–8.8 in order to maintain conditions as similar as possible to those of the water collected from the CIPs. The pH of the standards was adjusted by adding drops of 0.1 N sulfuric acid or sodium hydroxide. The prepared samples were analyzed by the FPA method. Metal analyses were completed by ICP analysis (APHA 1995).

RESULTS AND DISCUSSION

Lubricants and Rubber Gasket

Water Quality Analysis and Flavor-Profile Analysis (FPA)

Preliminary analysis of the lubricants showed an increase in pH with their addition to distilled or autoclaved water. The effect of two of the lubricants, BLL and AVA, was slight; however, the TJL showed a marked increase in pH with increasing concentrations of lubricant. The magnitude of the change in pH differed depending on the nature of the water in which TJL was dissolved. In distilled water of pH 6.6, 10 mg/L of TJL resulted in an increase to pH 7.5. However, in tapwater of pH 7.4, 20 mg/L only increased the pH to 7.6. This probably results from the presence of the natural buffering effect of tapwater. Autoclaving tapwater was observed to decrease the buffering effect of tapwater. The pH changes in the autoclaved tapwater changes the same way distilled water changed, and the pH was adjusted to pH 7–8. The TYTON® rubber gasket did not change the pH of any medium.

The effect of biological activity on the leaching tests was evaluated. Two sets of samples were tested for bacterial content. Heterotrophic plate counts by the method of Winnen et al. (1996), using base agar, were performed on the AVA and BLL samples to check for any significant biological regrowth during the analysis time period. The samples were used from the eighth day of analysis. Autoclaved tapwater was used as a background check to confirm that the samples were sterile initially. Results showed that little or no bacteria were present, which suggests that any contribution from biological activity would be insignificant. Thus, the autoclaving step was eliminated from subsequent tests.

The confirmatory test indicates that it is not necessary to use autoclaved water, although tapwater from the actual location where the material is to be used (or is being used) is recommended. The final protocol for lubricants and gaskets should be:

- All glassware used in the experiment should be cleaned by acid-washing for 24 h in an acid bath (1:1 HNO_3:HCl), rinsed with Milli-Q water, and oven-dried.
- Two-inch sections of the gasket or 10–20 mg/L portions of lubricant should be dissolved in dechlorinated tapwater and in chlorinated or chloraminated (as used at the utility) tapwater at 3 mg/L. The jars should be filled to the rim, excluding any headspace.
- Analytical tests should be completed for up to 10 days and the results compared to the data presented in the present testing procedure.

Figures 6.1a and b show the change in disinfectant concentration over time for each material tested. The lubricant TJL appears more reactive than the gasket, and BLL was slightly more reactive than AVA 10. The BLL and AVA sample solutions maintained a chlorine residual after 10 days. The reverse was the case with the chloraminated water, with AVA and BLL having a lower residual, while TJL maintained the initial concentration of 3 mg/L for 10 days.

TYTON® Rubber Gasket (TRG). The gasket leached up to 5 mg/L of TOC (Tables 6.1–3). Tables 6.1–3 showed that the gasket produced a distinct rubbery odor, with gasoline or kerosene odors being produced in disinfected waters. Figure 6.2 shows that the rubbery odor increased in intensity over time in the media without disinfectant, and with chlorine. The intensity of the rubbery odor in the chloramine medium was erratic, but persistent. The kerosene odor (intensity 1.5–4) was defined in chloraminated water, whereas a gasoline odor difficult to separate from the rubbery odor (with a persistent intensity of about 2.7) was defined in chlorinated water. The odor produced by this material had a high degree of consensus among the FPA panelists.

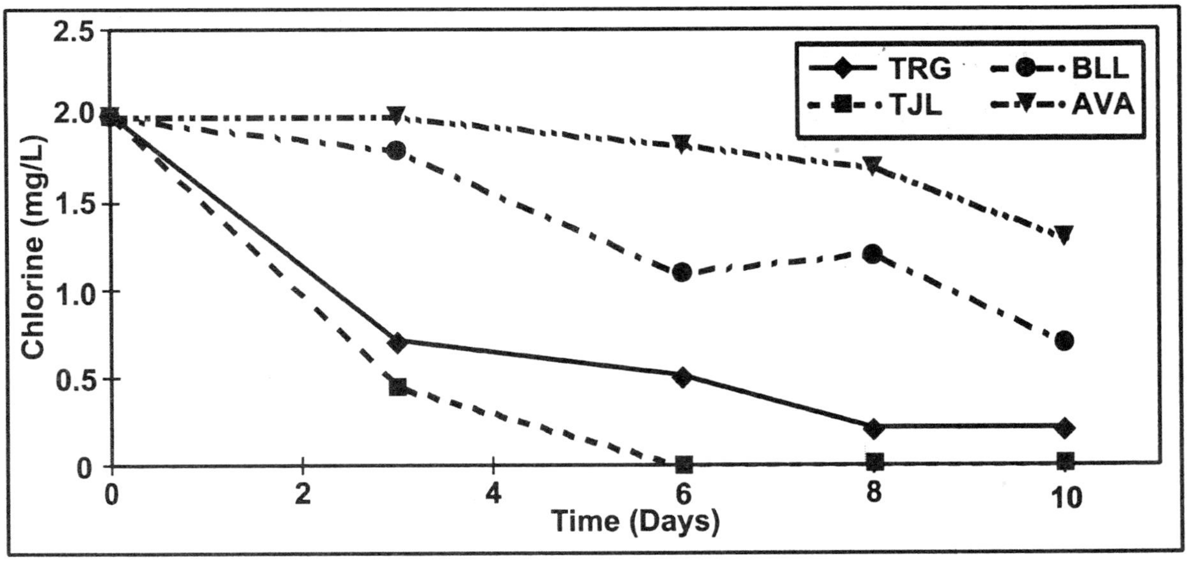

TRG - TYTON® rubber gasket
BLL - Blue Lube™ lubricant
TJL - TYTON® joint lubricant
AVA - Aquaflow VA-SP 10

Figure 6.1a Reduction in chlorine levels for different materials

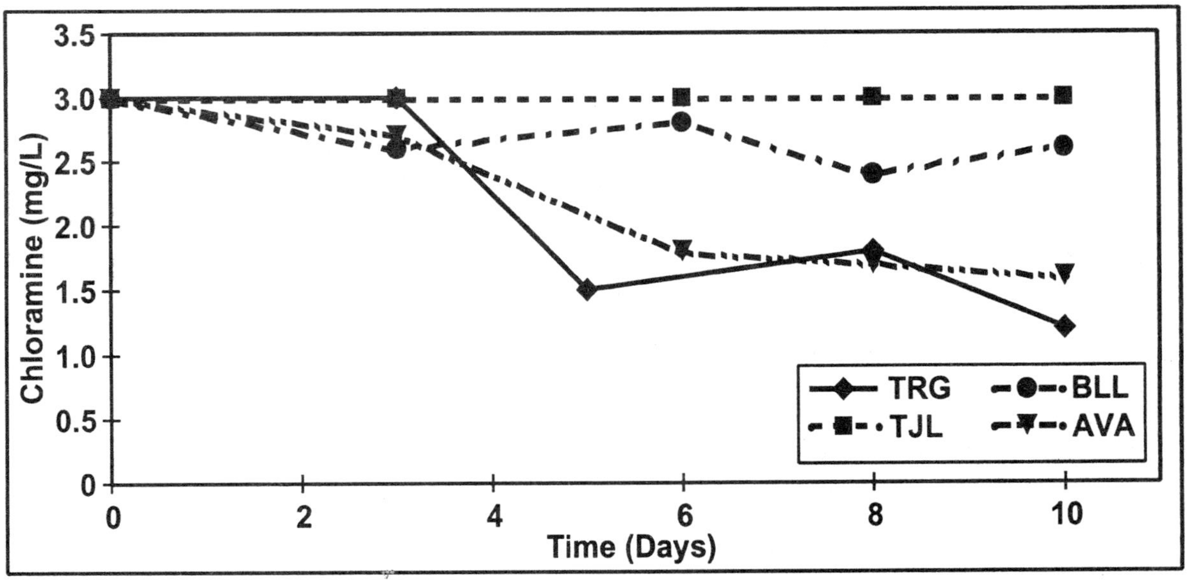

TRG - TYTON® rubber gasket
BLL - Blue Lube™ lubricant
TJL - TYTON® joint lubricant
AVA - Aquaflow VA-SP 10

Figure 6.1b Reduction in chloramine levels for different materials

Table 6.1

Water quality data of TYTON® rubber gasket (TRG) without disinfectant (temp. = 28°C)

Days	Cl_2 (mg/L)	TOC (mg/L)	UVA_{254}	Odor Intensities
Blank	-	1.0		odorless
3	-	3.3	0.0552	rubber (2.8±1.8), fruity (2.4±1.7)
6	-	3.6	0.0352	rubber (4.4±2.2), fruity (2.8±2.7)
8	-	4.0	0.0239	rubber (4.8±3.0)
10	-	4.3	0.0175	rubber (7.6±2.6)

Table 6.2

Water quality data of TRG in 3 mg/L chloramines (temp. = 28°C)

Days	NH_2Cl as Cl_2 (mg/L)	TOC mg/L	UVA_{254}	Odor Intensities
Blank	N/A			odorless
3	3.0	1.1	0.0023	rubber (4.3±1), kerosene (2.5±3.0)
5	1.5	2.5	0.0287	rubber (3±1.1), kerosene (2.0±2.3)
8	1.8	3.3	0.0504	rubber (5.0±2.2), kerosene (4.0±4.2)
10	1.2	4.6	0.0447	rubber (3.0±1.1), kerosene (1.5±1.9)

Table 6.3

Water quality data of TRG in 2 mg/L Cl_2 (temp. = 28°C)

Days	Cl_2 (mg/L)	TOC (mg/L)	UVA_{254}	Odor Intensities
Blank	N/A	1.0	0.0056	odorless
3	0.7	4.2	0.0513	gasoline/rubber (2.7±3)
6	0.5	5.0	0.0656	gasoline/rubber (2.7±3), sweet (2.0±1.8)
8	0.2	5.2	0.7832	rubber (5.3±1.0), gasoline (2.7±2.8)
10	0.2	6.0	0.1031	rubber (6.2±1.28)

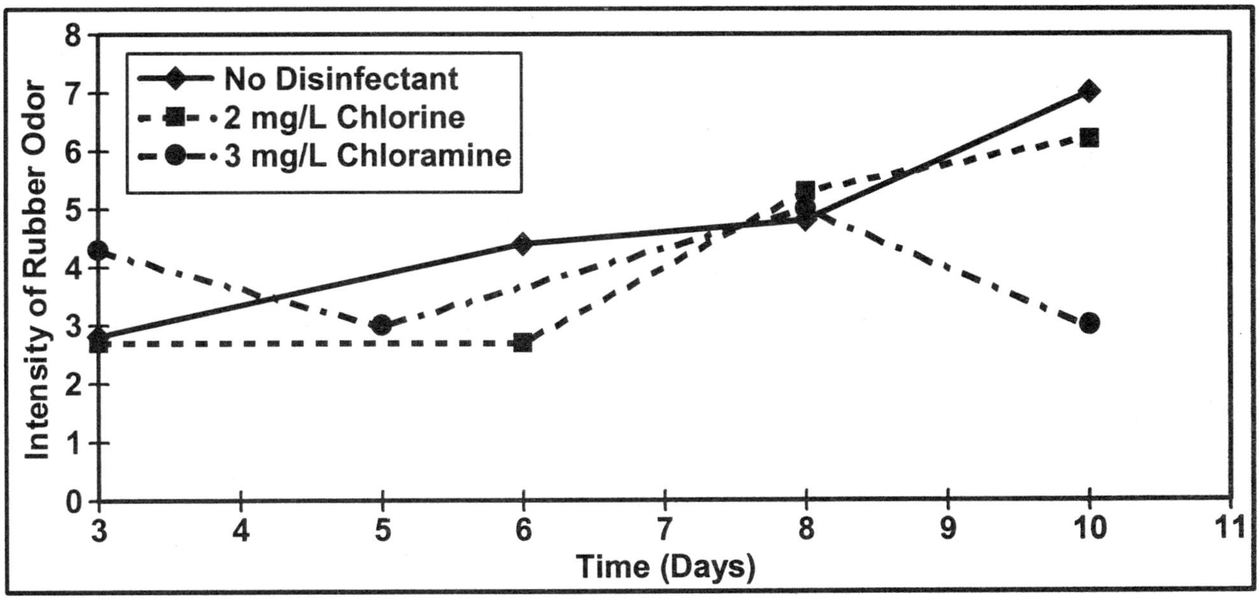

Figure 6.2 Trend in odor intensity of TRG

TYTON® Joint Lubricant (TJL). TJL has a distinct waxy/oily (rancid) odor (intensity of 5) and soapy odor (intensity of 5) when the lid of the product is opened after being opened and closed over a nine-month period. The most common odors perceived in the TJL water samples were the same waxy/oily (rancid) and soapy odors, with a rancid oil note associated with the waxy/oily odor (Tables 6.4–6). The waxy/oily odor was consistent in the medium without disinfectant at days 3, 6, and 10, although it was not detected at day 8. Generally, the dissolved TJL odor in water changed from waxy/wet cardboard to waxy/oily, and the soapy odor decreased with time in the chloramine and chlorine media as a rancid character became more associated with the waxy/oily note at days 6 and 8 for chlorinated and chloraminated water, respectively. Figure 6.3 plots the waxy/oily odor (rancid or not) versus time. The odor tends to build up to intensity 5 in nonchlorinated water. In chloraminated water, the odor intensity decreases from 3 to 2. In chlorinated water, the odor intensity decreases to background. Apparently, the chemicals causing this odor leach from the lubricant into water and can be

Table 6.4

Water quality data of TYTON® joint lubricant (TJL) without disinfectant (temp. = 28°C)

Days	Cl_2 (mg/L)	TOC (mg/L)	UVA_{254}	Odor Intensities
Blank	-	1.0	-	odorless
3	-	2.6	0.0320	wet cardboard (4.4±2.6), paint (3.2±3.0), waxy/oily (1.6±2.2)
6	-	1.8	0.0302	waxy/oily (4.0±28), wet cardboard (3.2±2.8), paint (2.4±2.2)
8	-	1.9	0.0212	citrus/sweet (3.6±0.9)
10	-	2.1	0.0263	waxy/oily (4.0±2.8), citrus/sweet (3.0±2.3)

Blank = Milli-Q water

Table 6.5

Water quality data of TJL in 3 mg/L chloramines (temp. = 28°C)

Days	NH_2Cl as Cl_2 (mg/L)	TOC (mg/L)	UVA_{254}	Odor Intensities
Blank	N/A	1.1		odorless
3	3	3.4	0.177	waxy/oily (3.0±1.1), soapy (3.0±3.8)
6	3	4.4	0.0262	waxy/oily (2.0±1.9), soapy (1.5±1.9), citrus/sweet (3.0±2.6)
8	3	4.7	0.0413	waxy/oily(rancid) (2.0±1.9), soapy (2.0±2.3)
10	3	4.3	0.0329	Soapy(2.5±2.5), Waxy/oily (2.0±2.8)

Blank = Milli-Q water

Table 6.6

Water quality data of TJL in 2 mg/L Cl_2 (temp. = 28°C)

Days	Cl_2 (mg/L)	TOC (mg/L)	UVA_{254}	Odor Intensities
Blank	N/A	1.0	0.006	odorless
3	0.45	5.0	0.063	Waxy/oily (5.7±1.3)
6	0	5.6	0.091	waxy/oily(rancid) (3±2), soapy (2±1.6)
8	0	3.3	0.086	soapy (1.6±1.9), citrus (1.4±1.7), sweet (0.8±0.8)
10	0	5.0	0.126	citrus (1.7±2), soapy (1.7±2.4), sweet (1±1.1)

Blank = Milli-Q water

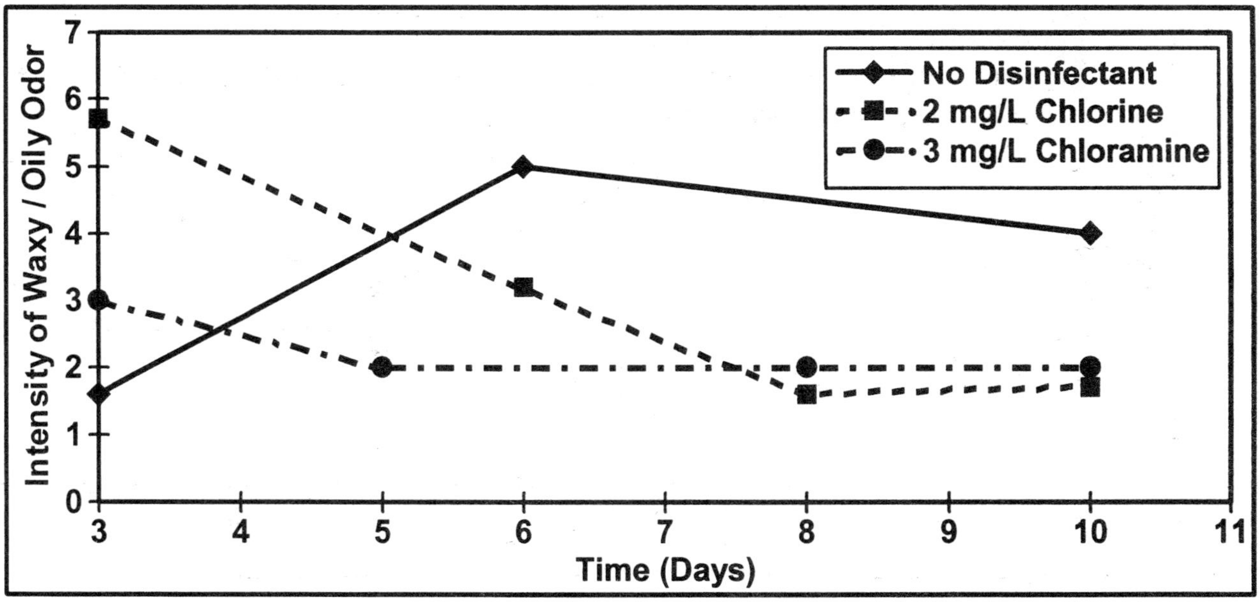

Figure 6.3 Trend in odor intensity of TYTON® joint lubricant (TJL)

eliminated by chloramination or chlorination with time. These samples always maintained a consistent soapy odor intensity of 1.5–3 when the water was disinfected. A sweet/citrus odor in the range of 0.8–3.6 appeared on days 8 and 10 for the samples in the chlorine and disinfection-free media and in the day 6 sample of the chloraminated sample.

Fadel et al. (1993) reported a wet paper odor (similar to that found in dechlorinated waters) and a chalky or wood putty odor from water in new mains using TJL. The chalky or wood putty odors were not obtained in this study. The amount of disinfectant, the amount of the TJL, and the study of pipes may be the causes of the differences between the Fadel et al. (1993) new water main study and this pipe study.

Blue Lube™ Lubricant (BLL) and Aquaflow VA 10 (AVA) Lubricant. BLL has a waxy odor (intensity 4) and soapy odor (intensity 4) when the can of the product is opened after being opened and closed over a nine-month period. Figure 6.4 and Tables 6.7–9 show that 20 mg/L of BLL has a sweet odor without disinfectant, and with chloramines it has an intensity of 1.4–2. Chlorine appears to oxidize something in BLL to a low kerosene and alcohol odor after 8 days with an intensity of 1–2. The odors of BLL were the lowest of the lubricants tested.

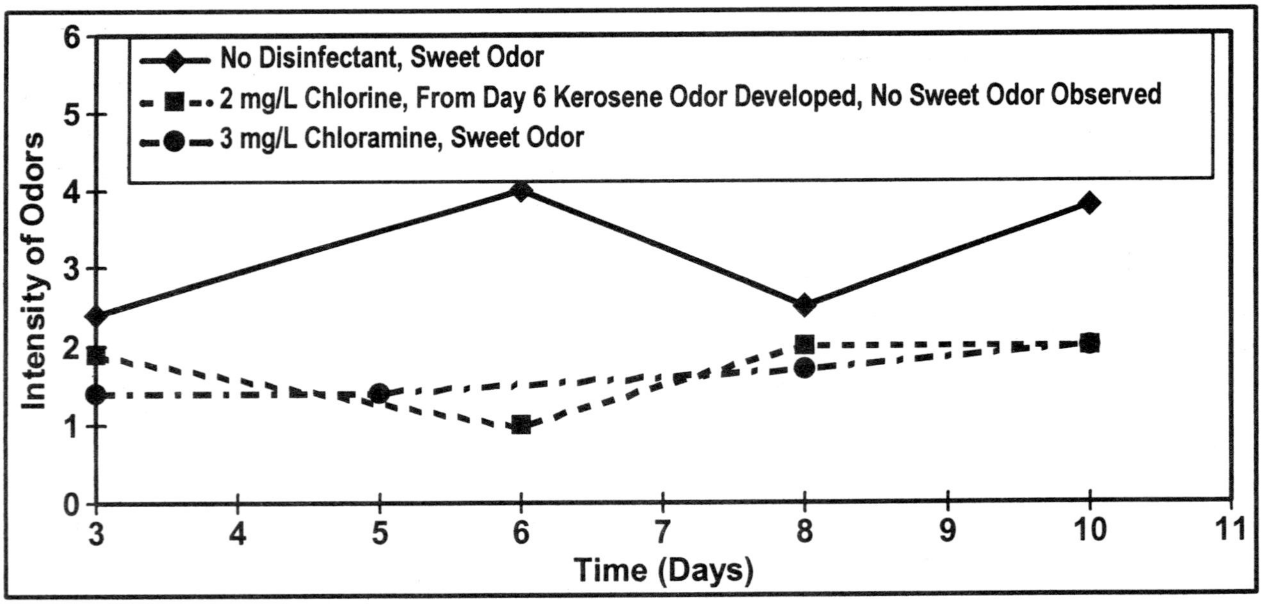

Figure 6.4 Trend in odor intensity for Blue Lube™ lubricant (BLL)

Table 6.7

Water quality data of Blue Lube™ lubricant (BLL) without disinfectant (temp. = 28°C)

Days	Cl_2 (mg/L)	TOC (mg/L)	UVA_{254}	Odor Intensities
Blank	-	1.0	-	Odorless
3	-	1.6	0.0423	sweet (2.5±1.0), paint (1.5±1.9)
6	-	1.7	0.0257	sweet (4±0), alcohol (1.5±1.9)
8	-	1.2	0.0247	sweet (2.5±1.9)
10	-	1.2	0.0339	sweet (3.8±2.1)

Table 6.8

Water quality data of BLL in 3 mg/L chloramines (temp. = 28°C)

Days	NH_2Cl as Cl_2 (mg/L)	TOC (mg/L)	UVA_{254}	Odor Intensities
Blank	N/A	1.1	-	odorless
3	2.6	2.0	0.0238	Sweet (1.4±0.9)
6	2.8	1.9	0.0191	sweet (1.4±0.9), earthy (1.0±1.0)
8	2.4	1.9	00218	Sweet (1.8±1.5)
10	2.6	2.2	0.0288	Sweet (2±1.2)

Table 6.9

Water quality data of BLL in 2 mg/L Cl_2 (temp. = 28°C)

Days	Cl_2 (mg/L)	TOC (mg/L)	UVA_{254}	Odor Intensities
Blank	N/A	1.0	0.0034	odorless
3	1.8	2.2	0.0299	earthy/musty (1.5±1.9)
6	1.1	1.8	0.0278	Sweet (1±1.1)
8	1.2	1.9	0.0251	alcohol (2±2.3), kerosene (2±2.8)
10	0.7	2.0	0.0355	Kerosene (2±1.6), musty (1.2±1.9), oily (rancid) (1±1.1)

The AVA product has a distinct pine-tar type of odor (intensity 6), a kerosene odor (intensity 4), a petroleum odor (intensity 4), and a background sweet odor (intensity 2) when the tube of the product is opened. However, Figure 6.5 and Tables 6.10 and 11 show that 10 mg/L of AVA also has a sweet odor (intensity 1.4–4), without disinfection or with chloramines, when dissolved in water.

In addition, the chloramines, and to a lesser extent the chlorination process, developed an astringency (intensity = 1.6–2.5) (Tables 6.11 and 12). Table 6.12 shows that the chlorination process produced a succession of odors that were then removed and replaced by another odor characteristic. Specifically, after an astringent mouthfeel occurred at <3 days and disappeared, an oily (rancid) odor occurred. The oily (rancid) odor appeared to be oxidized and lessened by day 8. Then, on day 10, phenol and pine odors were observed. The phenol and pine odors were quite different compared to the odors from chloramination treatment or the solution without disinfectant. When a higher concentration of about 40 mg/L AVA was dissolved in the water, the pine and kerosene odors were noticeable, each at an intensity of 6.

Both AVA and BLL had an odor that was more difficult to characterize compared to TJL. A sweet odor was common to both lubricants (see Figures 6.4 and 6.5) in samples derived from the chloramine and disinfectant-free media. The AVA samples in the chloramine medium had an additional astringent odor that was maintained over time. Most odors of BLL and AVA determined by FPA were of low intensities (2 or less), with only a few reaching a high of 4.0.

Table 6.10

Water quality data of AVA without disinfectant (temp. = 28°C)

Days	Cl_2 (mg/L)	TOC (mg/L)	UVA_{254}	Odor Intensities
Blank	-	1.1	-	odorless
3	-	2.1	0.0392	swampy (2.4±0.9), sweet (1.6±1.7)
6	-	1.5	0.411	sweet (4±1.4), tart (2±2)
8	-	1.1	0.0355	sweet (2.6±3.1)
10	-	1.0	0.0178	sweet (2.6±3.1), tart (2±2)

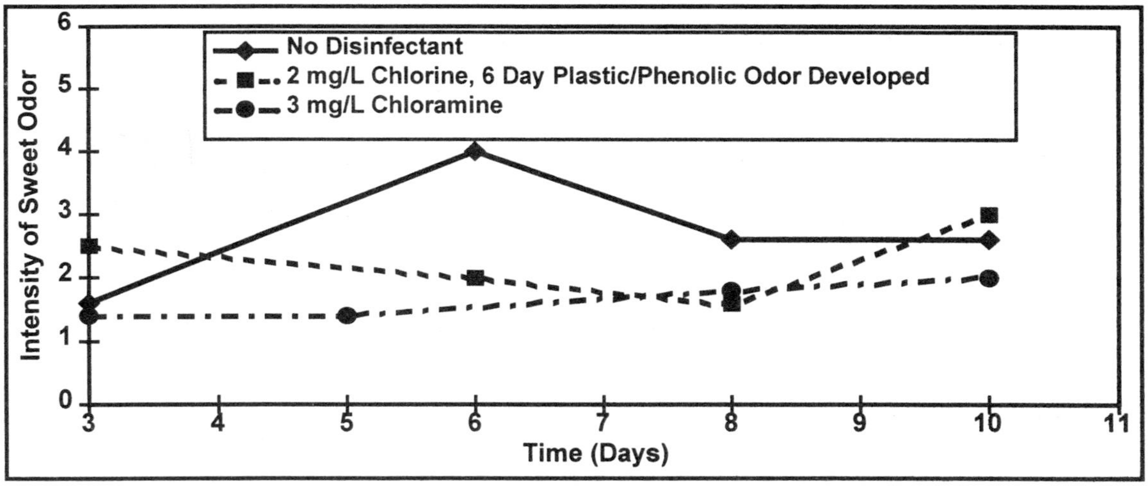

Figure 6.5 Trend in odor intensity of Aquaflow VA-SP 10 (AVA) lubricant

Table 6.11

Water quality data of AVA in 3 mg/L chloramines (temp. = 28°C)

Days	NH_2Cl as Cl_2 (mg/L)	TOC (mg/L)	UVA_{254}	Odor Intensities
Blank	N/A	1.1		odorless
3	2.7	1.2	0.0373	astringent (1.6±1.7), sweet (1.4±1.7)
6	1.8	1.5	0.0159	astringent (1.8±2.5), sweet (1.4±1.7)
8	1.7	1.9	0.0194	astringent (1.6±1.8), sweet (1.8±1.1)
10	1.6	2.5	0.0317	astringent (2.2±1.8), sweet (2.0±2.0)

Table 6.12

Water quality data of AVA in 2 mg/L Cl_2 (temp. = 28°C)

Days	Cl_2 (mg/L)	TOC (mg/L)	UVA_{254}	Odor Intensities
Blank	N/A	1.0	0.0034	odorless
3	2	2.2	0.0551	astringent (2.5±2.3)
6	N/A	1.3	0.0440	oily(rancid) (4±2.4), plastic (2±2)
8	1.7	0.8	0.0300	sweet (1.6±1.7), oily(rancid) (1.6±1.7)
10	1.3	0.8	0.0210	phenol (3±3.5), pine (3±3.5)

Conclusions of FPA Results on Lubricants

- The TRG without disinfectant and in chlorine had a rubbery odor that was of concern, with a kerosene background, in chlorinated water.
- The most common odors perceived in the TJL water samples were the same waxy/oily odors of the original product, which develops a rancid oil note as well as a soapy note. Chlorine disinfection decreased the odor to a sweet, citrus, and soapy odor at an intensity of 2.5 or less after 10 days. The dissolved TJL changed from an original odor, in water alone, of waxy/wet cardboard to waxy/oily (rancid) and soapy to sweet/citrus and soapy.
- The odor of AVA changed from astringent (0–<6 days) to oily (rancid) (>3–<10 days) to a pine and phenol odor (>8 days). All the odors were less than 3 on the FPA scale. The chloramine disinfection produced the astringency odor over 10 days of mixing.
- The BLL had a sweet odor without disinfectant, and with chloramine it had intensities of 1–4. Chlorine appeared to oxidize something in BLL to a kerosene and alcohol odor after >6 days, and the kerosene odor persists to >10 days with a low odor response (intensity of 1–2).
- BLL had odors that were the lowest of the lubricants tested. All three waters had the lowest odor intensities and the sweetest odor.

It was hypothesized that CLSA, GC/MS, and sensory GC of disinfection leaching studies may be able to identify the odors associated with the lubricants placed in disinfected water. Sensory GC and GC/MS have been found able to easily evaluate FPA data when the odor of the FPA intensity is above 4.0 on the seven-point FPA intensity scale of threshold (1) to 12 (Khiari, Barrett, and Suffet 1997). Thus, the chemicals causing the TRG's rubber odor, the TJL's waxy/oily (rancid) odor, and possibly the sweet odor of BLL are the most likely chemicals to be evaluated by sensory GC.

GC/MS and Sensory GC Analysis

Lubricant Dissolved in Solvent (Original Odor). TJL and BLL were dissolved in a solvent, and a GC/MS analysis was completed to try to identify any compound(s) associated with the original odors that could be present in subsequent aqueous leachate extracts. Figures 6.6a–c illustrate the GC/MS of the spell out dichloromethane (DCM) solvent used in these studies: TJL dissolved in DCM, and BLL dissolved in acetonitrile (as it would not dissolve in DCM). It should be remembered that:

- The GC/MS tentative identification level for the database percentage identification is 70%.
- The retention time limit for the sensory GC analysis is about 35 min, as burnt odors overcome any sensory evaluation after 40 min.

Figure 6.6a of the blank DCM did not show any GC peaks below 39 min. Figure 6.6b of the TJL had GC peaks at retention times of 17.1 and 39.7 min. Figure 6.6c of the BLL (dissolved in acetonitrile) did not show any GC peak below 39.4 min and showed two GC peaks at 39.4 and 39.7 min that were not column bleed. (Peaks from column bleed are noted as silanes and siloxanes on Figures 6.6a–c.) Only the GC peak at 39.7 min in TJL had a tentative identification above 70%. This peak was tentatively identified as a hydrocarbon, tetratriacontane. Sensory GC analysis could not associate any peak with the odor of the original lubricants.

Water Blank—Initial Study. Figures 6.7a–c are chromatograms for CLSA and GC/MS analyses of leaching protocol control blanks under the same experimental conditions as the leaching experiment, without the lubricants or gasket for experimental runs 1 and 2. Chlorine was used in the control blank. Figure 6.7a is a control Milli-Q water analysis with 2 mg/L chlorine after 10 days (first-run condition). Four of the eight peaks with a retention time less than 39 minutes were tentatively identified as common organic chemicals that appear to be present in Milli-Q water: 4-methyl-2-pentanone, toluene, 2-ethyl hexanol, and 2-decanone.

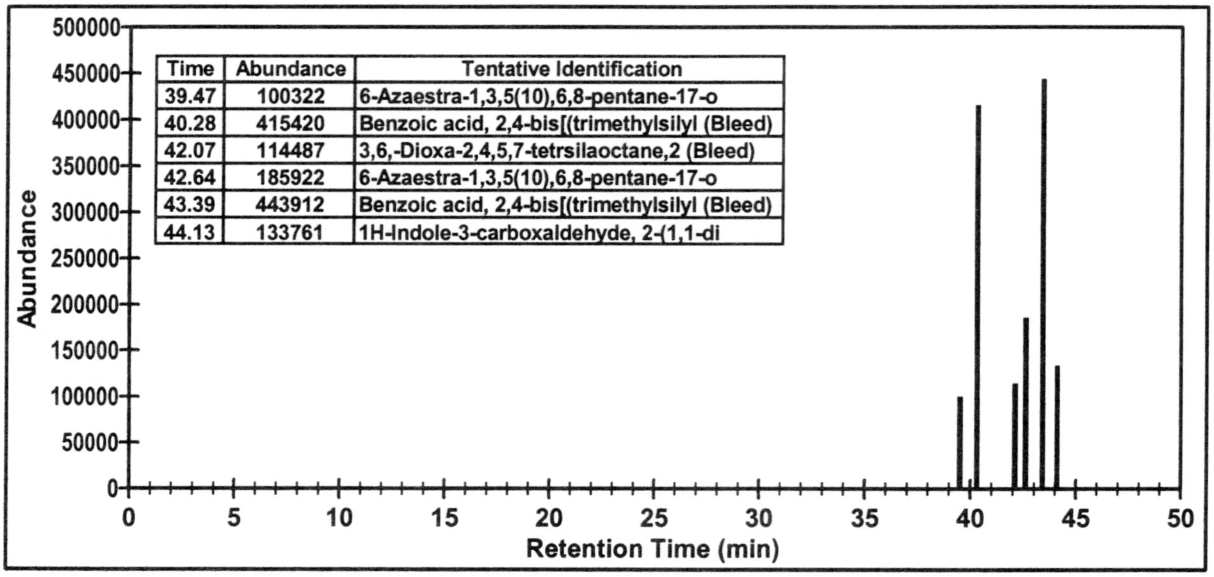

Time	Abundance	Tentative Identification
39.47	100322	6-Azaestra-1,3,5(10),6,8-pentane-17-o
40.28	415420	Benzoic acid, 2,4-bis[(trimethylsilyl (Bleed)
42.07	114487	3,6,-Dioxa-2,4,5,7-tetrsilaoctane,2 (Bleed)
42.64	185922	6-Azaestra-1,3,5(10),6,8-pentane-17-o
43.39	443912	Benzoic acid, 2,4-bis[(trimethylsilyl (Bleed)
44.13	133761	1H-Indole-3-carboxaldehyde, 2-(1,1-di

Figure 6.6a GC/MS of blank dichloromethane (DCM) solvent

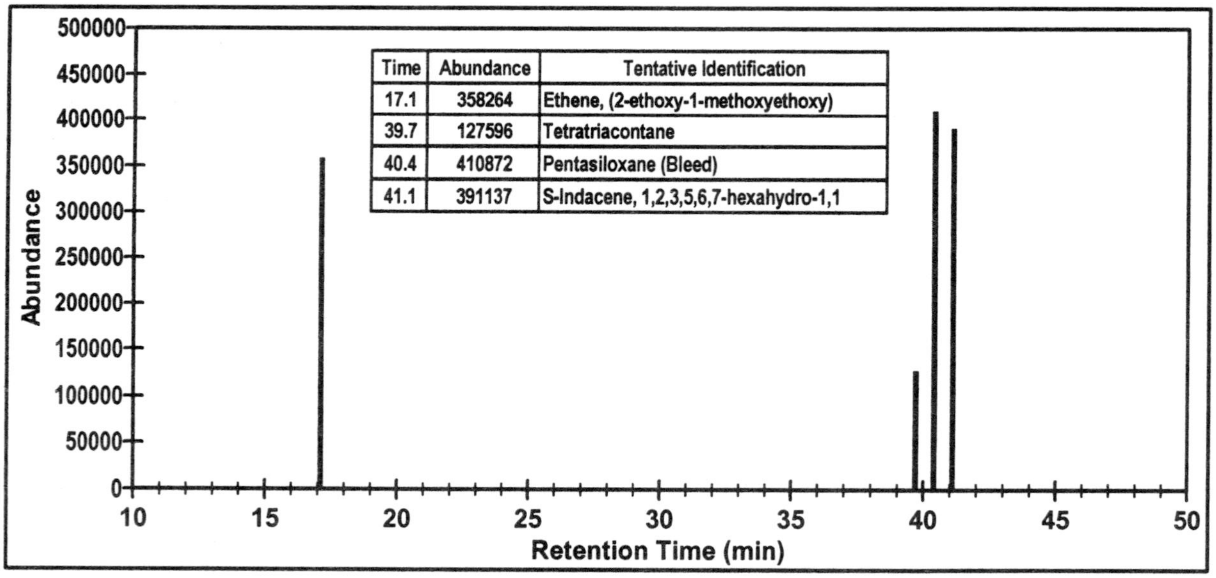

Time	Abundance	Tentative Identification
17.1	358264	Ethene, (2-ethoxy-1-methoxyethoxy)
39.7	127596	Tetratriacontane
40.4	410872	Pentasiloxane (Bleed)
41.1	391137	S-Indacene, 1,2,3,5,6,7-hexahydro-1,1

Figure 6.6b Gas chromatography/mass spectrometry (GC/MS) of TJL dissolved in DCM

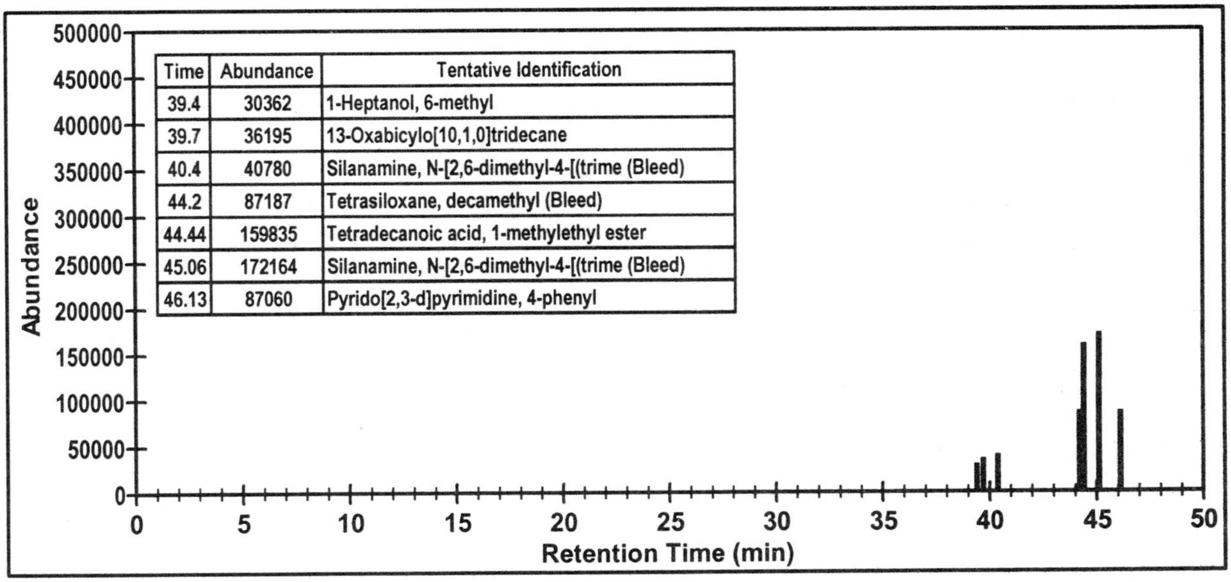

Time	Abundance	Tentative Identification
39.4	30362	1-Heptanol, 6-methyl
39.7	36195	13-Oxabicylo[10,1,0]tridecane
40.4	40780	Silanamine, N-[2,6-dimethyl-4-[(trime (Bleed)
44.2	87187	Tetrasiloxane, decamethyl (Bleed)
44.44	159835	Tetradecanoic acid, 1-methylethyl ester
45.06	172164	Silanamine, N-[2,6-dimethyl-4-[(trime (Bleed)
46.13	87060	Pyrido[2,3-d]pyrimidine, 4-phenyl

Figure 6.6c GC/MS of BLL dissolved in acetonitrile

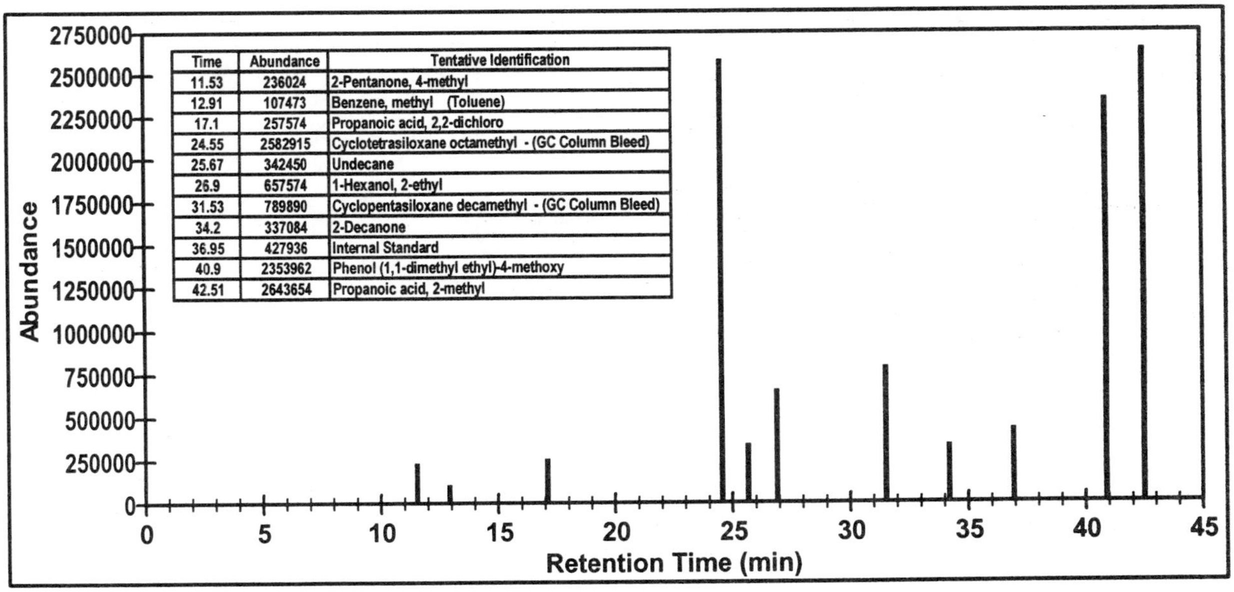

Time	Abundance	Tentative Identification
11.53	236024	2-Pentanone, 4-methyl
12.91	107473	Benzene, methyl (Toluene)
17.1	257574	Propanoic acid, 2,2-dichloro
24.55	2582915	Cyclotetrasiloxane octamethyl - (GC Column Bleed)
25.67	342450	Undecane
26.9	657574	1-Hexanol, 2-ethyl
31.53	789890	Cyclopentasiloxane decamethyl - (GC Column Bleed)
34.2	337084	2-Decanone
36.95	427936	Internal Standard
40.9	2353962	Phenol (1,1-dimethyl ethyl)-4-methoxy
42.51	2643654	Propanoic acid, 2-methyl

Figure 6.7a GC/MS of the control of autoclaved Milli–Q water, including chlorine without gasket or lubricants (first run)

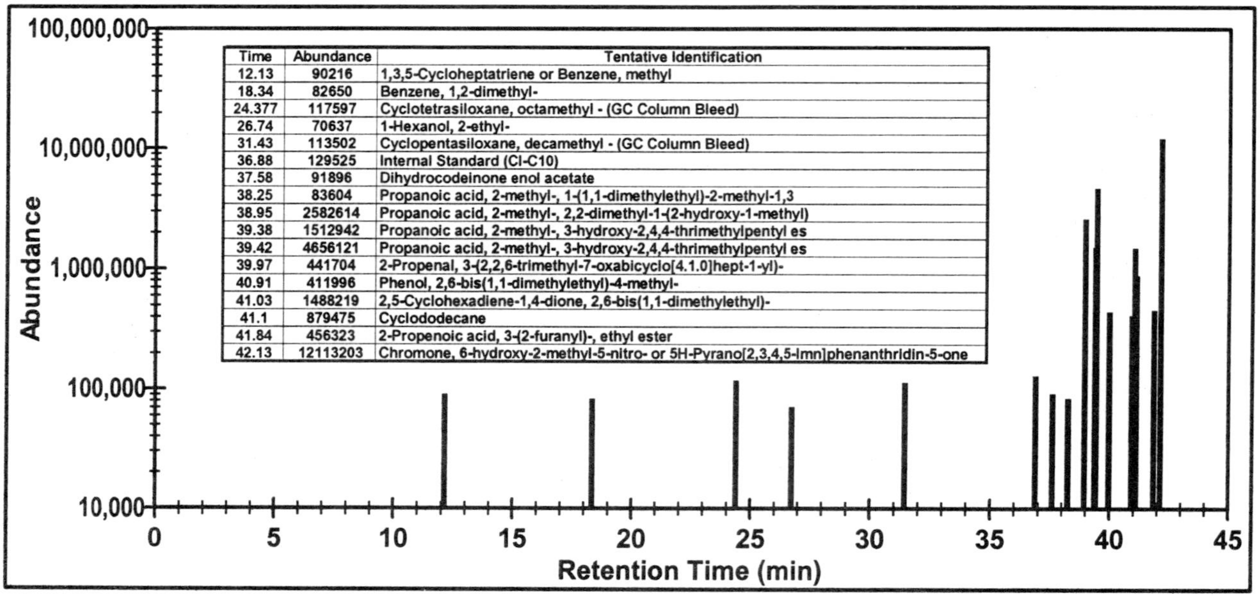

Time	Abundance	Tentative Identification
12.13	90216	1,3,5-Cycloheptatriene or Benzene, methyl
18.34	82650	Benzene, 1,2-dimethyl-
24.377	117597	Cyclotetrasiloxane, octamethyl - (GC Column Bleed)
26.74	70637	1-Hexanol, 2-ethyl-
31.43	113502	Cyclopentasiloxane, decamethyl - (GC Column Bleed)
36.88	129525	Internal Standard (Cl-C10)
37.58	91896	Dihydrocodeinone enol acetate
38.25	83604	Propanoic acid, 2-methyl-, 1-(1,1-dimethylethyl)-2-methyl-1,3
38.95	2582614	Propanoic acid, 2-methyl-, 2,2-dimethyl-1-(2-hydroxy-1-methyl)
39.38	1512942	Propanoic acid, 2-methyl-, 3-hydroxy-2,4,4-thrimethylpentyl es
39.42	4656121	Propanoic acid, 2-methyl-, 3-hydroxy-2,4,4-thrimethylpentyl es
39.97	441704	2-Propenal, 3-(2,2,6-trimethyl-7-oxabicyclo[4.1.0]hept-1-yl)-
40.91	411996	Phenol, 2,6-bis(1,1-dimethylethyl)-4-methyl-
41.03	1488219	2,5-Cyclohexadiene-1,4-dione, 2,6-bis(1,1-dimethylethyl)-
41.1	879475	Cyclododecane
41.84	456323	2-Propenoic acid, 3-(2-furanyl)-, ethyl ester
42.13	12113203	Chromone, 6-hydroxy-2-methyl-5-nitro- or 5H-Pyrano[2,3,4,5-lmn]phenanthridin-5-one

Figure 6.7b GC/MS for Milli–Q water blank (pH 7.3) for lubricant leaching experiment (second run)

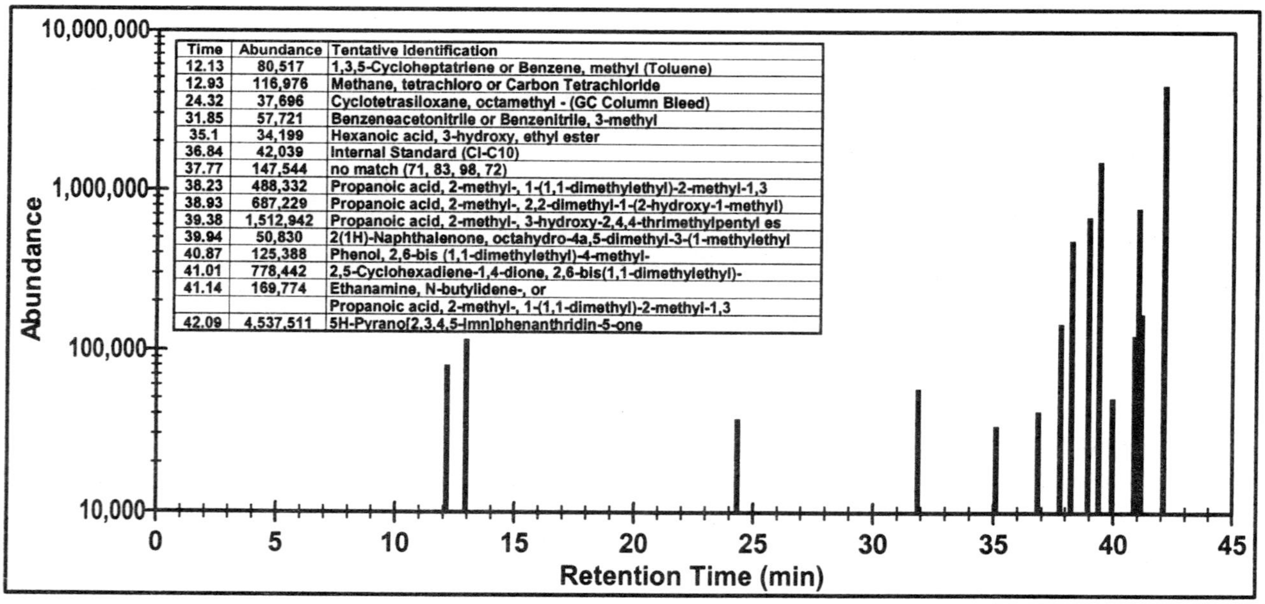

Time	Abundance	Tentative Identification
12.13	80,517	1,3,5-Cycloheptatriene or Benzene, methyl (Toluene)
12.93	116,976	Methane, tetrachloro or Carbon Tetrachloride
24.32	37,696	Cyclotetrasiloxane, octamethyl - (GC Column Bleed)
31.85	57,721	Benzeneacetonitrile or Benzenitrile, 3-methyl
35.1	34,199	Hexanoic acid, 3-hydroxy, ethyl ester
36.84	42,039	Internal Standard (Cl-C10)
37.77	147,544	no match (71, 83, 98, 72)
38.23	488,332	Propanoic acid, 2-methyl-, 1-(1,1-dimethylethyl)-2-methyl-1,3
38.93	687,229	Propanoic acid, 2-methyl-, 2,2-dimethyl-1-(2-hydroxy-1-methyl)
39.38	1,512,942	Propanoic acid, 2-methyl-, 3-hydroxy-2,4,4-thrimethylpentyl es
39.94	50,830	2(1H)-Naphthalenone, octahydro-4a,5-dimethyl-3-(1-methylethyl
40.87	125,388	Phenol, 2,6-bis (1,1-dimethylethyl)-4-methyl-
41.01	778,442	2,5-Cyclohexadiene-1,4-dione, 2,6-bis(1,1-dimethylethyl)-
41.14	169,774	Ethanamine, N-butylidene-, or
		Propanoic acid, 2-methyl-, 1-(1,1-dimethyl)-2-methyl-1,3
42.09	4,537,511	5H-Pyrano[2,3,4,5-lmn]phenanthridin-5-one

Figure 6.7c GC/MS for control for lubricant leaching experiment (Milli–Q water, pH 7.3, chlorine 2.8 mg/L) (second run)

At greater than 40 min, a high-molecular-weight phenol and a proprionic acid were tentatively identified. The other peaks that were present in the sample did not originate from gaskets and lubricants. The proprionic acid was subsequently identified to be in the water blank (see below).

Figures 6.7b and c are chromatograms for Milli-Q water blanks and control blanks, respectively, under confirmatory leaching run 2, where water was adjusted with 0.1 M orthophosphate buffer to pH 7.3 chlorinated with 2.8 mg/L chlorine, and shaken for 10 days, and the GC column was 60 m instead of 48 m. Autoclaved Milli-Q water was not used as in run 1, Figure 6.7a.

Figures 6.7a–c show a total of 19 different GC peaks (versus eight peaks, run 1) on the GC/MS total ion current chromatogram that are not noted as GC bleed or internal standards. At greater than 38 min, a series of three to four proprionic acid esters is tentatively identified on both Figures 6.7b and c. These types of compounds were identified in some of the samples during the initial study as well. These data indicate that these chemicals were in the water blanks and were not sample components. The chemicals in the blanks will be eliminated from all subsequent data analysis.

TJL Leached into Water. Figure 6.3 indicates that the waxy/oily odor intensity of TJL is leached from the lubricant and that chlorination and chloramination can break down the odor with time. Table 6.6 and Figure 6.3 show that chlorine is a more efficient oxidant, and the odor disappeared after 6 days of chlorination at 2 mg/L for run 1. A soapy odor that was part of the aroma of the original product appeared during the chlorination and chloramination processes.

Tables 6.13a–c show the results of the CLSA–GC/MS and sensory GC analyses after 10 days of leaching with chlorine, the first run with 10 mg/L TJL and 2.0 mg/L chlorine, the second run with 13 mg/L TJL and 2.8 mg/L chlorine, and the third run with 10–13 mg/L of TJL and 2.8 mg/L of chlorine. All extraneous GC peaks caused by column bleed and blanks have been eliminated. The first run (Table 6.13.a) shows that the waxy/oily/rancid odors were tentatively identified by sensory GC and GC/MS to consist of a mixture of heptanal, octanal, and nonanal. These data indicate that the odor of the original compounds in TJL was changed by chlorine, and the odor of the original product and the leaching product appear to be different.

Tables 6.13b and c confirm the CLSA–GC/MS and sensory GC analyses of run 1 after 10 days of leaching with chlorine. The waxy/oily/rancid odors were tentatively identified by sensory GC and GC/MS to consist of a mixture of aldehydes, as in the first run (heptanal, octanal, nonanal), plus other compounds: for example, in run 2 (Table 6.13b), ketones (e.g., 2-heptanone or 4- and 5-methyl-2-hexanone), an unsaturated alcohol (2-heptene-1-ol), and a borneol-type compound may also be part of the odorous mixture. Other odor notes (skunk and spoiled milk) observed by sensory GC analysis, in combination with the chemicals above, could give a rancid quality to the waxy/oily odor. Many other aldehydes, ketones, alcohols, hydrocarbons, and bicyclo compounds were tentatively identified by GC/MS, although the concentrations were too low to observe sensory GC descriptors. In all, 35 new compounds were observed as distinct GC peaks from the chlorination of TJL in run 2. The higher concentration of TJL (13 mg/L) and chlorine (2.8 mg/L) for the second run versus the first run at a TJL level of 10 mg/L and a chlorine level of 2 mg/L was needed to observe the reactions of TJL and chlorine.

The third TJL run had odor intensities similar to those of run 2. Table 6.13c shows that in the third run, the same 2-heptanone was also soapy/waxy and the bicyclo compound was waxy/wet cardboard instead of spoiled milk/waxy, as in run 2. The same or very similar compounds were noted, with the same odors of mushroom/sweet, apple, and skunky. This confirms the reproducibility of the sensory GC analysis and indicates that the process of reaction of the lubricant was occurring.

These data confirm the data of Fadel et al. (1993), who also observed these same aldehydes in TJL from water in new mains. The characterization of the odors by sensory GC (Tables 6.13a–c) indicates that heptanal (oily/waxy) is similar to that found in the odor threshold studies of Fadel et al. (1993), as well as researchers in Spain (see Chapter 4 of this work), who described the odor of heptanal as rancid walnut oil and rancid oil plus fish, and fruity, respectively. The characterization of the odors by sensory GC (Tables 6.13a–c) indicates that the octanal (waxy) odor is similar to that of the odor threshold studies discussed in Chapter 4, where octanal was described as rancid, fruity, citric, rancid oil, and sweet. The characterization of the odors by sensory GC (Tables 6.13a–c) indicates that nonanal (rancid oil) is similar to the odor threshold descriptions by Fadel et al. (1993) and in Chapter 4, with the odor described as hay-like and rancid, and fruity, respectively. The compounds tentatively identified by GC/MS in the sample are associated with the waxy odors. A soapy odor with an intensity of less than 3 has not been identified in these samples, except that the hexanal peak had a hint of thisodor. Apparently,

Table 6.13a

Gas chromatography/mass spectrometry (GC/MS) analysis of TJL (10 mg/L) leached into water for 10 days: Milli-Q water, pH 7.3, chlorine 2.0 mg/L, first run

Retention Time (min*)	Tentative Identification	Sensory GC Odor Description
10.5	Not Identified	N/A
13.2	Not Identified	N/A
14.5	Hexanal	Herbal/Grassy
20.2	Heptanal	Oily/Waxy
23.3	Not Identified	N/A
24	7-Octen-4-ol	Sweet
25.45	Octanal	Waxy
26.7	Nonanal	
27.9	Not Identified	N/A
28.2	Not Identified	N/A
30.08	Decanal	Sweet/Soupy
32.1	Not Identified	N/A
37.1	*Not Identified*	*N/A*
37.3	*Not Identified*	*N/A*
40.6	*Not Identified*	*N/A*
46.8	*Naphthalene*	*Burn Oil*
47.8	*Not Identified*	*N/A*
48.1	*Not Identified*	*N/A*
48.6	*Phenoxazine*	*Sweet/Doughy*

NOTES:

Italics correspond to a burnt sensory background of the sensory-GC as at 40 minutes the column oven is 270 °C. The column oven is at >150 °C after 27.5 minutes.

*—Retention time adjusted by adding 5 min to correspond to second GC/MS analysis.

Table 6.13b

GC/MS analysis of TJL (13 mg/L) leached into water for 10 days:

Milli-Q water, pH 7.3, chlorine 2.8 mg/L, second run

Ret. Time (min)	Tentative Identification	Identification %	Sensory GC Odor Description
14.26	Cyclohexanol, or hexanal, or 1-Penten-3-ol, 4-methyl, or 1-Hexen-3-ol	40, 39, 37, 37	
19.42	2-Heptanone or 2-Hexanone, 5-methyl	90, 64	Burnt/Waxy (4)
20.13	Heptanal; Hexanal, 4-methyl	64, 42	Burnt/Waxy (4)
23.12	2-Hepten-1-ol	28	Burnt/Waxy (4)
23.46	Benzaldehyde	83	
24.34	7-Octen-4-ol	53	Mushroom/Sweet(4-6)
24.81	2-Hexanone, 4-methyl	56	Mushroom/Sweet(4-6)
25.53	Octanal	47	Fresh Apple (4)
26.19	7-Oxabicyclo[2,2,1]heptane, 1-methyl-4-(1-methylethyl)-	72	Fresh Apple
27.1	Cineole	91	Apple(2)
28.18	2-Octenal, (E)-	33	
28.55	2,4-Heptadienal, (E,E)-	12	
29.65	11-Dodecen-2-one	38	
30.32	2-Nonen-1-ol	42	Green
31.4	Bicyclo[2,2,1]heptan-2-ol, 1,3,3-trimethyl	74	Burnt (2)
31.83	Bicyclo[4,1,0]heptan-3-ol, 3,7,7-trimethyl-[1S-(1.alpha., 3.	43	Skunk (2)
32.74	2-Nonenal, (E)-	64	Skunk (4)
33.55	Bicyclo[2,2,1]heptan-2-ol, 1,7,7-trimethyl- or Borneol, or Isoborneol	90, 58, 58	Spoiled Milk/Waxy (4)
34.41	3-Cyclohexene-1methanol, .alpha., alpha.4-trimethyl-	86	
35.49	Bicyclo[2,2,1]heptane, 7-methylene-2-(1-methylethylidene)	64	
36.71	2-Decenal, (Z)-	78	
37.02	3-Heptyn-1-ol	10	
37.61	*Cyclopentane, 1-methyl-2-methylene*	*45*	
38.18	*2,4-Decadienal, (E,E)-*	*91*	
38.51	*4-Octanol, 7-methyl*	*12*	
39.16	*2-Undecanethiol, 2-methyl*	*42*	
39.21	*1-Pentadecanol*	*43*	
39.7	*2-Undecanone, 6,10-dimethyl*	*64*	
39.8a	*Hexacosane*	*53*	
39.82a	*Octacosane*	*64*	
40.02a	*Oxirane, tetradecyl*	*47*	
40.31a	*Bicyclo[2,2,1]heptan-2-ol, 1,3,3-trimethyl*	*38*	
41.07a	*7-Hexadecene, (Z)-*	*91*	
41.32a	*2(1H)-Pyridione, 1-propyl*	*30*	
41.57a	*Eicosane*	*95*	

NOTES:

1. Italics correspond to a burnt sensory background of the sensory GC, as at 40 minutes the column oven is 270 °C. The column oven is at >150 °C after 27.5 min.
2. A tentative identification above 70% by the mass spectral data base is treated as a higher probability

Table 6.13c

GC/MS analysis of TJL leached into water for 10 days:

Milli-Q water, pH 7.3, chlorine 2.8 mg/L, third run

Ret. Time (min)	Tentative Identification	Identification %	Sensory GC Odor Description
14.15	Hexanal	38	
19.35	2-Heptanone or 5-methyl-2-Hexanone	90, 64	Soapy/Waxy(4)
20.09	4-methyl-Heptanal; Hexanal,	64, 42	
23.08	No match found		
23.4	Benzaldehyde	56	
24.05	Hexanoic acid	56	
24.27	7-Octen-4-ol	53	Mushroom/Sweet(6)
24.76	6-methyl-2-Heptanone	83	Mushroom/Sweet(6)
25.47	Octanal	47	Mushroom/Sweet(6)
26.14	7-Oxabicyclo[2,2,1]heptane, 1-methyl-4-(1-methylethyl)-	72	
27.04	Cineole	90	Apple(2-4)
28.14	2-Octenal, (E)-	64	Caramel (2-4)
28.5	2-Octen-1-ol, (E)	37	Caramel (2-4)
28.67	Cyclobutane, 1,2-diethyl-, cis-	36	
29.6	11-Dodecen-2-one	38	Caramel (2-4)
30.27	Nonanal	45	Green/Citrus (2)
31.14	Bicyclo[2,2,1]heptan-2-ol, 1,3,3-trimethyl	74	
31.79	Cyclopropane, trimethylmethylene or 2,4-Heptadiene, (E,E)-	43	Painty(4-6)
32.7	2-Nonenal, (E)-	59	Skunk(4-6)
33.51	Bicyclo[2,2,1]heptan-2-ol, 1,3,3-trimethyl-, (1S-endo)-	90	Waxy/Wet Cardboard
34.36	3-Cyclohexene-1methanol, .alpha., alpha.4-trimethyl-	78	
35.06	2,4-Nonadienal, (E,E)-	86	
36.12	1H-Pyrazole, 3-ethyl-4,5-dihydro-	46	
36.68	2-Decenal, (Z)-	78	
37.57	*2,4-Decadienal, (E,E)-*	*80*	
38.16	*2,4-Decadienal, (E,E)-*	*91*	
38.49	*2-Heptanal, (E)-*	*35*	
39.13	*2-Undecenal*	*86*	*Oily/Castor Oily(6)*
39.78	*Nonadecane*	*80*	*Oily/Castor Oily(6)*
39.87	*4H-Naphthol[1,2-b]pyran-4-one, 5,6-dimethoxy-2-methyl-*	*46*	
40.62	*Oxabicyclo[3.2.1]octan-7-one, 1,5-dimethyl-8-[2-[3-(1-methyl*	*50*	
41.04	*7-Hexadecene, (Z)-*	*87*	
41.26	*1-Phenanthrenecarboxilic acid, 1,2,3,4,4a,9,10,10a-octahydro*	*93*	
41.52	*Phenol, 2,6-bis(1,1-dimethylethyl)-*	*53*	

NOTES:

1. Italics correspond to a burnt sensory background of the sensory GC as at 40 min the column oven is at 270 °C. The column oven is at >150 °C after 27.5 min.
2. A tentative identification above 70% by the mass spectral data base is treated as a higher probability for future identification.

the sensitivity of the CLSA/sensory GC method is too low or the method does not efficiently extract the soapy odor from aqueous solution.

In summary, a series of straight-chained C-6 to C-10 aldehydes and a mixture of other aldehydes, ketones, alcohols, hydrocarbons, and bicyclo compounds are implicated as causing the waxy/oil/rancid odor of the TJL in chlorinated and chloraminated water. Fadel et al. (1993) identified by GC/MS that the aldehydes were coming from the leaching of TJL, but they identified the odor from water in new mains with lubricant as wet paper, chalky, or wood putty odors. These results match the results of the beginning of the study, where no oxidation has occurred at low dosages. After disinfection, the odor changes to a sweet, citrus, and soapy odor after 10 days at an intensity of 2.5 or less. The dissolved TJL changes from an original odor in water alone of waxy/wet cardboard to waxy/oily (rancid) and soapy to sweet/citrus and soapy.

The sweet-type odors include the tentatively identified aldehydes (octanal, E-2-octenal, and nonanal) in runs a and b. Other compounds that were common to both runs: one ketone (6-methyl-2-heptanone or 4-methyl-2-heptanone) and three alcohols (7-octen-4-ol, E-2-octene-1-ol, and cineole). These compounds are logical candidates for producing these sweet odor types.

AVA 10 Lubricant Leached into Water . The initial odor of the AVA 10 (pine, kerosene petroleum oil, and sweet) quickly disappears upon dissolving in water and/or reaction with disinfectants in water (Tables 6.10–12). Tables 6.10 and 11 show that 10 mg/L of AVA 10 has a sweet odor (intensity = 1.4–4) without disinfection or with chloramines. In addition, the chloramines and, to a lesser extent, the chlorination process with 2 mg/L chlorine, develop an astringency (intensity = 1.6–2.5). The chlorination process produces oily (rancid), phenol, and pine odors. The highest FPA intensity was 4 for sweet, which is near the FPA level that enables GC/MS and sensory GC analyses to be successful.

Table 6.14 shows the results of CLSA–GC/MS and sensory GC data analyses after 10 days of leaching with chlorine. The original odor of the AVA 10 has disappeared. Table 6.14 shows that the sensory GC analysis could not identify any odors for the six compounds from AVA 10 extracted for a solution of 10 mg/L AVA and 2.0 mg/L chlorine.

BLL Leached into Water. BLL has a distinct waxy odor (intensity 4) and soapy odor (intensity 4) when the can is opened. However, Tables 6.7–9 and Figure 6.4 show that 20 mg/L of BLL has a sweet odor without disinfectant, and with chloramine it has an intensity of 1-4 Chlorine (2 mg/L) appears to oxidize something in BLL to a kerosene and alcohol odor after

Table 6.14

GC/MS analysis of AVA 10 leached into water for 10 days:

Milli-Q Water,pH 7.3, chlorine 2.0 mg/L, first run

Retention Time (min*)	Tentative Identification	Identification %	Sensory GC Odor Description
28.33	Oxirane, 2-ethyl-2-methyl	43	
30.45	Cyclopentane, 1,3,-dimethyl-trans	33	
33.65	Isooctane, (ethenyloxy)-	28	
35.64	1-Pentadecene, 2-methyl-	9	
41.12	*2-Oxonanone*	*9*	
45.07	*Phenethylamine, N-methyl-beta., 3, 4,-t*	*14*	

NOTES:

Italics correspond to a burnt sensory background of the sensory GC as at 40 min the column oven is at 270 °C. The column oven is at >150 °C after 27.5 min.

*—Retention time adjusted by adding 5 min to correspond to second GC/MS analysis.

8 days. However, the odor response is low, with an intensity of 1–2. This response is below the levels of sensitivity (FPA intensity of 3–4) that have been necessary for the identification of odorants by GC/MS and sensory GC analyses. Figure 6.6c shows the GC/MS analysis of BLL dissolved in acetonitrile. An ester of a fatty acid of tetradecanoic acid is indicated as the major GC/MS peak, with only a 58% identification. Sensory GC analysis could not identify these or any odors from the BLL extracts of the chlorine or chloramine solutions of the first leaching experiment. These data indicate that the chemical concentrations are below threshold.

Table 6.15 shows the results for CLSA–GC/MS and sensory GC analyses after 10 days of leaching with chlorine solution from the confirmatory second set of leaching experiments with nearly double the concentration of lubricant and disinfectant: 23 mg/L of BLL and 2.8 mg/L of chlorine. Only the same mushroom/sweet compounds identified in TJL as ketones (6-methyl-2-heptanone or 4-methyl-2-heptanone) and alcohols (7-octen-4-ol, E-2-octene-1-ol, and cineole), in addition to a possible acid, 2-ethyl hexanoic acid, were identified with sweet odors. A bicyclic compound or camphor was observed for a waxy/plastic odor with over a 95% confidence level.

Table 6.15
GC/MS analysis of BLL leached into water for 10 days:
Milli-Q water, pH 7.3, chlorine 2.8 mg/L, second run

	Tentative Identification	Identification %	Sensory GC Odor Intensity
14.17	57, 82, 72, 67 no match found		
17.53	91, 69, 57, no match found		
19.16	3-Heptanone	74	
20.1	Heptanal; Hexanal, 4-methyl	72, 45	
22.96	No match found		
23.5	Pentane, 1,1-diethoxy-	10	
24.23	Hexanoic acid, 2-ethyl-, or 7-octen-4-ol	46, 53	Mushroom/Sweet
24.76	2-Heptanone, 6-methyl-	83	Mushroom/Sweet
25.8	Octanal	47	
26.26	7-Oxabicyclo[2,2,1]heptane, 1-methyl-4-(1-methylethyl)-	80	
27.08	Cineole	86	
28.53	2,4-Decadienal, (E,E)	16	
28.85	Heptanoic acid	42	
29.78	Bicyclo[2.2.1]heptan-2-one, 1,3,3-trimethyl-, (1R)-	87	
30.6	Hexanoic acid, 2-ethyl-	78	
31.17	Bicyclo[2,2,1]heptan-2-ol, 1,3,3-trimethyl	74	
31.86	117, 90, 89, 62—no match		
32.47	Bicyclo[2.2.1]heptan-2-one, 1,7,7-trimethyl-, (or Camphor)	96, 95	Waxy/Plastic(4)
32.75	60, 69, 73, 83 91, 57—no match		
33.28	Bicyclo[2,2,1]heptan-2-ol, 2,3,3-trimethyl-, or 3-Cyclohexen-1-ol, 4-methyl-1-(1-methylethyl)-	17, 35	
33.53	Borneol or Bicyclo[2.2.1]heptan-2-ol, 1,7,7-trimethyl-, (1S-endo)-	70, 58	
34.57	3-Cyclohexen-1-ol, 3-methyl-	40	
36.56	Nonanoic acid	53	
37.01	6-Nonenal, (Z)-		
38.43	*3-Piperidione, 1-methyl-*	*64*	
38.5	*Bicyclo[2.2.1]heptane-2-carboxaldehyde, 3-methyl*	*12*	
39.99	*Nonadecanol*	*53*	
40.6	*2,4-Heptadiene, 2,6-dimethyl-*	*60*	
42.07	*Pyrene, 4-decyl-1,2,3,6,7,8-hexahydro-*	*35*	

NOTES:

1. Italics correspond to a burnt sensory background of the sensory GC, as at 40 min the column oven is at 270 °C. The column oven is at >150 °C after 27.5 min.
2. A tentative identification above 70% by the mass spectral data base is treated as a higher probability for future identification.

Other odors were a fatty acid odor of 2,4-decadienal (E,E) and a cardboard/fatty acid odor of octanal. No kerosene or alcohol odor was observed.

The compounds identified from leaching BLL at higher concentrations are of a similar type to those observed for the TJL. These data indicate that a similar scenario of chlorination reactions is occurring with these compounds in the water distribution system, but the effect is much lower or slower for BLL.

TRG Leached into Water. Tables 6.1–3 and Figure 6.2 show that leaching of the TRG produced a distinct rubbery odor with gasoline or kerosene odors being produced in disinfected waters. Figure 6.2 shows that the rubbery odor increased in intensity levels as high as 6 to 7.6 over time in the waters without disinfectant, and with chlorine. Thus, constant diffusion of the rubber compounds from the gasket material to the water contributes to the increase in odor intensities over time. The intensity of the rubbery odor in the chloramine medium was erratic, but persistent. Results for this material have the highest degree of consensus among the FPA panelists.

Tables 6.16.a and b show the results of two independent tests of the CLSA–GC/MS and sensory GC analysis of TRG material leached with chlorine for 10 days. The first test shows that the compounds tentatively identified with gasoline rubber odors were butane, 1,1-oxybis (4-chloro)—commonly called bis-(2-chlorobutyl) ether—and 2,5-dimethylpiperazine, with odors of gasoline/rubber and burnt rubber, respectively. The GC/MS confirmation levels were 47% and 72%, respectively. These compounds were purchased and did not have the odor perceived by sensory GC analyses. The odors of the bis-(2-chlorobutyl) ether and 2,5-dimethylpiperazine were green apple/citrus and shoe polish, respectively. Thus, the real compounds causing rubber/gasoline odor are different from these compounds. Table 6.16b did not show that the second analysis had the same tentatively identified peaks as the first analysis, and only the 1,2,3-trimethylbenzene peak was identified with a burnt rubber odor. The GC/MS identifications indicate the presence of two other low-molecular-weight alkyl benzenes and three esters of butanoic acid without sensory GC odors. At present, it is concluded that the CLSA–GC/MS and sensory GC analysis did not identify the chemicals associated with the rubbery odor or with the gasoline or kerosene odors.

Table 6.16a
GC/MS analysis of TRG leached into water for 10 days:
Milli-Q water, pH 7.3, chlorine 2.8 mg/L, first run

Ret. Time, (min)	Tentative Identification	Odor
2.7	Not Identified	N/A
2.98	Not Identified	N/A
3.2	Not Identified	N/A
4	Not Identified	N/A
6.97	Pentanone, 4-methyl (Blank)	N/A
7.6	Not Identified	N/A
12.88	Not Identified	N/A
16.8	Not Identified	N/A
18.3	Not Identified	N/A
21.4	Not Identified	N/A
21.8	Butane, 1,1-oxybis (4-chloro)	Gasoline/Rubber
28.4	1-Hexanol, 2-ethyl- (Blank)	Sweet
30.2	Not Identified	N/A
31.6	Not Identified	N/A
33.2	Not Identified	N/A
34	Not Identified	N/A
43.3	Not Identified	N/A
34.6	Not Identified	N/A
35	Not Identified	N/A
35.6	Not Identified	N/A
35.7	Not Identified	N/A
36.8	Not Identified	N/A
37.8	Not Identified	N/A
38.4	Ethanone, 1-(4-(hydroxy-1-methyl))	N/A
39	Not Identified	N/A
39.2	Internal Standard	N/A
39.4	Not Identified	N/A
39.6	Not Identified	N/A
41.5	Piperazene, 2,5-dimethyl-, trans	Burnt Rubber
42	*Benzene, trimethyl(1-methylethyl)-*	*N/A*
43	*Propanoic acid, 2-methyl-, 3-hydroxy-2,4 (Blank)*	*Waxy*
43.2	*Not Identified*	*N/A*
43.5	*Not Identified*	*N/A*
43.7	*Not Identified*	*N/A*
45	*Not Identified*	*N/A*
45.98	*Phenol, (1,1-dimethyl)-4-methoxy)*	*Sweet/Alcohol*
46.6	*Not Identified*	*N/A*

NOTES:

1. Italics correspond to a burnt sensory background of the sensory GC, as at 40 min the column oven is at 270 °C. The column oven is at >150 °C after 27.5 min.
2. A tentative identification above 70% by the mass spectral data base is treated as a higher probability for future identification.

Table 6.16b

GC/MS analysis of TRG leached into water for 10 days

Milli-Q water, pH 7.3, chlorine 2.0 mg/L, second run

Ret. Time (min)	Tentative Identification	Odor Identification & Intensity	Identification %
12.65	Benzene, methyl (Blank)		83
14.65	Not found		
15.45	Arsenous acid, tris(trimethylsilyl) (GC column bleed)		50
18.07	Benzene, 1,2-dimethyl		53
18.58	Benzene, 1,2-dimethyl		87
23.56	1,3-cyclopentadiene, 5-(1-methylpropy	Paint (4)	47
25.28	Benzene, 1,2,3-trimethyl	Burnt Rubber (4)	72
26.83	1-Hexanol, 2-ethyl (Blank)	Sweet	50
28.41	2,4-Cyclohexadien-1-one, 3,5-bis(1,1-		38
28.76	Butanedioic acid, (phenylmethylene)-,	Astringent (4)	59
29.7	1H-Trindene, 2,3,4,5,6,7,8,9-octahydr-		45
34.12	2-Undecanone, 6,10-dimethyl		38
35.31	Not found		
37.79	Butanoic acid, 1-methyloctyl ester		25
38.22	Butanoic acid, 2,3-dihydroxypropyl ester		38
38.92	Propanoic acid, 2-methyl-, 2, 2-dimethyl (Blank)		83
39.35	Butanoic acid, 1-methylpropyl ester		50
39.65	2-Undecanone		38
40.95	*2,5-Cyclohexadiene-1,4-dione, 2,6-bis*		*99*
44.84	*1,2-Benzenedicarboxylic acid, butyl 2*		*58*
45.68	*1,2-Benzenedicarboxylic acid, butyl 2*		*93*
46.29	*Phenethylamine, N-methyl-beta., 3,4-t*		*38*
47.09	*6-Azaestra-1,3,5(10),6,8-pentaen-17-o*		*27*

NOTES:

1. Italics correspond to a burnt sensory background of the sensory GC, as at 40 min the column oven is at 270 °C. The column oven is at >150 °C after 27.5 min.
2. A tentative identification above 70% by the mass spectral data base is treated as a higher probability for future identification.

BLL is the least offensive "odor former" of the three lubricants, followed by AVA 10 and TJL. Because no GC peaks of low-molecular-weight organic compounds are present on the GC/MS chromatograms of the lubricants dissolved in a solvent, low-molecular-weight gas-chromatographable compounds are not present in the original product at high concentrations. The AVA 10 and BLL lubricants appear less reactive than the TJL with chlorine and do not form as much odor. However, the reaction scenario for the TJL appears to be followed at a significantly lower level of odor production for the AVA 10 and BLL. For example, Table 6.15 shows that the aldehyde, ketone, alcohol, and carboxylic acid products of BLL are similar to the products of TJL (Table 6.13). In fact, the common reaction products of BLL and TJL include, among others: 2-heptanones, heptanal, 7-octene-4-ol, and cineole. It must be realized that in an actual pipe, insoluble lubricant and soluble lubricant will be present. Therefore, dissolution will occur continuously with time in a pipe with flowing water. In the laboratory protocol equilibrium experiments, the original odor of the insoluble product will disappear with time and the residual odor will be accentuated.

Generally, when 10–13 mg/L TJL is dissolved in water, the odors change from a waxy/wet cardboard to the waxy/oily and soapy odors and decrease with time in the chloramine and chlorine media as a rancid character becomes more associated with the waxy/oily note at days 6 and 8 for chlorinated and chloraminated water, respectively. Also, after 3 days, a citrus/sweet odor appears and a soapy odor is consistently present in most samples.

In chlorinated water in the water distribution system near the use of the TJL, the odor intensity of waxy/oily would decrease to background after about a week. Apparently, chlorine can react with the lubricant, and low-molecular-weight waxy/oily (rancid) chemicals not initially present are formed, as shown by GC/MS analyses (Table 6.13). The waxy/oily (rancid) odor can then react with chlorine and be removed with time. These samples would always maintain a consistent soapy odor intensity of 1–3 when the water was disinfected for 10 days. A sweet/citrus odor in the range of 1.5–3 would appear on days 8 and 10 in the samples in the chlorine and disinfection-free media only.

The apparent chemical mechanism occurring is that the fat and soap components of the lubricants are reacting with chlorine. Chlorine reacts as a hypochlorous acid electrophile with organic compounds, by three different mechanisms: addition of chlorine to double and triple bonds, substitution by displacement of another atom, and oxidation. The first two mechanisms produce chlorinous products. It has been observed that oxidation accounts for most of the chlorine demand of natural waters and waste treatment effluents (Jolley et al. 1978 and Morris 1975), and this is apparently occurring for the lubricants used for drinking water pipe applications.

The lubricants studied are made of soaps, which are salts of a fatty acid. Chlorination of the lubricants does not produce chlorinated products, which indicates that few if any substitution and addition reactions are occurring. Specifically, TJL is stated to contain potassium oleate (C-9 to C-10 olefinic, C-18 fatty acids). As shown by the data from this study, 2–3 mg/L of chlorine reacts with 10–20 mg/L of the TJL and produces aldehydes, ketones, alcohols, and hydrocarbons in the C-6 to C-20 range. These are the same compounds that are found when unsaturated fatty acids in food products are oxidized (Kochhar 1996). This process is known as oxidative rancidity and is one of the major causes of food spoilage. In food products, the terms soapy, painty, rancid, and oily refer to common odors observed during this oxidation.

The mechanism for oxidative rancidity that has been described involves the abstraction of hydrogen by a peroxy radical from the alpha methylene group of a fat molecule. This is a free-radical reaction mechanism of hydroperoxy formation. A free-radical chain reaction occurs with initiation at the site of the olefinic group of fatty acids. The initiation step is followed by propagation and termination steps. For oleic acid, the hydrogen abstraction on C-8 and C-11 produces two allylic radicals. These radicals can react with oxygen to produce mixtures of C-8 to C-11 secondary and tertiary products such as aldehydes, ketones, lactones, acids, alcohols, furans, esters, and hydrocarbons. Oleate is reported to produce more C-8 and C-11 *cis* and *trans* compounds than C-9 and C-10 compounds. Aldehydes and hydrocarbons appear to increase with the level of oxidation (Snyder, Frankel, and Selke 1985). Of the fatty acids, oleate (one double bond) is much more stable than other fatty acids of the same carbon chain size, but fatty acids with more double bonds, such as linoleric (two double bonds) and linoleic acids (three double bonds), are less stable than other fatty acids of the same carbon chain size.

The odors of C-7 to C-9 aldehydes have been described by investigators as different odors and as mixtures of odors. The following are results reported in Chapter 4, and by Soderhjelm and Eskelinen (1985), and by Fadel et al. (1993)—shown as A, B, and C, respectively.

- Heptanal
    - A. Rancid walnut oil, rancid oil plus fish, and fruity
    - B. Oil
    - C. Rancid walnut oil
- Octanal
    - A. Rancid, fruity, citric, rancid oil, sweet
    - B. Rancid fat
    - C. Not reported
- Nonanal
    - A. Rancid oil, hay-like, rancid, fruity
    - B. Fatty
    - C. Hay-like

The odors observed by the odor panel and during sensory GC analyses generally matched these odors.

It is hypothesized that the combined concentration of the aldehydes, ketones, alcohols and hydrocarbons in the C-6 to C-20 range effects the odor perceived. Weber-Fechner curves, where the FPA of each compound versus its concentration is plotted, need to be investigated. Any change of odorant character as a function of concentration must be carefully defined. It should be remembered that C-7 to C-9 aldehydes have been indicated as the cause of fruity (citrus)/sweet odors from ozonation of natural organic matter in drinking water (Anselme, Suffet, and Mallevialle 1988). Apparently, type and concentration of the chemicals present in a mixture are involved with the overall odor.

Conclusion—Evaluation of Lubricants and Gasket

BLL is the least offensive odor former of the three lubricants, followed by AVA 10 and TJL. In chlorinated water in the water distribution system near the point of use of the TJL, the odor intensity of waxy/oily would decrease to background after about a week. Apparently, chlorine can react with the lubricant and low-molecular-weight paint, and waxy/oily (rancid) odors form after 3 days. The waxy/oily (rancid) odors can then react with chlorine and be removed with time and the citrus/sweet odor forms after 8 days. A soapy odor was found consistently in most samples.

The AVA 10 and BLL appear less reactive than the TJL with chlorine and do not form as much odor. However, the reaction scenario for the TJL appears to be followed at a significantly lower level of odor production for AVA 10 and BLL.

GC/MS and sensory GC analysis have tentatively identified a series of straight-chained C-7 to C-10 aldehydes with odors of TJL off-flavors; however, the true odor of the sample was not matched by the known odors of the chemicals. Also, many other aldehydes, ketones, alcohols, and hydrocarbons in the C-6 to C-20 range in chlorinated and chloraminated water were found after chlorinating water with TJL dissolved in it. The study verifies the results of Fadel et al. (1993), indicating that aldehydes are present and part of the odor mixture, although the FPA panel describes the samples differently. TJL produces many aldehydes, ketones, alcohols, and hydrocarbons in the C-6 to C-20 range, and a combination of these in the ratios present are probably causing the odor in the distribution system. The mechanism of odor production is proposed to be the same oxidative rancidity reaction that is the major cause of food spoilage. A free-radical chain reaction occurs with initiation at the site of the olefinic group of the fatty acids of oleic acid as occurs in food spoilage. Further studies are needed to determine the specific combination of chemicals and concentrations that are causing the odor.

At present, for the TRG, it is concluded that the CLSA–GC/MS and sensory GC analysis did not identify the chemicals associated with the rubbery, gasoline or kerosene odors. Further chemical analyses are required. A more polar isolation method appears to be needed for the study.

Leaching of PVC Pipes

The PVC primer and the cement, alone and together with PVC white and PVC gray pipes, were analyzed for production of tastes and odors in drinking water under the worst water quality conditions that could cause the most leaching. Free-chlorine residues were removed by the addition of a stoichiometric amount of sodium sulfite before analyses.

- pH = 6.5
- NaCl = 500 mg/L
- Dissolved oxygen = 8 mg/L, with
- Disinfectant
  (a) Cl_2 concentration = 2.7–3.0 mg/L or NH_2Cl concentration = 3.0 mg/L
  (b) Leaching time = 2 and 7 days

FPA, water quality, and GC/MS–sensory GC profiles for each were determined from a seven-part experiment:

- Blank (Milli-Q water at pH = 7)
- Control (all chemicals added without a pipe, primer, or cement)
- Pipe only
- Pipe + primer + cement
- Pipe + cement
- Pipe + primer
- Water + primer

Tables 6.17a–b and 6.18a–c report the FPA and water quality data for the leaching of white PVC pipe, by experiments 1–7. Tables 6.19a–b and 6.20a–c report the FPA and water quality data for the leaching of gray PVC pipe, by experiments 1–7.

Figure 6.8 indicates that the FPA intensity of the odor notes for the seven experimental set-ups did not increase significantly after the two consecutive days of leaching of white pipe. This small difference in odor intensity could be attributed to the low solubility of the PVC

Table 6.17a

FPA results of the white PVC pipe for seven different set-ups:

After 2 days of leaching

Experiment Set-Up	Description of Odor	Intensity, Average ± SD	Notes/Comments
Blank	Sweet/Earthy	1.2 ± 1.1	N/A
Control	Sweet/Earthy	1.3 ± 1.0	N/A
Pipe only	Sweet	1.0 ± 1.0	plastic
Pipe + Primer + Cement	Glue/Varnish	4.0 ± 2.5	burnt plastic, sweet, acetone
Pipe + Cement	Glue/Varnish	6.4 ± 1.5	sweet phenolic, thinner
Pipe + Primer	Glue/Varnish	5.8 ± 1.3	sweet phenolic, hydrocarbon
Water + Primer	Glue/Varnish	6.6 ± 1.7	paint

Table 6.17b

FPA results of the white PVC pipe for seven different set-ups:

After 7 days of leaching

Experiment Set-Up	Description of Odor	Intensity, Average ± SD	Notes/Comments
Blank	Sweet/Earthy	1.3 ± 1.0	N/A
Control	Sweet/Earthy	1.3 ± 1.0	N/A
Pipe only	Sweet	0.8 ± 1.0	plastic
Pipe + Primer + Cement	Glue/Varnish	4.0 ± 1.8	hydrocarbon, sweet phenolic, acetone
Pipe + Cement	Glue/Varnish	6.7 ± 1.8	sweet phenolic, hydrocarbon
Pipe + Primer	Glue/Varnish	6.0 ± 1.2	sweet phenolic, acetone
Water + Primer	Glue/Varnish	6.4 ± 1.6	acetone

Table 6.18a

Water quality data for white pipe experiments:

Before the leaching

Experiment Set-Up	Conductivity (μmhos/cm)	pH	O_2 (mg/L)	Cl_2 (mg/L)	TOC (mg/L)	$UVA_{230/254}$	T °C
Blank	0.01	7.4	8.0	0.0	0.2	0.036/0.015	25
Control	1040	6.5	8.0	2.8	0.2	0.024/0.007	25
Pipe only	1100	6.5	8.0	2.8	0.2	0.032/0.008	25
Pipe + Cement + Primer	1150	6.5	8.0	2.8	0.2	0.041/0.009	25
Pipe + Cement	1050	6.5	8.0	2.8	0.2	0.035/0.009	25
Pipe + Primer	1100	6.5	8.0	2.8	0.2	0.045/0.009	25
Water + Primer	1100	6.5	8.0	2.8	0.2	0.044/0.009	25

Note: The initial concentrations of primer and cement used were 1.33 g/L of primer on pipe, 4.0 g/L of cement on pipe, 4.26 g/L of both cement and primer on pipe, 0.75 g/L of primer in water.

Table 6.18b

Water quality data for white pipe experiments:

After 2 days

Experiment Set-Up	Conductivity (μmhos/cm)	pH	O_2 (mg/L)	Cl_2 (mg/L)	TOC (mg/L)	$UVA_{230/254}$	T °C
Blank	N/A	7.4	8.0	0.0	0.2	0.036/0.015	28
Control	1050	6.5	6.5	2.8	0.3	0.024/0.007	28
Pipe only	1100	7.3	6.0	1.7	1.3	0.032/0.008	28
Pipe + Cement + Primer	1150	7.5	6.0	0.0	231	0.110/0.034	28
Pipe + Cement	1050	7.5	6.5	0.0	216	0.144/0.053	28
Pipe + Primer	1100	7.0	6.0	0.8	246	0.072/0.60	28
Water + Primer	850	5.0	6.0	0.9	280	0.052/0.108	28

Table 6.18c

Water quality data for white pipe experiments:

After 7 days

Experiment Set-Up	Conductivity (μhos/cm)	pH	O_2 (mg/L)	Cl_2 (mg/L)	TOC (mg/L)	$UVA_{204/210/230/254}$	T °C
Blank	N/A	7.5	8.0	0.0	0.3	0.012/0.002/ 0.0001/0.02	28
Control	1050	6.4	6.0	2.8	0.3	0.083/0.004/ 0.002/0.003	28
Pipe only	1100	7.4	6.0	1.7	1.4	0.155/0.035/ 0.006/0.005	28
Pipe +Cement + Primer	1150	7.5	6.0	0.0	241	2.065/1.325/ 0.121/0.038	28
Pipe +Cement	1050	7.5	6.0	0.0	225	2.411/2.186/ 0.174/0.083	28
Pipe + Primer	1100	7.0	6.0	0.8	248	0.152/0.016/ 0.010/0.605	28
Water +Primer	850	5.0	6.0	0.9	283	0.747/0.503/ 0.072/0.118	28

Table 6.19a

FPA results for gray PVC pipe for five different set-ups:

After 2 days of the leaching

Experiment Set-Up	Description of Odor	Intensity, Average ± SD	Notes/Comments
Blank	Sweet/Earthy	0.9 ± 1.02	earthy, chalky
Control	Earthy/Chalky	4.0 ± 0.0	N/A
Pipe only	Sweet/Earthy	1.3 ± 1.0	glue
Pipe + Primer + Cement	Paint/Varnish	4.2 ± 2.5	primer, glue
	Sweet	1.4 ± 1.3	
Pipe + Cement	Paint/Varnish	4.2 ± 1.1	sweet phenolic, glue, cement
Pipe + Primer	Paint/Solvent	2.0 ± 1.4	earthy, glue
	Sweet	2.2 ± 1.1	

Table 6.19b

FPA results for gray PVC pipe for five different set-ups:

After 7 days of the leaching

Experiment Set Up	Description of Odor	Intensity, Average ± SD	Notes/Comments
Blank	Sweet/Earthy	2.2 ± 1.1	earthy
Control	Earthy/Chalky	2.6 ± 1.3	sweet
Pipe only	Earthy/Sweet	1.0 ± 1.0	chalky, musty
Pipe + Primer + Cement	Paint/Varnish Sweet	6.6 ± 1.3 2.4 ± 0.9	glue, cement, primer
Pipe + Cement	Paint/Varnish	6.4 ± 2.6	sweet phenolic, glue
Pipe + Primer	Paint/Varnish Sweet	7.6 ± 2.2 2.8 ± 3.1	acetone, glue

Table 6.20a

Water quality parameters before 2 and 7 days of leaching

Experiment Set-Up	Conductivity (μmhos/cm)	pH	O_2 (mg/L)	Cl_2 (mg/L)	TOC (mg/L)	$UVA_{230/254}$	T °C
Blank	0.01	7.4	8.0	0.0	0.3	0.0004/0.0005	25
Control	1100	6.5	8.0	2.7	0.3	0.0004/0.0005	25
Pipe only	1200	6.5	8.0	2.7	0.3	0.0004/0.0005	25
Pipe + Cement + Primer	1150	6.5	8.0	2.7	0.3	0.0004/0.0005	25
Pipe + Cement	1100	6.5	8.0	2.7	0.3	0.0004/0.0005	25
Pipe + Primer	1120	6.5	8.0	2.7	0.3	0.0004/0.0005	25
Water + Primer	1120	6.5	8.0	2.7	0.3	0.0004/0.0005	25

Note: The initial concentrations of primer and cement used were 0.043 g/L of primer on pipe, 0.143 g/L of cement on pipe, 0.155 g/L of both cement and primer on pipe, 0.75 g/L of primer in water.

Table 6.20b

Water quality parameters after 2 days of leaching

Experiment Set-Up	Conductivity (μhos/cm)	pH	O_2 (mg/L)	Cl_2 (mg/L)	TOC (mg/L)	$UVA_{210/230/254}$	T °C
Blank	0.6	6.4	7.0	0.0	0.5	0.007/0.06/0.0006	28
Control	1200	6.3	7.0	2.9	0.4	0.010/0.04/0.002	28
Pipe only	1100	7.4	7.0	2.5	0.4	0.010/0.03/0.0037	28
Pipe +Cement + Primer	1175	6.8	7.0	2.0	52.0	1.522/0.005/0.008	28
Pipe + Cement	1075	7.0	7.0	2.0	40.9	1.490/0.08/0.007	28
Pipe + Primer	1250	6.7	7.0	2.3	5.2	0.108/0.09/0.006	28

Table 6.20c

Water quality parameters after 7 days of leaching

Experiment Set-Up	Conductivity (μhos/cm)	pH	O_2 (mg/L)	Cl_2 (mg/L)	TOC (mg/L)	$UVA_{210/230/254}$	T °C
Blank	0.6	6.4	6.5	0.0	0.5	0.007/0.004/0.000 2	28
Control	1220	6.3	6.5	3.0	0.7	0.063/0.008/0.002	28
Pipe only	1200	6.5	6.5	2.4	0.7	0.032/0.006/0.003	28
Pipe + Cement + Primer	1150	7.0	6.5	0.5	130.8	2.207/0.102/0.020	28
Pipe + Cement	1180	6.4	6.5	1.2	15.5	0.065/0.010/0.008	28
Pipe + Primer	1150	7.0	6.5	0.9	90.8	1.908/0.070/0.014	28

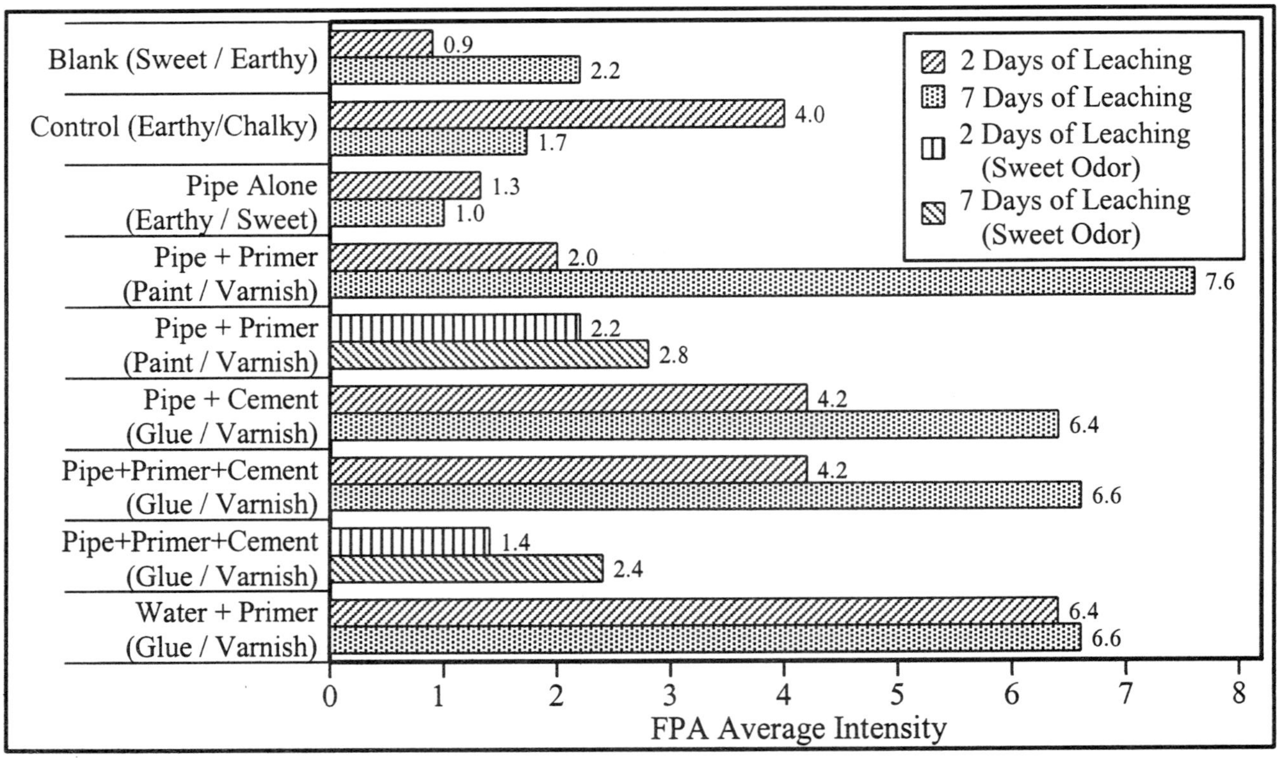

Figure 6.8 FPA of white pipe leached for 2 and 7 days versus intensity for seven experimental set-ups

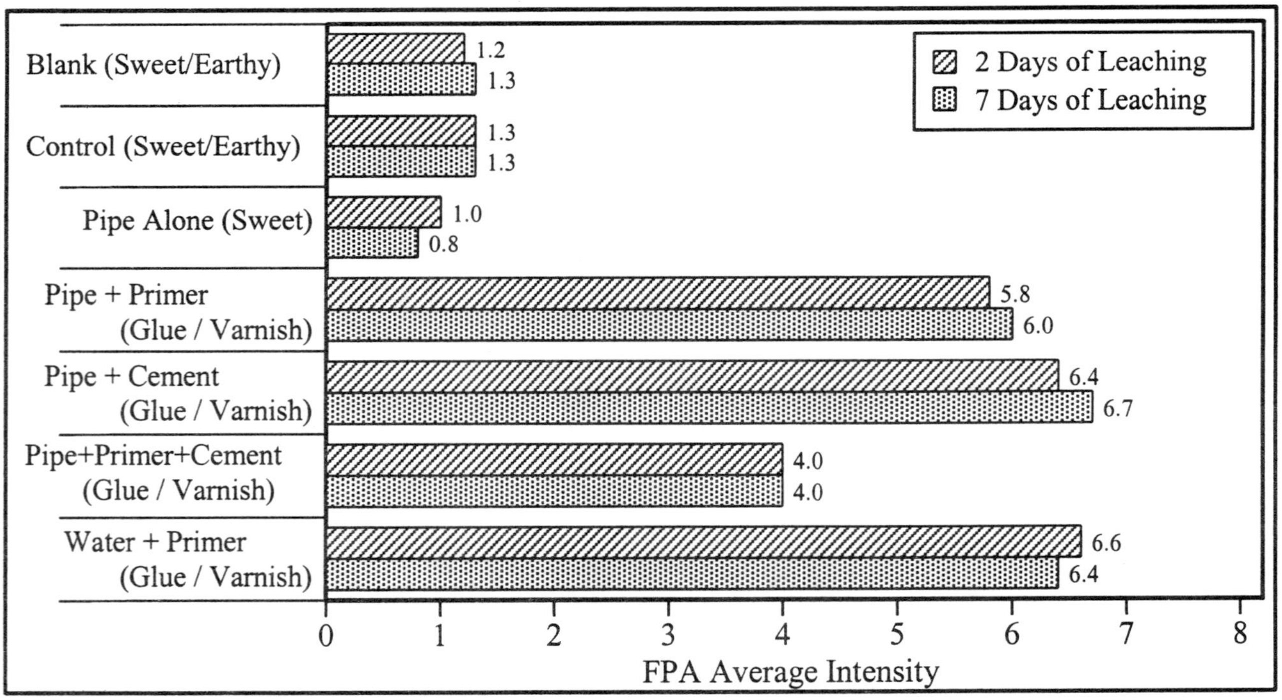

Figure 6.9 FPA of gray pipe leached for 2 and 7 days versus intensity for seven experimental set-ups.

cement and the primer in water. On the contrary, Figure 6.9 shows that there were many differences between 2 and 7 days of leaching the gray pipe for the seven experimental set–ups. Figures 6.8 and 9 also clearly show that the primer and cement increase odor intensities to 6–7 and are the causes of off-odors. Table 6.20 shows that the chlorine residual decreased in the presence of primer and cement. Table 6.21 presents the CLSA–GC/MS, sensory GC, and FPA data from the various experiments and shows that the primer and the cement are the causes of off-flavors and sensory GC peaks in distributed drinking water.

PVC Pipes Leached Alone

CLSA–GC/MS, sensory GC, and FPA data indicate that no odor was produced from the leaching of white or gray PVC pipes under the worst water quality conditions (Tables 6.17a, 19a, and 21). The exposed crosscut surfaces from cutting the PVC pipes did not contribute to the production of the odor in water. These data indicate that the PVC material is made of an identical composition throughout the pipe.

Evidence includes:

- The FPA panel identified a sweet/earthy odor with an intensity of 2 for the white or gray PVC and also for the control/blank.
- No odor was observed from the sensory GC analysis of both the control/blank study and the PVC pipes leaching study.
- The GC/MS chromatograms were identical for both the control/blank study and the PVC pipes leaching study. The siloxane compounds reported in Table 6.21 are known components of GC column bleed or bleed from the GC system itself. The proprionic acids and other compounds were identified in the blank water.

PVC Pipes Leached in the Presence of Primer and Cement

The PVC primer and cement container labels indicate that the following constituents are present in these products:

- Primer—acetone, tetrahydrofuran, methyl ethyl ketone, and cyclohexanone

Table 6.21

Summary of Chemicals Produced by Leaching of Primer, Cement, or Primer and Cement Together in Water and with Grey and White Pipe. Organic Chemicals Extracted by CLSA

Compound Name	Primer and Cement on PVC Pipe		Primer on PVC Pipe		Cement on PVC Pipe		Primer in Water (alone)	ID Range %	Odor Description	
	White	Gray	White	Gray	White	Gray			White	Gray
Cycloheptane or Cyclopentanone, 2-methyl	X	X	X	X	X	X	X	36-78		
2-Octanoic acid, methyl ester					X	X		37-40	Sweet/ Phenolic	Sweet/ Phenolic
Acetamide, 2-(2,4-dichloro-phenoxy)-N,N-dimethyl	X	X		X				39-45	Sweet/ Phenolic	Sweet/ Phenolic
Acetoacetic acid, 1-thio, S-sec-butyl ester	X	X		X				28-38	Medicinal	Medicinal
Phosphetane, 1-[(1,1-dimethylelthyl)thio]-2,2,3,4,4-pentamethyl	X			X				24-43		
Phenol, 2,6-bis(1,1-dimethylethyl)-4-methyl	X			X				33-37		
Phenol. Bis(1,1-dimethyl ethyl)	X			X				15-22		
Benzofuran, 4-Nitro	X			X				46-50	Sweet	Sweet
Cyclohexanol, 4-(1,1-dimethylethyl)	X			X				17-37	Sweet/ Phenolic	Sweet/ Phenolic
2-Heptadecanone, 1-[(2,2-dimethyl-1,3 dioxalan-4-yl) methoxy		X	X					11-25	Sweet	Sweet
Cyclopetanone, oxime		X	X				X	22-32	Phonolic/ Sweet	Phenolic/ Sweet
Benzaldehyde, 3-hydroxy		X	X					33-46	Sweet	Sweet
Benzoic acid, 2,5-dichloro-,3,6-dimethyl-, methyl ester					X	X		21-34	Earthy/ Sweet	Earthy/ Sweet
2H-tetrazole-5-carboxylic acid, 2-phenyl							X	33		

- Cement—acetone, tetrahydrofuran, methyl ethyl ketone, methylpyrolidone, and PVC resin

The FPA odor panel identified the original product to have the same odor as the products dissolved in water—i.e., a varnish/glue odor that had a sweet phenolic character. Acetone and methyl ethyl ketone have sweet odors with fruity and sharp characters (Mallevialle and Suffet 1987). The GC/MS chromatogram of the primer and cement dissolved in methylene chloride showed the primer to contain cycloheptane at a retention time of 20 min. The cement was shown to contain methylpyrolidone as 1-methyl-2-pryrolidone (retention time 27 min—86% tentative identification) and 2-methyl cyclopentanone (retention time 20 min—78%).

Column and system bleed make up the remaining components. The compounds on the label of the PVC primer and cement (methyl ethyl ketone, acetone, tetrahydrofuran and cyclohexanone) are not all observed by the CLSA–GC/MS method used. This is because these compounds are more volatile compounds than the CLSA method can isolate and/or these compounds appear in the solvent peak of the GC chromatographic analysis and cannot be analyzed by the GC/MS used.

Sensory GC analysis for the leaching experiments is shown in Table 6.21. Reaction products of the original primer and cement (i.e., acetone, tetrahydrofuran, methyl ethyl ketone, methylpyrolidone, cylohexanone, and cycloheptane) are found. Some of these compounds are tentatively identified by GC/MS associated with the odors of the primer and cement in water solution—i.e., sweet, sweet phenolic, and medicinal (see Table 6.21).

Table 6.21 shows that three individual chemicals were identified as reaction products from the primer, and two other chemicals were identified as individual reaction products of the cement by CLSA–GC/MS. In the presence of the primer, CLSA–GC/MS, sensory GC, and FPA data indicate that the original chemicals (acetone, tetrahydrofuran, methyl ethyl ketone, and cyclohexanone) and one other tentatively identified sweet phenolic odor of cyclopentanone oxime contribute to the high intensity (6) of the glue/varnish odor. However, the compound identification percentile from the GC/MS database for the oxime was very low (below 35%). The oxime is a contributing component to the high intensity (6) of the overall sweet type glue-varnish odor identified by the FPA panel. However, although there definitely are reaction products of the primer, their identities have not been determined.

In the presence of the cement (on the pipe), CLSA–GC/MS, sensory GC, and FPA data indicate that the original chemicals of the product (acetone, tetrahydrofuran, methyl ethyl ketone, methylpyrolidone), the tentatively identified sweet phenolic odor of the methyl ester of 2-octanoic acid, and the methyl ester of a chlorinated benzoic acid contribute to the high intensity (6) of the glue/varnish odor (see Table 6.21) this does not correspond with Table 6.21. However, the compound identification percentile from the GC/MS database for these two chemicals was below 40%. Therefore, these compounds can only be considered initially as tentatively identified at best. There certainly are reaction products of the cement, but their identities have not been determined.

A solvent joint for PVC pipes is made by spreading a combination of cement and primer on the surfaces to be joined. They react chemically by dissolving the surface of the pipe. Therefore, some chemicals might be formed as a result of the fusing process. Table 6.21 indicates that the primer formed seven compounds with the PVC pipe, three with sweet phenolic odorous GC peaks. These compounds were identified as a dichloroacetamide, 4-nitrobenzofuran, and a cyclohexanol, but with very low percent identifications (<50%). Therefore, although there definitely are reaction products caused by spreading the primer, their identities have not been determined.

It should be noted that the concentrations of the cement and primer on the pipes were 0.004 to 0.04 g/cm^2. The conditions that were used to coat the pipe might cause a greater or lesser amount to be applied than is actually applied in the field. There is no control over how much solvent gets into the distribution system because the installation instructions do not indicate how much of the solvent and cement should be applied in order to achieve a sufficiently strong joint between the pipes. The instructions state that solvent, cement, and primer shall be applied deliberately, but without delay. Therefore, the concentration of the cement and primer in the water distribution system depends on the size of the joints and on the quantity of the solvents applied. The *Piping Handbook* (Nayyar 1992) illustrates solvent–cement joints for plastic pipes. It can be concluded that contamination of the water by the primer and cement, and their reaction with the pipe, produces compounds that enter the water distribution system if the solvents are not applied with extra care. The performance of workers installing pipe should be monitored for quality control.

It should also be pointed out that chlorination of acetone (dimethyl ketone) and methyl ethyl ketone to chloroacetones (e.g., 1,1,1-trichloroacetone)—although not part of this study—is highly likely (e.g., Suffet et al. 1976, Gurol et al. 1983). These acetones are on the disinfection by-product list of the U.S. Environmental Protection Agency.

In conclusion, PVC gray or white pipes alone do not cause odors in the distribution system; however, the primer and the cement, alone and when applied to gray or white pipes, can cause odor problems of the glue/varnish type. The chemicals applied are a primary source of the odor, and reaction products formed during the bonding process; reactions with chlorine appear to add to the odors observed.

Leaching of Cast-Iron Pipes (CIPs)

Guo, Toomuluri, and Eckert (1998) describe the process of making DIP with cement–mortar lining. Cement–mortar is a mixture of cement, sand, and water in suitable proportions. Cement–mortar-lined DIP and gray CIP were analyzed for production of tastes and odors in drinking water under the worst water quality conditions that could cause the most leaching. Two-foot pieces of gray CIP and DIP were disinfected with 20 mg/L chlorine solution and shaken for 24 h. This is near the recommended chlorine concentration of 25 mg/L required for disinfecting repaired water mains (AWWA 1999). The disinfected pipe was rinsed carefully with Milli-Q water to remove all the chlorine before addition of the leaching solution. The leaching conditions were set as:

- pH = 6.5
- NaCl = 500 mg/L
- Dissolved oxygen = 8 mg/L, with
- Disinfectant
  - (a) Cl_2 concentration = 2.7–3.0 mg/L or NH_2Cl concentration = 3.0 mg/L
  - (b) Leaching time = 7 days

Each pipe was filled with 2 L of leaching solutions. The pipes were closed with a Teflon-taped, aluminum-wrapped silicon stopper. Each pipe was shaken for 7 days and analyzed by the methods used for plastic pipes, with the addition of ICP analysis for metals (APHA 1995). Any free-chlorine residuals after 7 days were removed by the addition of a stoichiometric amount of sodium sulfite.

Leaching of DIPs with Cement–Mortar Lining

Tables 6.22a and b show that the oxygen concentration decreased, and the chlorine and chloramine residuals disappeared, while the pH, conductivity, UVA_{254}, and TOC levels increased over 7 days of leaching. Table 6.23a shows that the taste of the water had a rubber/greasy and a metallic/dry taste at intensities of 6–7 and 4.8–6, respectively. Table 6.24b shows that the odor of the water was similarly rubbery, with wet paper and soapy characteristics. Tables 6.24a and b show the CLSA–GC/MS and the SDE–GC/MS results, respectively, and the associated sensory GC analysis of both the CLSA and SDE samples after 7 days of chlorination.

Organic Chemicals Leached from DIPs with Cement–Mortar Lining. As discussed above sensory GC analysis is possible up to about 35–39 min on the GC/MS column used in this study. The temperature of the outlet of the column is <150°C (after 35 min). At 40 min the column outlet is at a maximum temperature of 270°C, and a burnt odor overcomes most other odors at the sensory port. The sensory GC data did not detect a rubbery odor. However, a few peaks correspond to a wet paper odor:

- 3-Hexen-1-ol (z)—CLSA with a 80% identification.
- Benzaldehyde—SDE with a 78% identification, although benzaldehyde is known to have the odor of sweet almonds (Mallevialle and Suffet 1987).
- Methyl pentyl proprionic acid—SDE with a low 33% identification.

Table 6.22a

Water quality data for DIP with cement–mortar lining: Before leaching

Experiment Set-Up	Conductivity (µmhos/cm)	pH	O_2 (mg/L)	Cl_2 (mg/L)	TOC (mg/L)	$UVA_{230/254}$	T °C
Pipe+Milli-Q water	0.1	6.5	8.0	0.0	0.2	0.035/0.015	25
Pipe + Cl_2	1150	6.5	8.0	2.9	0.2	0.035/0.015	25
Pipe + NH_2Cl	1150	6.5	8.0	3.0	0.2	0.035/0.015	25

Table 6.22b

Water quality data for DIP with cement–mortar lining: After 7 days of leaching

Experiment Set-Up	Conductivity (μmhos/cm)	pH	O_2 (mg/L)	Cl_2 (mg/L)	TOC (mg/L)	$UVA_{200/210/254}$	T °C
Pipe + Milli-Q water	000.6	8.5	5.5	0.0		2.236/0.928/0.033	28
Pipe + Cl_2	1150	9.2	5.5	0.0		2.294/1.181/0.0472	28
Pipe + NH_2Cl	1150	9.6	5.5	1.1		2.183/0.760/0.050	28

Table 6.23a

FPA taste results for DIP-lined pipe: After 7 days of leaching

Experiment Set-Up	Description of Taste	Intensity Average ± SD	Description of Taste	Intensity, Average ± SD	Other Notes
Pipe + Milli-Q water	Metallic/Dry	6.2 ± 1.6	Rubber/Grease	6.0 ± 1.0	dirt, iron
Pipe + Cl_2	Metallic/Dry	4.8 ± 3.6	Rubber/Grease	7.0 ± 0.8	oily, iron
Pipe + NH_2Cl	Metallic/Dry	4.8 ± 1.3	Rubber/Grease	6.0 ± 1.2	cement, iron

Table 6.23b

FPA odor results for DIP-lined pipe: After 7 days of leaching

Experiment Set-Up	Description of Odor	Intensity, Average ± SD	Other Notes
Pipe+Milli-Q water	Rubber /Wet paper/Soapy	5.5 ± 1.8	solvent, grease
Pipe + Cl_2	Rubber /Wet paper/Soapy	4.7 ± 2.0	solvent
Pipe + NH_2Cl	Rubber /Wet paper/Soapy	4.8 ± 2.2	solvent

Table 6.24a
GC/MS of DIP with cement–mortar lining leached in water and extracted by CLSA

Peak #	RT min	Tentative Identification	ID %	Sensory Analysis	Odor Value
1	11.51	2-Hexanone	64		
2	12.34	*Azuleno[4,5-b]furan-2(3H)-one*	64		
3	12.88	Benzene, methyl (Blank)	49		
4	14.17	2-Hexanone, 4-hydroxy-3-propyl	64	Flower (lilac)	4
5	17.92	3-Hexen-1-ol, (z)	80	Wet paper	4
6	18.26	Benzene, 1, 2-dimethyl	64		
7	19.62	2-Heptanone	90		
8	20.66	*Ethanol, 2-butoxy-*	64	Onion	4
9	22.38	2-Heptanone, 4-methyl-	64		
10	22.67	3-Octanol, 6-ethyl-	50		
11	23.27	2-Heptanone, 6-methyl-	62	Solvent	4
12	24.07	1-Heptanol	64		
13	25.10	2-Heptanol, 6-methyl	86	Sewage	2
14	27.02	1-Hexanol, 2-ethyl	45		
15	27.37	Cineole	86		
16	27.84	2,7-Octanedione	40		
17	28.94	Cyclopropane, pentyl	83	Licorice	4
18	29.87	2-Nanone	57		
19	30.29	Undecane (Blank)	74		
20	32.73	Bicyclo(2,2,1)heptan-2-one	96		
21	33.06	*Bicyclo[3,1,0]hexan-2-one, 5-(1-methyl)*	93	Fruity	2
22	33.34	Cyclopropane, 1-ethyl-2-heptyl	78		
23	33.87	*Cyclohexanol, 5-methyl-2-(1-methylethyl)*	89		
24	34.21	2-Decanone (Blank)	90	Spices/coffee	6
25	35.31	Hexane	22		
26	36.44	2-Cyclohexene-1-one	22		
27	37.04	1-Dodecanol	83	Sugar candy	4
28	37.58	Undecanone	72		
29	37.82	*Formamide, N,N-dibutyl-*	46		
30	37.85	Not found			
31	38.23	Benzene, trimethyl(1-methylethyl) (Blank)	59		
32	38.74	Octadecanoic acid, 2-methyl,	35		
33	38.96	Ether, methyl 1-tetradecenyl	43		
34	39.40	Propanoic acid, 2-methyl, 3-hydroxyl (Bl)	78	Burnt	6
35	39.70	2-Undecanone	56		
36	40.45	*Butane, 1,1'-[oxybis(2,1-ethanedylox*	87	Wet paper	2

RT = Retention time
Odor value—from FPA
Italics—Similar compound in both CLSA and SDE (6 compounds)

Table 6.24b
GC/MS of DIP with cement–mortar lining leached in water and extracted by SDE

Peak #	RT (min)	Tentative Identification	ID %	Sensory analysis	Odor Value
1	12.32	*Azuleno[4,5-b]furan-2(3H)-one*	74		
2	14.33	Not found			
3	17.05	3-Heptanol, 5-methyl-	36		
4	17.60	6-Hepten-3-ol,4-methyl-	50		
5	18.25	Not found			
6	18.60	Not found			
7	20.17	Cyclopentanone, 2-methyl-	59		
8	20.77	*Ethanol, 2-butoxy*	64		
9	23.78	Benzaldehyde	78	Wet paper	4
10	27.08	Propanoic acid, 2-methyl-, pentyl (Blank)	33	Wet paper	4
11	28.86	1,2-Propanedione, 1-phenyl	64	Fragrant apricot	4
12	29.74	Benzoic acid, 4-hydroxy	12		
13	30.24	Heptadienal	8		
14	30.49	Hexane, 2,2,3,3-tetramethyl	9		
15	31.57	Bicyclo[5,1,0]octane, 8-methylene	25		
16	32.30	9-Octadecenal	45		
17	32.69	2,3-Pentanediol, 2,4-dimethyl	56		
18	33.05	*Bicyclo[3,1,0]hexan-2-one, 5-(1-methyl)*	94		
19	33.32	Ether, heptyl	36		
20	33.58	Ethanone, 1-(4-methylphenyl)-	47		
21	33.86	*Cyclohexanol, 5-methyl-2-(1-methylethyl)*	64		
22	33.97	Ethanol, 2-(2-butoxyethoxy)	64		
23	34.18	Ethanone, 1-C3-methylphenyl)	87	Burning sensation	6
24	35.28	Not found			
25	35.48	S-Triazolo[4,3-a]pyridine, 7-methyl	37		
26	36.00	1,2-Benzisothiazole	49		
27	36.41	Ethanol, 2-(9-octadecenyloxy)-	10		
28	36.91	3,4-Hexadien-2-ol, 3-isopropenyl-2,5-	35		
29	36.95	Nonane, 1-chloro-	83		
30	37.80	*Formamide, N,N-dibutyl*	87		
31	38.82	9H-Xanthen-9-one, 1,3,8-trihydroxyl-	59		
32	38.96	Propanoic acid, 2-methyl-, 2-ethyl-3-(Bl)	37		
33	39.39	Propanoic acid, 2-methyl-, 3-hydroxy (BL)	64	Alcohol/Sour	4
34	40.10	Benzene, (2-nitroethenyl)	49		
35	40.33	Morphinan, 7,8-didehydro-3-methoxy	18		
36	40.44	*Butane, 1,1'-[oxybis(2,1-ethanediylox*	50		

RT = Retention time
Odor value—from FPA
Italics—Similar compound in both CLSA and SDE (6 compounds)

The CLSA and SDE analyses had only six compounds in common, indicating that each is a unique isolation technique. The rubber/wet paper/soapy odor is not clearly defined by the sensory GC data generated. CLSA and SDE identified a series of nine alcohols, 15 ketones, two acids, two aldehydes, and six hydrocarbons. However, the specific cause of the off-odor cannot be directly correlated with a specific chemical for the DIP with the cement lining, although numerous unique compounds were identified (see Table 6.24).

The rubber gasket and the lubricants are used to hold DIPs together. The lubricants produce the same type of mixture of chemicals as the DIP. Therefore, the leachate of the cement–mortar-lined DIP adds to the odor of the lubricants. The rubber gasket produces different types of odors. In summary, the data indicate that a complicated mixture of chemicals from the pipes and lubricants is generating odors. Careful flushing of the pipes and minimum use of lubricants are an obvious recommendation. Also, a coated rubber or a differently formulated material should be developed to minimize the rubbery odor of the gasket.

Metallic Taste Leached from DIPs with Cement–Mortar Lining. In a comprehensive leaching study of cement–mortar linings of DIPs, Guo, Toomuluri, and Eckert (1998) found that significant increases of chromium, cadmium, and barium over those in the water at the faucet were leached out in 14 days. The metal levels leached in the present study are shown in Table 6.25. The ICP analysis found very low concentrations of soluble iron in water samples of DIP with cement lining. Figure 6.10 shows the Weber-Fechner curve and indicates that the iron (II) sulfate concentration of the leached water should be approximately 6 mg/L (iron = 1.2 mg/L) for

Table 6.25

ICP results for DIP with cement–mortar lining (in μg/L)

Pipe	Fe	Al	B	Cr	Mn	Ni	Sr	Pb
Pipe+Milli-Q water	1	987	103	17	<1	2	178	<1
Pipe + Cl_2	1	732	173	48	<1	1	316	1
Pipe + NH_2Cl	1	447	137	10	<1	1	169	1

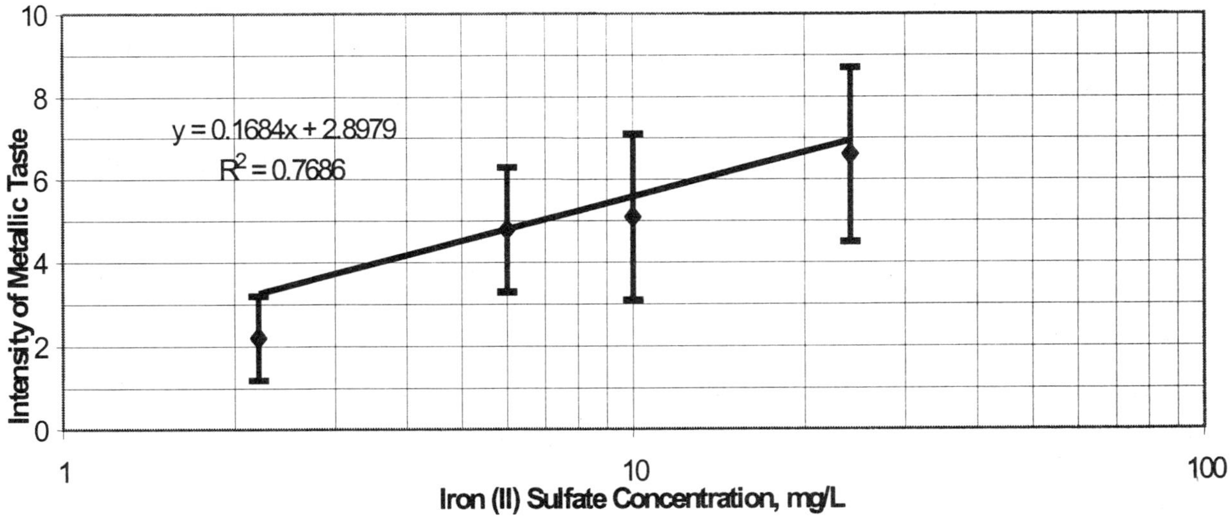

Figure 6.10 Plot of different concentrations of iron (II) sulfate versu intensity of metallic/astringent taste in Milli-Q water

an intensity of 5. The maximum allowable (aesthetic secondary standard in California) iron concentration in drinking water is 0.3 mg/L. This concentration is the near the taste threshold of the metallic taste. Therefore, the concentrations of the iron from the leached DIP were below drinking water standards.

Matia (1995) studied the main mouthfeel descriptors for iron salts by FPA and found that metallic, pungent, and rough were the primary descriptors. The taste threshold concentration for iron was found to be 0.12 mg/L in mineral water, with 500 mg/L of total dissolved salts. However, iron can be detected by taste in distilled water at a concentration of 0.05 mg/L (Ainsworth et al. 1981). For water with sulfate and low mineral content, the iron should be reduced to 50 μg/L to remove the taste from the water (Chedal 1987). White (1986) reported the treatment of iron by chlorine. Inorganic iron oxidation occurs within a pH range of 4 to 10, and the optimal pH is 7.0. These data indicate that iron may indeed be part of the metallic taste, but obviously it would not cause all the metallic taste.

In this study, the aluminum concentration was above 450 μg/L, which is twice the California secondary standard of 200 μg/L. The distribution-system taste-and-odor wheel indicated that an aluminum concentration of 4.4 mg/L and above causes astringent/dry taste. The aluminum concentration in the leached samples was five times less than the threshold value for the metallic/astringent/dry taste. Thus, according to the distribution-system taste-and-odor wheel the aluminum alone did not contribute completely to the metallic/astringent taste in the water sample. The metallic/astringent taste with intensity above 5 could result from aluminum in combination with other salts in the water samples. Aluminum concentrations of <100–200 μg/L in drinking water s (>200 μg/L) of aluminum in drinking water used for kidney dialysis causes neurological disorder in hemodialysis patients (Alfrey 1993). The amount leached in this study was two to four times that value in a static sample build-up test.

Leaching of Gray CIPs

Gray CIP, which has been used for water mains and is usually used for sewage collection systems, was tested for comparison to cement–mortar-lined DIP. Table 6.26 shows that chlorine and chloramines disappeared, while UVA_{254} and TOC levels increased over 7 days of leaching. Table 6.27 shows that the water had a wet paper taste and a kerosene/rubber odor in Milli-Q water. Dramatically, the wet paper taste of the water that was not disinfected tasted metallic/bitter, with intensities of 4.8–5.1 after disinfection, while the kerosene/rubber odor became buttery/sweet, with intensities of 3.7–4.8 after disinfection.

Organic Chemicals Leached from CIPs. Surprisingly, CLSA–GC/MS and sensory GC analysis of the chlorinated water after 7 days did not show any of the tastes or odors identified in the FPA analyses (Table 6.28). High concentrations of ten C-3 and C-4 alkyl-substituted benzene-containing compounds were detected in leached water and identified by CLSA–GC/MS analysis. The benzene-containing compounds were eluted from the GC with retention times between 21.0 and 31.0 min. Thus, there is a possibility that these alkyl benzenes, although not sufficiently odorous individually, could present a combined odor or that these compounds might be covering the other odorous compounds present. The fact that disinfected gray CIP leached these compounds indicates that oily hydrocarbons are probably used to oil the pipes when they are manufactured.

Table 6.26a

Water quality data for cast-iron pipe (CIP):

Before leaching

Experiment Set-Up	Conductivity (μmhos/cm)	pH	O_2 (mg/L)	Cl_2 (mg/L)	TOC (mg/L)	$UVA_{230/254}$	T °C
Pipe+Milli-Q water	0.1	7.5	8.0	0.0	0.3	0.039/0.012	25
Pipe + Cl_2	1150	6.5	8.0	2.8	0.3	0.039/0.012	25
Pipe + NH_2Cl	1150	6.5	8.0	2.7	0.3	0.039/0.012	25

Table 6.26b

Water quality data for CIP:

After 7 days of leaching

Experiment Set-Up	Conductivity (μmhos/cm)	pH	O_2 (mg/L)	Cl_2 (mg/L)	TOC (mg/L)	$UVA_{200/210/254}$	T °C
Pipe+Mill-Q water	1100	5.2	5.5	0	1.5	0.862/0.427/0.020	28
Pipe + Cl_2	1050	5.7	5.5	0	1.5	0.675/0.238/0.015	28
Pipe + NH_2Cl	1100	7.1	5.5	0	1.3	0.793/0.361/0.008	28

Table 6.27a

FPA results for CIP for three different set-ups after 7 days of the leaching:

FPA taste results

Experiment Set-Up	Description of Taste	Intensity Average ± SD	Description of Taste	Intensity, Average ± SD	Other Notes
Pipe + Milli-Q water	Wet paper	2.5±1.8	Sweet	2.1±2.2	metallic, dry, oily, soupy, earthy
Pipe + Cl_2	Metallic/bitter	5.1±2.0	N/A	N/A	Sweet, musty, bloody
Pipe + NH_2Cl	Metallic/ dry/bitter	4.8±2.6	N/A	N/A	Moldy, sweet, paint

N/A = not analyzed

Table 6.27b

FPA results for the CIP for three different set-ups after seven days of the leaching

b. FPA odor results

Experiment Set-Up	Description of Odor	Intensity, Average ± SD	Description of Odor	Intensity, Average ± SD	Other Notes
Pipe+Milli-Q water	Kerosene/Rubber	2.7±2.4	Sweet	1.4±1.5	Musty, earthy, plastic
Pipe + Cl_2	Buttery/Sweet	3.7±2.0	N/A	N/A	Petroleum, rubber, grassy
Pipe + NH_2Cl	Buttery/Sweet	4.8 ±2.6	N/A	N/A	Dust, earthy, rancid

N/A = Not analyzed

Table 6.28

CLSA–GC/MS and sensory GC analysis for CIP for three different set-ups after 7 days leaching

Peak Number	Retention Time, min	Tentative Identification	Intensity	Description	Possible Source
Peak 1	13.28	Cyclotrisiloxame, hexamethyl	N/A	N/A	Column bleed
Peak 2	21.04	N/A	1	Sweet fragrant	Pipe
Peak 3	21.23	N/A	1	Faint fragrant	Pipe
Peak 4	21.67	Benzene, 1,2,3-trimethyl	N/A	N/A	Pipe
Peak 5	22.12.	Benzene, 1-ethyl-2-methyl	N/A	N/A	Pipe
Peak 6	22.27	Cyclotetrasiloxane, octamethyl	N/A	N/A	Column bleed
Peak 7	22.89	Benzene, 1,2,3-trimethyl	N/A	N/A	Pipe
Peak 8	22.92	Benzene, 1-ethyl-3-methyl	N/A	NA	Pipe
Peak 9	23.69	Benzene (1-methylpropyl)	N/A	N/A	Pipe
Peak 10	24.06	N/A	2	Dry, chalky	Pipe
Peak 11	24.24	Benzene, 1,2,3-trimethyl	N/A	N/A	Pipe
Peak 12	25.95	Benzene, 2-ethyl-1,4-dimethyl	N/A	N/A	Pipe
Peak 13	27.22	Benzene, 2-ethyl-1,4-dimethyl	N/A	N/A	Pipe
Peak 14	28.80	Benzene, 1,2,4,5-tetramethyl	N/A	N/A	Pipe
Peak 15	29.16	Cyclopentasiloxane, decamethyl	N/A	N/A	Column bleed
Peak 16	30.13	Benzene, 1,2,3,4-tetramenthyl	N/A	N/A	Pipe
Peak 17	35.84	2H-1,4-Bensodiazepin-2-one,7-chloro-1,3,-dihydro-5-phenyl-1	N/A	N/A	Blank

N/A = Not analyzed

Metallic Taste of CIPs. The metallic/bitter taste was identified from the leached water samples by the FPA panel with minimum and maximum intensities of 4.8–5 in disinfected waters. The ICP analysis shown in Table 6.29 indicated the high concentration of iron in all three experimental set-ups. The highest concentration of iron (2.15 mg/L) was found in water for the pipe leached with free chlorine. The FPA panel identified the highest metallic/bitter taste with an intensity of 5.2. Analysis of the Weber-Fechner curve, Figure 6.10, indicates that an intensity of 5 in water should be observed for iron (II) sulfate levels of 7 mg/L (iron = 2.5 mg/L). This roughly corresponds to the iron concentration (2.2 mg/L) found by the ICP analysis. Thus, iron appears to be a major component of off-tastes in the samples. The amount of aluminum leached from gray CIP is below the detection limit. Only manganese levels were raised significantly. In summary, the results of leaching of gray CIP and cement–mortar-lined DIP are completely different.

Table 6.29

ICP results for CIP cast-iron pipes (in μg/L)

Pipe	Fe	Al	B	Cr	Mn	Ni	Sr	Pb
Pipe + Milli-Q water	189	<9	5	<1	27	1	<1	1
Pipe + Cl_2	2146	<9	7	<1	139	11	<1	<1
Pipe + NH_2Cl	687	<9	1	<1	82	5	<1	<1

Summary

Protocols were designed and shown to work well for lubricants, rubber gaskets, metal pipes, and plastic pipes with and without primer and cement. The protocols were developed in Milli-Q water. Manufacturers of distribution-system materials should be required to have their components tested by a certifying organization using protocols such as those developed for this study.

- Lubricants and gaskets, primer, cement, and PVC pipe—Complete the leaching test individually or together in an appropriate group in a dark glass bottle with a Teflon cap. Shake for an appropriate time. Test lubricant with 10–25 mg/L portions of lubricant at room temperature. (Use acid-washed and oven-dried glassware.)
- Metal pipes—Fill a pipe section of 2 ft by 3–4 in., seal it with a Teflon plug, and shake overnight.
- Leaching solution—pH and general water quality parameters should be those expected in the distribution system (ambient).
- Add NaCl to bring the concentration of total dissolved solids to 500 mg/L. Chlorinate or chloraminate the water to a 3 mg/L residual.
- Monitor chlorine or chloramine removal and water quality until the disinfectant dissipates or a maximum of 10 days is reached.
- Monitor by odor, flavor, and taste tests and, if possible, by CLSA–sensory GC analysis and GC/MS after the disinfectant is removed or at a maximum of 10 days.

Odors produced by each material were characterized by an odor panel under the testing protocol. Table 6.30 shows a summary table of leached components after exposure to the leaching solution. The successive odors are shown that are formed by further oxidation of the leached components. In one case, for TJL, the reactions actually continue after the primary disinfectant cannot be successfully measured in the solution.

Table 6.30

Summary of taste-and-odor changes for distribution-system material

Material	Disinfectant (Disinfectant Removal)	Odor Characteristic Observed	Time Odor Observed (Days)	FPA Intensity: Low (1–3) Med (4–5) High (>6)
Lube—TJL (TYTON Joint)	Choramines >10 days	a) Waxy/oil(rancid*) b) Soapy*	<3–>10 <3–>10	Low (<3) Low (<3)
	Chlorine <6 days@	a) Waxy/oil(rancid*) to a') Sweet b) Soapy*	>3–<8 <8–>10 <6–>10	Med (3–6) Low (<1) Low (<2)
Lube—AVA 10 (Aquaflow)	Choramines >10 days	a) Astringent* b) Sweet	<3–>10 <3–>10	Low (<2) Low (<2)
	Chlorine >10 days	a) Astringent* to a') Oily(rancid*) to a") Pine*, Phenol*	<3 >3–<10 <8–>10	Low (2.5) Med (4) to Low (2) Low (3)
Lube—BLL (Blue Lube™)	Choramines >10 days	Sweet	>10	Low (<2)
	Chlorine >10 days	a) Sweet to a') Alcohol to Oily(rancid*) b) Kerosene*	>3–<6 >6–<8, >8 >6–>10	Low (1) Low (2) Low (2)
Gasket—TRG (TYTON Joint)	Choramines >10 days	Rubber, Kerosene*	<3–>10, <3–>10	Med (Avg 4) Low(Avg 2.5)
	Chlorine >10 days	Gasoline*/Rubber	<3–>10	Med (2.7–6.2)
PVC Pipe (G&W), Primer + Cement	Chlorine <2 days-White >7 days—Grey	Glue/Varnish(Paint/Varnish) Sweet	<2–>7 <2–>7	Med (4) Low (1–2)
PVC Pipe (G&W) Cement only	Chlorine <2 days-White >7 days—Grey	Glue/Varnish(Paint/Varnish)	<2–>7	Med (4) to High (6)
DIP—Pipe (Ductile Iron/ Cement–Mortar)	Choramines >7 days	Metallic/Dry* a) Rubber/Grease* to a')Rubber/wet paper/soapy	<2–>7	Med (4.8) High (6) Med (5)
	Chlorine 2-<7 days	Metallic/Dry Rubber/Grease* to Rubber/wet paper/soapy		Med (5) High (7) Med (5)
CIP—Pipe (Cast-Iron)	Choramines <7 days	Metallic/dry/bitter* Buttery/Sweet*	<7 >1	Med (5) Med (5)
	Chlorine <7 days	Metallic/bitter* Buttery/Sweet*	<7 >1	Med (5) Med (4)

Notes: chlorine—2 mg/L, chloramines—3 mg/L; FPA—1 (threshold), 2, 4, 6, 8, 10, 12.
*Formed by chlorine oxidation; a, a', a'"—chemicals form successive odors and reaction continues after disinfectant is gone.

Some of the chemicals leached from these materials were tentatively identified when odors, flavors, and tastes were perceived. The off-odors produced by material in the distribution system by the general testing protocol are as follows:

- Lubricants: Apparently chlorine can react with the lubricant and low-molecular-weight compounds, with paint and waxy/oily (rancid) odors forming after 3 days. The waxy/oily (rancid) odor can then react with chlorine and be removed with time, and a citrus/sweet odor forms after 8 days. A soapy odor was found consistently in most samples.
- Gasket—TRG: rubber/kerosene or gasoline/rubber.
- PVC pipe (gray and white) + primer + cement: glue/varnish (paint/varnish).
- DIP with cement–mortar lining: rubber/grease to rubber/wet paper/soapy.
- Gray CIP: buttery/sweet.

The chemicals associated with off-odors leaching from water distribution materials were evaluated.

- The lubricants and DIP with cement–mortar lining produce odors that are continually oxidized by chlorine or oxygen with time. The mechanism of oxidative rancidity, one of the major causes of food spoilage, is the apparent mechanism of oxidation. In food products, odor terms such as soapy, painty, rancid, and oily are the common odors observed during this oxidation. Apparently, the dissolved lubricants are being oxidized in succession—astringent, oily (rancid), pine, and phenol odors. BLL, is the least easily oxidized, is the least offensive odor former of the three lubricants, followed by AVA 10 and TJL. The AVA 10 and BLL lubricants appear less reactive than the TJL lubricant with chlorine and do not form as much odor.
- The rubber gasket used with the lubricants and DIP can add a rubber odor as well as a gasoline/kerosene odor to the mixture, as leaching occurs after installation in a water distribution system. Although 20 new compounds were associated with the chemical oxidation of the gasket alone, no confirmatory

identifications were made to define the causes of the odors. Also, a coated rubber or a differently formulated material should be developed to minimize the rubbery odor of the gasket.

- The white and gray PVC pipes alone did not produce an odor during the leaching tests.
- The PVC pipe primer and cement and the products of their reactions with chlorine appear to cause the primary odor of glue/varnish (paint/varnish) when PVC pipes are connected in the water distribution system. The chemical causes are the initial chemicals (acetone, tetrahydrofuran, methyl ethyl ketone, cyclohexanone, methylpyrolidone, cycloheptane, and 2-methyl cyclopentanone) plus seven of the 11 reaction products of the primer and cement with chlorine in the leaching solution. The seven reaction products determined by sensory GC analysis had a sweet or sweet/phenolic odor. However, all the GC/MS chemical identifications are of low probability for positive identification.
- Careful flushing of the pipes and minimum use of lubricants are an obvious recommendation.
- People who apply the lubricants in the field should be properly instructed in the application of the proper amount of lubricant.

Taste-and-Odor Problems in Actual Distribution Systems

Case Study I

The following case study illustrates the occurrence of a persistent and intense taste-and-odor problem associated with numerous organic compounds after installation of a new main in the distribution system. This kind of problem, although observed over the years, was never documented with FPA and CLSA until the months of January and February 1994 in the Philadelphia area, where such a problem occurred. The event started on January 6, 1994, when a new main was installed. There was only one residential customer to provide flow through the new main. This was not intended. Cold weather caused postponement of the completion of the main installation, and the hook-up of other residential customers, until warmer weather. As a result, this situation was prolonged and continued to cause problems for nearly a month.

After a customer complained, water samples from the house and from the blow-off were taken and analyzed by sensory and chemical analyses. FPA described both samples as having tastes and odors of an oil-base paint-type. Follow-up samples taken on January 14 still had the same odor and a metallic, bitter taste. CLSA followed by GC/MS identified a series of substituted benzenes and naphthalenes in both samples. Continuous monitoring of the site showed that the odors were reduced with time and had disappeared completely by February 7. The GC/MS identification obtained from the CLSA extract showed no significant levels of alkyl benzenes to be present, as observed by the absence of alkyl benzene and naphthalene peaks. Based on similar instances reported in the past, it was assumed that the origin of the taste-and-odor problem was TJL, the vegetable-based soap used to lubricate the pipe-joint gaskets. Consequently, TJL was allowed to soak in water that was subsequently extracted by CLSA followed by GC/MS determination. The GC/MS total ion chromatogram of the resulting CLSA extract showed that no significant levels of substituted benzenes were found, as noted by the lack of characteristic alkyl benzene peaks.

It is now suspected that the problem may have been caused by improper curing of the interior asphaltic coating of the pipe, rather than by the gasket lubricant, as previously suspected. Both the interior pipe coating and the gasket lubricant can cause taste-and-odor problems.

Case Study II

In winter 1995, taste-and-odor problems associated with installation of new pipe occurred in two separate locations. Although the first problem resulted in a customer complaint, the second problem was identified before any complaint was reported. Both circumstances were very similar. FPA described the samples as having a paint/putty type of odor. Chemical analysis identified alkyl benzenes and naphthalene with xylene isomers (dimethylbenzenes) as being the most prominent.

It is suspected that the problem was caused by leaching from the bituminous coating on the inside of the pipe. To verify this assumption, some water was allowed to sit in contact with a new pipe for several days. The resulting water sample was analyzed for chemical content and odor and found to be very similar to the samples that had been causing the problem in the field. It was interesting to note that in the first problem, when the customer complained, the odor decreased after flushing and reappeared 10 days later at an even higher intensity. Further flushing took care of the problem, and no further complaints were reported.

Practical Application of Results of Historical Case Studies Presented

Initial migration of additives from newly installed synthetic material is sometimes accompanied by the appearance of intense taste-and-odor problems and may trigger extensive consumer complaints. The application solvents for coatings have a tendency to leach taste-and-odor–causing chemicals, such as alkyl benzenes, when improperly cured. When investigating a taste-and-odor problem characterized by plastic, medicinal, solvent, or any type of chemical-like descriptors, the investigators should always inquire whether new pipes have been recently installed or internal coatings have been recently applied. Such a field investigation will often allow the successful determination of the origin of the problem.

Pretesting of Products and Systems

The bench-scale studies and a number of case studies highlighted the need to include an approval (clearing) procedure before new pipes and other materials are used and before pipes or recently coated reservoirs are released for use. The product should be tested and approved for use by a recognized authority before the new main is installed or coating applied. The recognized authorities should include taste and odor testing in their approval procedures. Ideally, the same conditions that exist in the specific system should be used. This means that water of similar quality, with the same disinfectant at a similar concentration and a similar contact time, should be used. These tests, when applied systematically, should minimize the consequences of the potential water quality changes and reduce consumer complaints.

CHAPTER 7
TASTE-AND-ODOR PROBLEMS ASSOCIATED WITH DESIGN AND OPERATIONS

INTRODUCTION

The following case studies report the investigation of various taste-and-odor problems in real distribution systems. No bench-scale experiments could be performed to simulate these kinds of problems. Instead, these case studies highlight the complexity of the distribution system and the difficulty in identifying the exact cause of a problem. The following case studies are used to describe issues with corrosion, operations, and a main break.

TASTES AND ODORS ASSOCIATED WITH CORROSION

This case study was intended to illustrate taste-and-odor problems caused by small-diameter galvanized pipes used in some distribution systems. The samples studied were obtained from a water utility in the Pacific Northwest. The utility had been using 2-in. diameter galvanized pipes in their distribution system. Galvanized zinc pipes have been and still are used in some service lines in distribution systems. Galvanized pipes are made by coating steel or iron pipes with zinc to minimize rusting.

Although the water utility has a very active program of pipe replacement, some areas of the system have yet to be changed. Some pipes are more than 100 years old. Although most customer complaints were related to "red water" rather than tastes and odors, it was thought worthy to evaluate the organoleptic properties of the water.

Two sets of samples were taken from the distribution system. Each set consisted of two samples: one taken from the blow-off on the galvanized pipe studied and a second sample, for use as a control, from the nearest fire hydrant on the supply line (within three to four blocks). Table 7.1 shows the types of samples and their origins. All four samples were subjected to general water quality analysis, trace metals analysis, and flavor-profile analysis (FPA). In addition, the samples were extracted by closed-loop stripping analysis (CLSA), followed by gas chromatography/mass spectrometry (GC/MS), for their organic content.

Results

The results obtained are presented in Table 7.2 for the general water quality parameters, Table 7.3 for trace metals analysis, and Table 7.4 for FPA.

Table 7.1

Description of case-study samples for Pacific Northwest

	A	B (control for A)	C	D (control for C)
Main size diameter (in.)	2	8	2	20
Installation date	1933	N/A	1967	1902
Main type	Galvanized	Ductile iron	Galvanized	Cast iron
Estimated water age (h)	200	200	130	130
Sample point	Blow-off	Hydrant	Blow-off	Hydrant

Table 7.2

Water quality parameters of Pacific Northwest samples

	A	B (control for A)	C	D (control for C)
Alkalinity (mg/L)	10	10	10	10
Conductivity (μmhos/cm)	26.3	26.5	28.1	27.5
Hardness (mg/L)	6	7	6	6
Turbidity (NTU)	0.73	0.46	35	3.9
pH	7.23	7.26	7.21	7.22
Calcium (mg/L)	1.7	1.7	1.7	1.7
Magnesium (mg/L)	0.6	0.6	0.6	0.6

Table 7.3

Trace metals analysis

Trace metals (μg/L)	A	B (control for A)	C	D (control for C)	Secondary MCL (State and/or Federal)
Aluminum	23.1	22.8	1520	51.8	50 to 200
Arsenic	ND	ND	0.12	ND	50*
Barium	0.89	1.3	8.5	1.2	2000*
Chromium	ND	ND	0.81	ND	100*
Copper	0.67	1.1	6.1	1.9	1000 1.3**
Iron	22.8	40.6	2970	293	300
Lead	0.39	0.92	3	1	15**
Lithium	ND	0.91	1	ND	
Manganese	3.8	5	135	25.9	50
Nickel	ND	0.12	1.9	0.16	100*
Strontium	12.4	12.4	17.3	12.3	
Zinc	0.73	0.46	35	3.9	5000

MCL: Maximum contaminant level   ND: Non-detected   *Primary MCL   **Action level

Table 7.4

Chlorine residuals and FPA results

	A	B (control for A)	C	D (control for C)
Total Chlorine Residual (mg/L)	1.09	1.17	1.38	1.42
FPA				
Odor	Chlorinous 4 Flowery 2	Chlorinous 4	Chlorinous 6 Plastic 4	Chlorinous 4 Plastic 2
Flavor	Sweet 2 Bitter 2		Sour 1 Bitter 2	Bitter 2
Mouthfeel	Metallic 2 Astringent 4	Metallic 4 Astringent 4	Astringent 6	Metallic 4 Astringent 4

Discussion

The unusually high concentrations of trace metals (especially aluminum and iron) obtained for sample C do not represent the concentrations of metal in solution. As indicated by the high turbidity number of 35 NTU, sample C contained a large amount of precipitate. The sample was acidified to dissolve the precipitate prior to inductively coupled plasma/mass spectrometry (ICP/MS) analysis. This explains the fact that these results do not correlate with the taste intensity perceived during FPA analysis. However, the metallic taste in the other samples appears to parallel the increase in iron concentration obtained by chemical analysis.

Preliminary tests performed with iron (II) sulfate ($FeSO_4 \bullet 7H_20$) indicate that this compound may be associated with a metallic mouthfeel, as discussed in Chapter 6. Similarly, taste experiments performed with aluminum sulfate indicate that this compound may be associated with an astringent mouthfeel. Experiments on the taste characteristics of zinc chloride indicate that there is a link with the sour descriptor. Zinc chloride has been added to the flavor wheel (Figure 8.1) as a standard for the astringent mouthfeel.

In addition to the chlorinous odor, a flowery odor was reported in the sample. A number of short- and intermediate-chain nitriles, with the number of carbon atoms in the chain ranging from six to 11, were identified by GC/MS analysis of the CLSA extract of this sample. These nitriles were reported as having sweet/citrus/flowery odors. The sensory characteristics of nitriles have been described in Chapter 4 of this study. Nitriles have been reported (White 1992, LeCloirec 1985) to form from oxidation of aldehydes and appear to increase with chloramine residual and detention time.

GC/MS identification of the CLSA extracts of samples C and D showed the presence of a large number of substituted alkyl benzenes that may explain the plastic descriptors. The origin of these alkyl benzenes could not be identified in this specific case study because additional information on the other components of the distribution system was not available. Alkyl benzenes have been associated with leaching of pipe materials and lubricants in case studies, some of which are discussed in Chapter 6.

Summary

Except for sample C, the metallic taste seems to be related to the level of iron measured in the water, the astringent mouthfeel to aluminum sulfate and zinc chloride, and the sour taste to zinc chloride. The flowery odor appears to be related to nitriles, and the plastic descriptors may be related to alkyl benzenes. This case study is a "real world" example of what was observed in bench-scale experiments and discussed in previous chapters.

TASTES AND ODORS ASSOCIATED WITH OPERATIONS

Background

A taste-and-odor episode occurred in southern California in which a water utility received 15 consumer complaints on September 16, 1998. The problem was widespread and covered six pressure zones in the distribution network. The episode lasted until September 28, 1998. During that time, numerous samples were collected at consumer taps, the treatment plant effluent, and fire hydrants in the area of the complaints. Consumers consistently described the water as having a tingling, bitter, and numbing sensation in the mouth and on the tongue.

Results

Table 7.5 shows the results obtained from FPA analysis. The tingling sensation described by consumers was confirmed by trained panelists for most of the complaint sites sampled on September 17, 1998. When the same samples were analyzed the following day, this tingling sensation could no longer be perceived. This problem was not found in the treatment plant effluent, which confirmed that the problem originated in the distribution system. Because the complaints were mainly associated with the tingling sensation, it was believed that the cause was the inorganic composition of the water. Metal analyses (Table 7.6) were performed on most

Table 7.5

Sensory analysis results for taste-and-odor episode in California.

Sampling date	Analysis date	Flavor-Profile Analysis			Organic Analysis
		Odor	Flavor	Comments	
Site 1a (fire hydrant near Site 1b)					
9/17/98	9/17/98	Rubbery	Bitter almond extract Stagnant Metallic Soapy Drying Astringent Tingling sensation	Not acceptable (very strong taste intensities)	CLSA–GC/MS
9/17/98	9/18/98	Floral/Almond Oily Metallic	Chlorine Astringent	Acceptable	
Site 1b (consumer tap)					
9/17/98	9/17/98	Rubbery	Bitter almond extract Soapy Drying Astringent Tingling sensation	Not acceptable (strong taste intensities)	CLSA–GC/MS
9/17/98	9/18/98	Chlorine Plastic Floral/Almond	No flavor perceived	Acceptable	
Site 2 (consumer tap)					
09/17/98	09/17/98	Rubbery	Metallic Bitter (almond extract) Astringent Rubbery Tingling sensation	Not acceptable (strong taste intensities)	CLSA–GC/MS
Site 3 (consumer tap)					
09/17/98	09/17/98	Rubbery	Chalky Metallic Drying Tingling sensation	Not acceptable (strong taste intensities)	CLSA–GC/MS

(continued)

Table 7.5 (continued)

Sampling date	Analysis date	Flavor-Profile Analysis		Comments	Organic Analysis
		Odor	Flavor		
Site 4 (consumer tap)					
09/18/98	09/18/98	Chlorine Almond Nose burn	No flavor perceived	Acceptable	
Site 5 (consumer tap)					
09/18/98	09/18/98	No odor	Chlorine Grassy Drying	Acceptable	
Site 6 (consumer tap)					
09/18/98	09/18/98	Chlorine	Chlorine Chalky Drying Oily	Acceptable	
Site 7 (consumer tap)					
09/18/98	09/18/98	Chlorine Chalky	Chalky Drying Oily	Acceptable	
Site 8 (consumer tap)					
09/18/98	09/18/98	Plastic	Chalky Drying Plastic Bitter	Acceptable	
Treatment plant effluent					
09/18/98	09/18/98	Chlorine Plastic	Chlorine Plastic Drying Bitter	Acceptable	CLSA–GC/MS

(continued)

Table 7.6

Trace metal analyses, (μg/L)

Metal	Reporting limit	Site 1a	Site 1b	Site 2	Site 3	Site 4	Site 5	Site 6	Site 7	Plant effluent
Aluminum	5	22	13	8	17	NA	7	ND	11	ND
Antimony	2	ND	ND	ND	ND	NA	ND	ND	ND	ND
Arsenic	0.5	0.6	ND	0.8	0.5	NA	0.7	0.6	0.6	0.5
Barium	5	35	17	50	28	NA	36	32	31	32
Beryllium	0.1	ND	ND	ND	ND	NA	ND	ND	ND	ND
Cadmium	0.1	ND	ND	ND	ND	NA	ND	ND	ND	ND
Chromium	2	ND	ND	ND	ND	NA	ND	ND	ND	ND
Copper	10	ND	16	14	27	NA	11	ND	ND	ND
Iron	20	334	24	24	ND	NA	ND	ND	ND	ND
Lead	1	3	ND	ND	ND	NA	ND	ND	ND	ND
Lithium	10	ND	ND	ND	ND	NA	ND	ND	ND	ND
Manganese	5	27	ND	ND	ND	NA	ND	ND	ND	6
Molybdenum	2	ND	ND	2	ND	NA	ND	ND	ND	ND
Nickel	2	ND	ND	ND	ND	NA	ND	ND	ND	ND
Selenium	5	ND	ND	ND	ND	NA	ND	ND	ND	ND
Silver	10	ND	ND	ND	ND	NA	ND	ND	ND	ND
Strontium	20	172	155	193	165	NA	172	174	170	170
Thallium	1	ND	ND	ND	ND	NA	ND	ND	ND	ND
Zinc	20	ND	ND	ND	ND	NA	ND	ND	ND	ND

ND = not detected; NA = not available

samples. When comparing the results for customer sites where the water was not acceptable (Sites 1b, 2, and 3) to the sites where the water was acceptable (Sites 5, 6, and 7), it was observed that values for aluminum, barium, copper, iron, and strontium are slightly higher for at least one or two of the sites. The levels of aluminum and of iron and copper may be contributing to the astringent mouthfeel and the metallic taste, respectively. No zinc was measured above the reporting limit in these samples. Selected samples were analyzed by CLSA and, in some cases, simultaneous distillation-extraction (SDE). Table 7.7 shows the GC/MS identifications. Benzaldehyde and butylated hydroxytoluene (BHT) were found in the complaint samples and could account for the bitter descriptors.

Discussion

The taste-and-odor problem affected several pressure zones of the water utility distribution area. One potential contributing factor was suspected to be a shutdown at a pumping station. During this time, water to customers was being supplied by an alternate treatment plant. The new configuration resulted in a reversal of flow in a 24-in mortar-lined clay pipe that was approximately 40 years old. It was thought that this may have stirred up sediment in the tanks and transmission lines.

Additionally, during the repair work, a 24-in. cast-iron gate valve was found closed when it should have been open. The usual flow in that area was approximately 1800 gpm. Once the valve was opened, flow increased to approximately 3000 gpm, and the velocity increased through lines that had been receiving low-velocity water for approximately 15–25 years. It was suggested that the elevated aluminum concentrations may have been caused by an accumulation of flocculating agents. The reversal of flow may have agitated these settled solids and contributed to the bitter and metallic tastes reported in many of the complaints. The problem subsided after extensive flushing.

Table 7.7

GC/MS identification of samples from California episode

Compound	Site 4 (9/17/98)	Site 1a [a] (9/17/98)	Site 3a (fire hydrant)[b] (9/18/98)	Treatment Plant Effluent[a] (9/18/98)	Taste-and-Odor Characteristics	Use
Dibromochloromethane			x	x		
Bromoform			x	x		
Benzaldehyde	x[c]	x[c] (40 μg/L)	x[c]		Bitter almond	
Benzonitrile			x		Almond oil	Solvent
Caprinitrile			x[c]			
Nonanal	x	x			Fruity/Rancid	
Dodecanol	x	x	x		Soapy	Wetting agent
Butylated hydroxy-toluene (BHT)	x[c]	x[c]	x[c]		Gasoline/Bitter	Additive in HDPE pipes
Trichloroethyl phosphate			x			Plasticizer, additive in lubricants

x: Found in sample

HDPE: High-density polyethylene

[a]Extracted by CLSA

[b]Extracted by SDE

[c]Identification confirmed

TASTES AND ODORS ASSOCIATED WITH A MAIN BREAK

On July 24, 1996, a fracture occurred in water pipes in a Barcelona street, next to which there was an electricity transformer. The water leak flooded the underground room where the transformer was located, causing the transformer's cooling oil to leak out.

Despite measures taken and cleaning of the pipe works after repair, a small amount of oil unavoidably entered the distribution system. In spite of company advice and warnings to consumers, this led to complaints, particularly from residents in the buildings next to where the fracture occurred. Analysis of 26 samples found the significant hydrocarbon content in the water to be between 2 and 290 µg/L. However, it was subsequently observed that the odor of oil was noticeable only in samples with a hydrocarbon content equal to or greater than 91 µg/L. Only a chlorine odor was reported in a sample with a hydrocarbon concentration of 47 µg/L. FPA of one of the affected samples revealed a rubber-like odor, which was not detected in the control samples.

The study revealed a degraded heavy oil, characterized by the presence of heavy, straight-chain hydrocarbons forming a distribution of numbers of carbon atoms ranging from 16 through 23, with 21 as the most common number of carbon atoms observed. Other hydrocarbons were also found, including branched-chain and unsaturated compounds.

Another taste-and-odor event (another case study) occurred in early December 1997 in Barcelona, Spain. Numerous complaints of "bad odors" were reported in water from newly built restrooms in a commercial building. Four water samples from different locations within the building and from the main reservoir serving the building were collected on December 15:

- New office in level 1
- Office in level 0
- Restroom in level 0
- Main reservoir in level 1

Although the "bad odor" could not be properly described, its intensity was highest in a new office in level 1 of the building. Water from the reservoir did not show any odor. The relative odor intensities were as follows:

New office (level 1) >> Office (level 0) > Restroom (level 0) >> Main reservoir (level 1)

Results

Table 7.8 shows the volatile organic compounds (VOCs) identified by GC/MS as obtained from the CLSA extracts of the four samples.

New Office—Level 1

The chromatogram of the "new office" sample (Figure 7.1) was markedly different from the other three samples, not only in identified compounds but also in concentration levels. Solvents, antioxidants, detergents, and hydrocarbons from lubricant oils were among the main identified compounds. This sample presented an unresolved chromatographic mixture (UCM) of compounds, shown on the chromatogram in the last 20 min of the run. The UCM of compounds corresponded to lubricant oils. Antioxidants and 2-ethyl-1-hexanol, which were present at high concentrations in this sample, were not detected in the other samples.

Office—Level 0

This sample (Figure 7.2) exhibited a UCM of compounds less intense than that shown in Figure 7.1, but of similar intensity to that in Figure 7.3. The 2-ethyl-1-hexanol concentration (Table 7.8) was approximately one-tenth of that in the sample from the "new office."

Restroom—Level 0

The chromatographic profile of the sample taken from the restroom (Figure 7.3) showed a UCM of compounds characteristic of lubricant odors.

Main Reservoir—Level 1

The chromatographic profile (Figure 7.4) did not show a UCM of compounds characteristic of lubricant oils.

Table 7.8

Gas chromatography/mass spectrometry (GC/MS) identification of closed-loop stripping analysis (CLSA) extracts for the four water samples

Compound	Concentration (ng/L)			
	New office (level 1)	Office (level 0)	Restroom (level 0)	Main reservoir (level 1)
Toluene	0.28	0.13	0.24	0.17
Ethylbenzene	0.06	0.03	0.04	0.03
m + *p*-Xylene	0.05	0.08	0.14	0.1
o-Xylene	0.05	0.03	0.05	0.03
Styrene	0.07	<0.01	<0.01	<0.01
m + *p*-Ethyltoluene	0.04	0.03	0.04	0.03
1,2,4-Trimethylbenzene	0.05	0.03	0.04	0.04
2-Ethyl-1-hexanol	3.85	0.42	0.02	<0.01
Naphthalene	0.24	0.02	0.02	0.02
Aldehydes (total)	0.36	0.28	0.17	0.13
2,2,4-Trimethyl-1,3-pentanediol monoisobutyrate (two isomers)	0.03	0.14	0.03	0.03
2,2,4-Trimethyl-1,3 pentanediol di-isobutyrate	0.64	0.12	0.21	0.12
Normal hydrocarbons C_{12}–C_{20} (total)	0.40	0.33	0.18	0.15
Siloxanes (total)	0.40			
2,6-Di-*t*-butyl-*p*-cresol	0.48			
2,6-Di-*t*-butyl-*p*-quinone	1.50			

C_{12}–C_{20} = carbon atom chain numbers ranging from 12 to 20

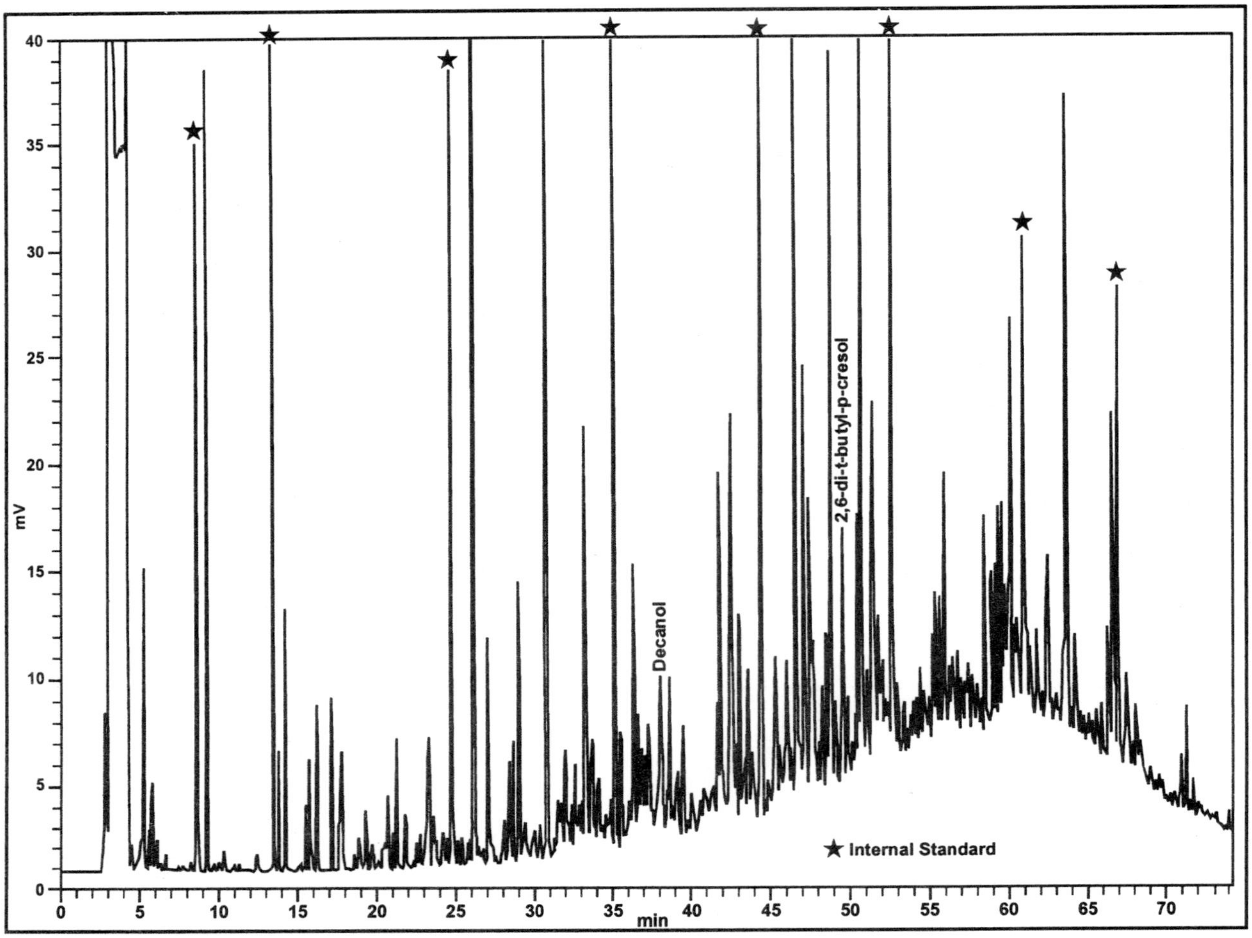

Figure 7.1 Gas chromatography/mass spectrometry (GC/MS) results for closed-loop stripping analysis (CLSA) extract of "Level 1/new office" sample

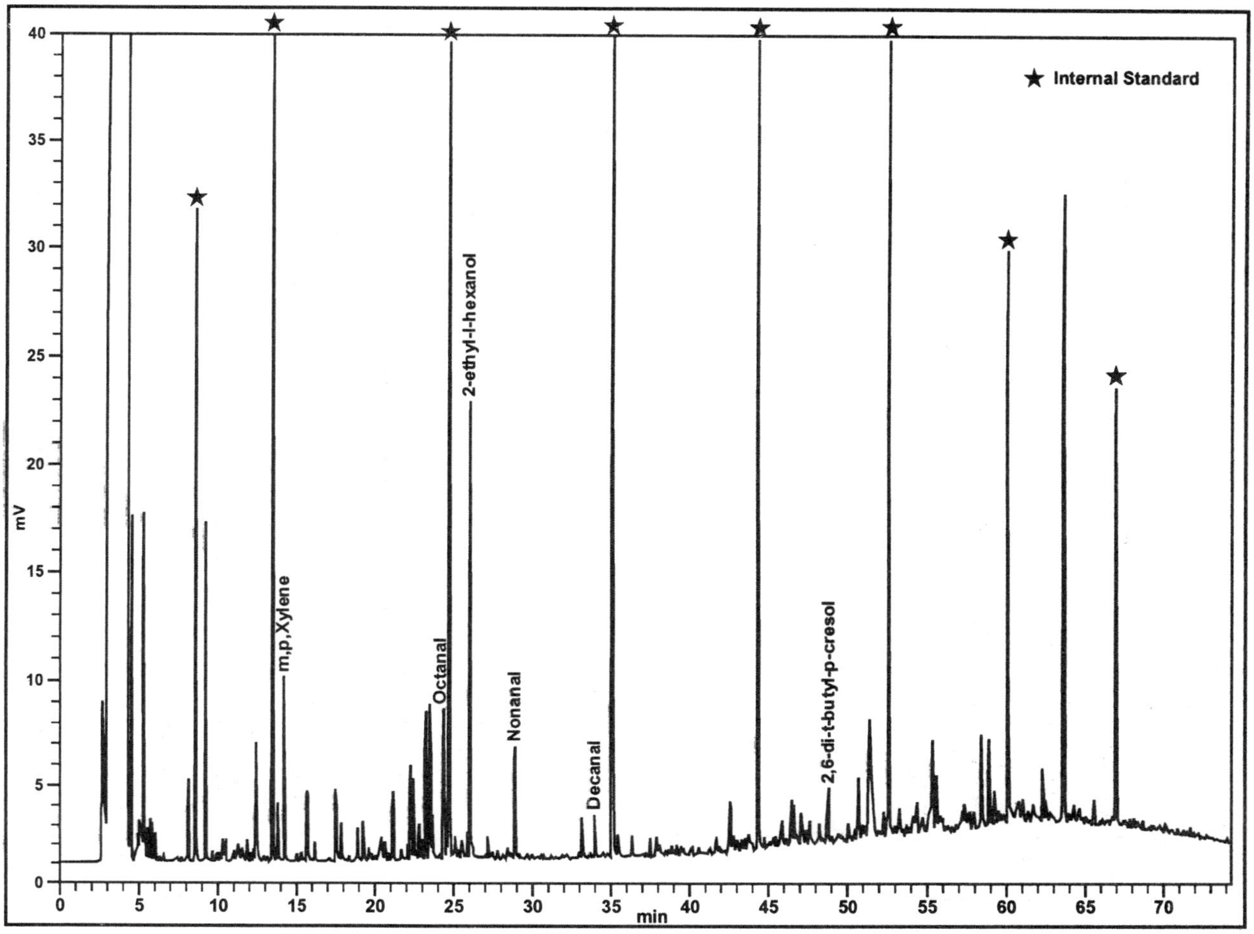

Figure 7.2 GC/MS results for CLSA extract of "Level 0/Mr. C. office" sample

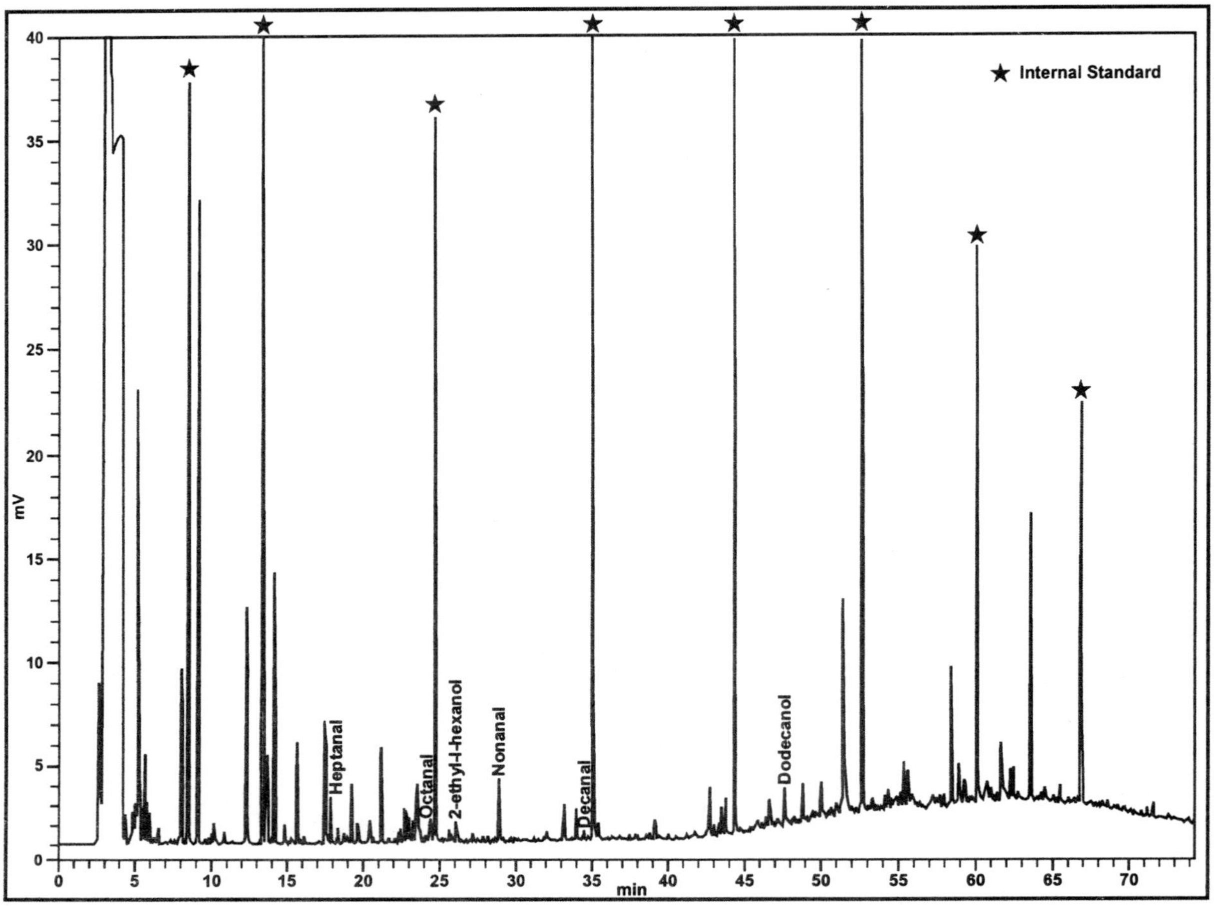

Figure 7.3 GC/MS results for CLSA extract of "Level 0/ladies' restroom" sample

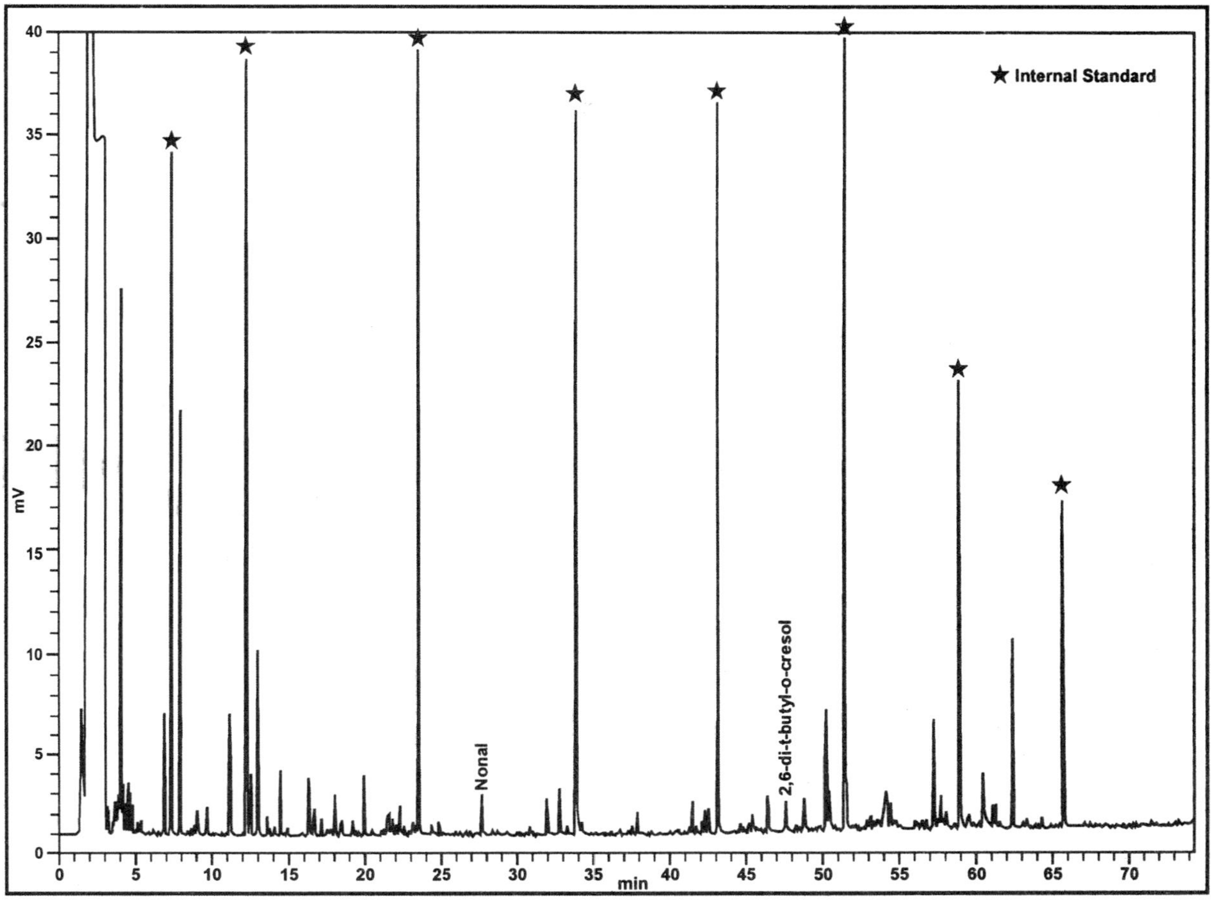

Figure 7.4 GC/MS results for CLSA extract of "Level 1/main reservoir" sample

The results for the semi-VOCs identified by liquid-liquid extraction (LLE) and GC/MS are discussed below.

New Office—Level 1

Base/neutrals fraction. The chromatographic profile (Figure 7.5) presented a large number of nitrogenated compounds, such as *N*-alkylamines (i.e., *N*-dibutyl, *N*-dibutylethyl, and several other unidentified amines); *N,N*-dichlorohexylurea and surfactants (nonylphenols with an ethoxylation degree of $n = 0–4$); solvents (styrene, 2-ethyl-1-hexanol, acetophenone, α,α-dimethyl benzyl alcohol, and caprolactam); and miscellaneous compounds (benzothiazole, 1,2-benzoisothiazolone, etc.).

Acidic fraction. The sample contained $C_6–C_{18}$ fatty acids with a predominance of the even-numbered carbon atoms. Of these fatty acids, palmitic acid (C_{16}) was the most abundant. Other compounds, such as diacids (C_9) and phenylacetic and phenylpropionic acids (degradation products of alkylbenzenesulfonates), were also identified.

Office—Level 0

Base/neutrals fraction. The chromatogram showed the presence of hydrocarbons, nonylphenols, and several other compounds such as styrene, 2-ethyl-1-hexanol, acetophenone, α,α -dimethyl benzyl alcohol, benzothiazole, and *N*-dibutylamine.

Acidic fraction. Even $C_{14}–C_{18}$ fatty acids were observed. As was the case with the "new office" sample, palmitic acid (C_{16}) was the most abundant.

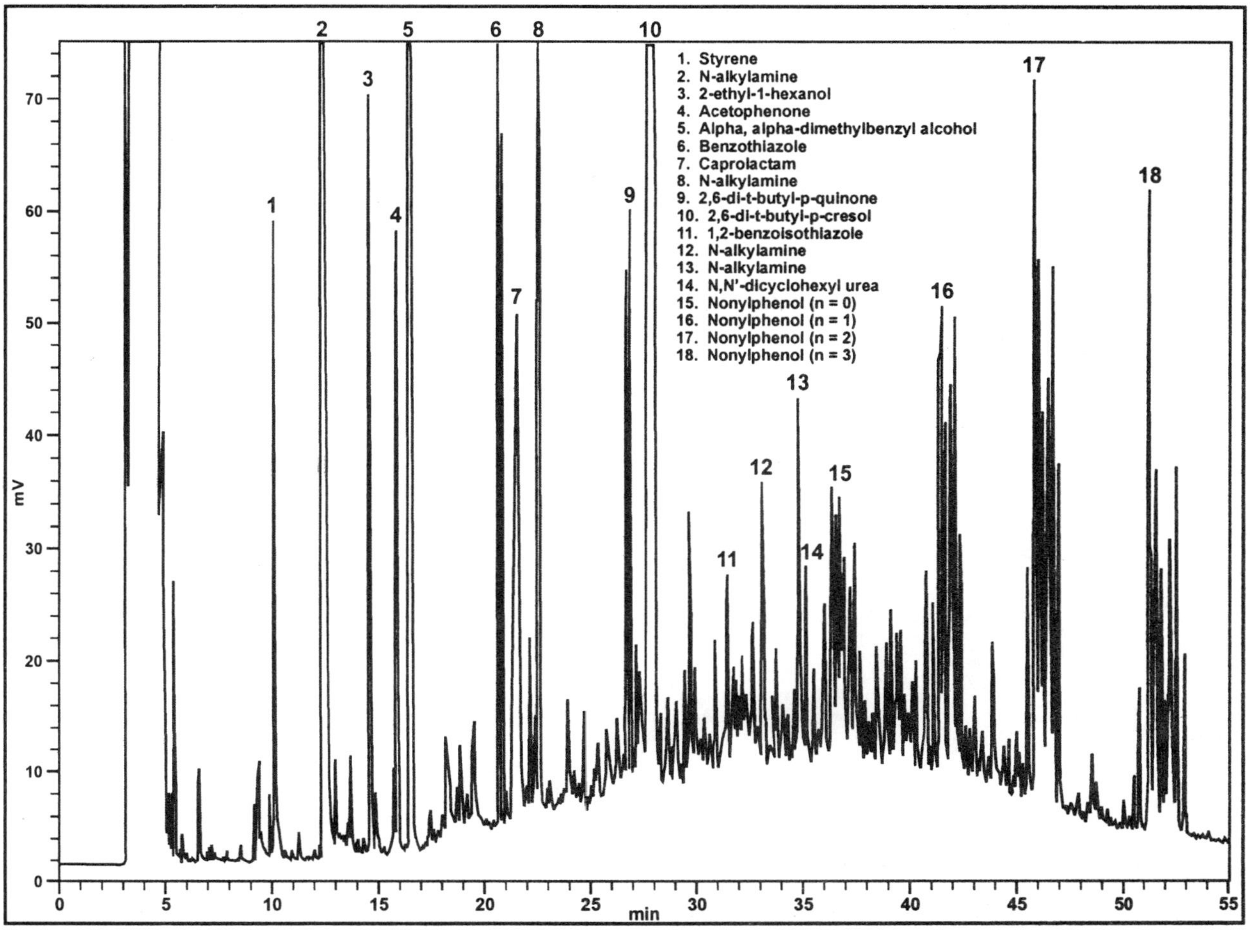

Figure 7.5 GC/MS results for base/neutral fraction of "Level 1/new office" sample

Restroom—Level 0

Base/neutrals fraction. Some of the compounds already identified in the "new office" sample, such as styrene, *N*-dibutylamine, and a characteristic UCM, were identified. These compounds were at lower concentrations than in the previous sample. These low concentrations could explain why this sample had a less offensive odor than the other two.

Acidic fraction. Even C_{14}–C_{18} fatty acids were observed. Once again, palmitic acid (C_{16}) was the most prevalent of the fatty acids identified in the sample.

Main Reservoir—Level 1

Base/neutral fraction. No compound of interest was identified in this sample.

Acidic fraction. Only traces of fatty acids were found.

Summary

The analysis of the acidic and base/neutral extracts from the "new office" sample showed the presence of a large variety of compounds, but little information was obtained from the analysis of the other three samples. In the "new office" sample, some of the compounds, such as *N*-alkylamines, appeared to be the most likely cause of the offensive odors. Several alkylamines are known to produce objectionable, ammonia-like odors. Their exact origin was not identified in this case study. Other compounds identified in the samples may have originated from the materials used during the installation of the copper water mains (e.g., cutting oils, grease, pipe joint lubricants).

CHAPTER 8

PRACTICAL APPLICATIONS FOR UTILITIES

INTRODUCTION

This study highlighted certain factors that need to be considered in order to minimize tastes and odors in the distribution system and reduce consumer complaints. As with tastes and odors originating in source waters or treatment plants, there are actions that can be taken to correct or prevent distribution-generated tastes and odors, thus increasing customer satisfaction. The first step is to understand where potential changes may occur and how they can be avoided.

TASTE-AND-ODOR INVESTIGATION PROCEDURES

Although it is not possible for water utilities to avoid all tastes and odors in their systems, many of them can be anticipated, and preventive or mitigative measures can be taken. When quality changes cannot be anticipated, and taste-and-odor complaints are received by a water utility, it is important to conduct a thorough investigation to identify the origins and chemical causes of taste- and odor-related consumer complaints. Tracking down the problem is often simple, and experienced water utility personnel may correct it easily. However, because of the complexity of distribution systems and the intricacy of their pipe networks, it is sometimes difficult to find the source of the change and decide on a simple solution. Only a systematic investigative procedure can resolve this type of situation.

An investigative procedure was developed to help water utilities understand and respond efficiently to taste- and odor-related consumer complaints. Appendix B presents the field investigation form and questionnaire.

SOURCES OF TASTES AND ODORS IN DISTRIBUTION SYSTEMS

In the previous sections of this report, numerous causes and associations of potential tastes and odors in the distribution system have been described. They can be summarized as follows:

- Replacement, repair, or coating of pipes or reservoirs
- Disinfectant residuals that are too high or too low
- Biological activity in the water or on surfaces contacted by the drinking water
- Changes in water quality
- Disinfection by-product (DBP) formation and continuing reactions
- Blending factors

Replacement, Repair, or Coating of Pipes and Reservoirs

Some of the bench-scale and case studies demonstrated the potential of chemical leaching from distribution system materials as a causative factor in taste-and-odor production. It is generally assumed that if the material has been used in the past without creating water quality problems, it will then be safe to use under all conditions. However, case studies have shown that a defective material, or even a slight change in the product formulation, can lead to the leaching of chemicals associated with tastes and odors. In other cases, it was seen that leached compounds can react further with the residual disinfectant, producing additional taste-and-odor changes at the consumer's tap.

Migration of additives from newly installed synthetic material is sometimes accompanied by the appearance of intense tastes and odors. When coatings have not been adequately cured, their application solvents have a tendency to leach taste- and odor-causing compounds such as alkyl benzenes.

The bench-scale studies and a number of case studies highlighted the need to include an approval (clearing) procedure before new pipes and other materials are used and before pipes or recently coated reservoirs are released for use. The product should be tested and approved for use by a recognized authority before the new main is installed or coating applied. The recognized authorities should include taste and odor testing in their approval procedures. Ideally, the same conditions that exist in the specific system should be used. This means that water of similar quality, with the same disinfectant at a similar concentration and a similar contact time, should be used. These tests, when applied systematically, should minimize the consequences of the potential water quality changes and reduce consumer complaints.

Disinfectant Residuals

Disinfectant residuals should be evaluated regularly throughout the entire distribution system, and the system should be adjusted for optimal customer satisfaction while still maintaining adequate microbial control. Even small changes in residual concentrations (free chlorine) can provoke consumer complaints. A regular flushing program may be necessary to maintain good water quality. The system should be designed, maintained, and operated so as to carry the lowest total chlorine residual possible, at a consistent level throughout the system, at all times.

Biological Activity

Biological activity in the distribution system is inversely related to disinfectant level. As the residual decreases, biological activity increases. Removing dead ends, if possible, and instituting a regular flushing program may help minimize biological activity. Use of a different disinfectant and corrosion control may be necessary in certain situations.

Water Quality

It is very important for water utilities to be aware of changes in the water quality they provide to their consumers. If the water is obtained from a wholesaler, the latter should notify the water utility when blends are changed. Results should be provided for flavor-profile analysis (FPA) of current and projected blends. When the use of a new well is anticipated, FPA should be performed on the product water prior to distribution. The water tested should be representative of the water that will ultimately be delivered to customers. This means that the water from the well should be of consistent quality, the untreated water should be tested before and at the end of bench- or pilot-scale treatment, and the treated water should be tested at appropriate detention times. Analyses for bromide, iodide, ammonia, and sulfur at low detection limits, together with analyses for the usual regulated chemicals, should be conducted on the untreated and treated water.

DBP Formation and Continuing Reactions

The sensory characteristics of a large number of DBPs are listed in the present report. This compilation could be used to estimate potential taste-and-odor changes. By comparing the results of the chemical analyses with the odor threshold concentrations of these compounds, it should be possible to anticipate the taste-and-odor potential of a water sample. It may be possible to optimize treatment to minimize DBP formation while maintaining adequate disinfection for microbial control.

Blending

When it is necessary to use blends of water of different qualities, it is recommended that bench-scale experiments, such as the ones presented in Chapter 5, be performed. These experiments will help determine the optimum blend that can be used without impairment of the sensory characteristics of the water distributed.

DISTRIBUTION-SYSTEM TASTE-AND-ODOR WHEEL

Figure 8.1 presents the distribution-system taste-and-odor wheel. This wheel stems from the original flavor wheel of Mallevialle and Suffet (1987) but has been updated and modified for the specific taste-and-odor compounds associated with distribution systems. The inner circle shows the four categories of basic tastes that pertain to water, one category of mouthfeel and nosefeel, and eight categories of odors. The middle circle from the center shows the common descriptors for each category, and the outer circle lists the chemical compounds that can be used as reference standards for the specific descriptors. All compounds designated by an asterisk have been actually confirmed to cause tastes or odors in the distribution system. Compounds such as methyl butanal and the iodomethanes have actually been identified in some of the case studies presented in this report.

Table 8.1 complements the taste-and-odor wheel by identifying the possible sources (column 2) of the different taste- and odor-causing agents and by suggesting appropriate corrective actions (column 3) to be taken to eliminate or prevent tastes and odors. The taste-and-

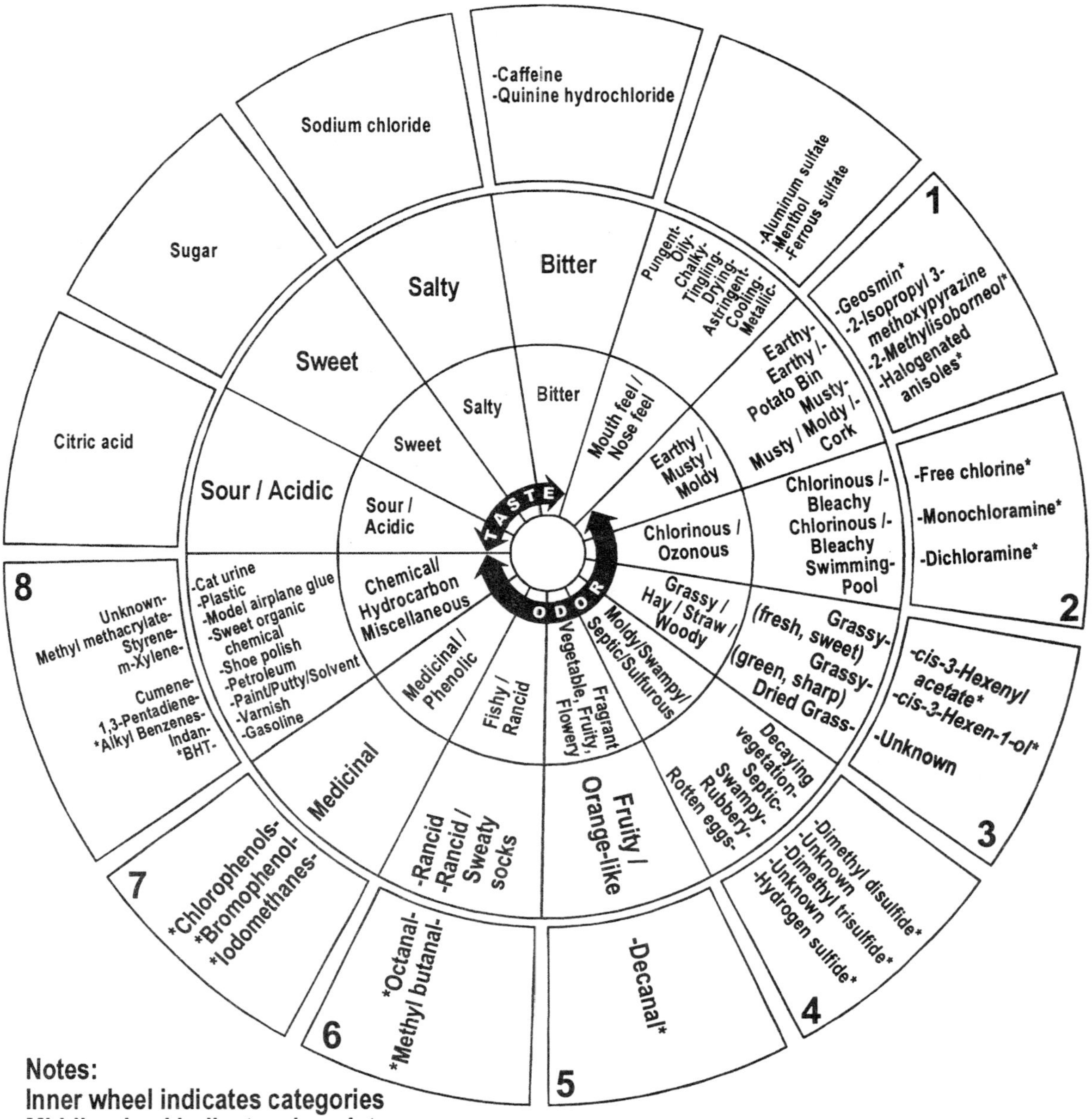

Notes:
Inner wheel indicates categories
Middle wheel indicates descriptors
Outer wheel indicates reference standards
Distribution system has not been positively confirmed to be the source of the compounds in italics
*Presence confirmed in water

Figure 8.1 Distribution-system taste-and-odor wheel

Table 8.1

Sources and possible corrective actions

Category	Possible Source	Possible Corrective Actions
Sweet	• Open reservoirs, biological activity	• *Cover reservoir. Breakpoint-chlorinate. Clean regularly.*
Salty	• Cross-connections • Treatment breakage at reverse osmosis membrane • Well contamination	• *Eliminate cross-connections. Survey for potential cross-connections and install appropriate backflow prevention devices.*
Sour/Acidic	• Hot water systems • Poorly maintained point-of-use (POU) device • Cross-connections	• *Raise water temperature in water heaters. Flush water heaters annually. Replace sacrificial anodes in water heaters.* • *Replace POU device filters more regularly.* • *Eliminate cross-connections. Survey for potential cross-connections and install appropriate backflow prevention devices.*
Bitter	• High pH from corrosion of cement lining • Metal leaching • Cross-connections (e.g., backflow from carbonation)	• *Fill, hold, and flush to waste newly installed or relined cement-lined pipelines left to stand. Periodically flush.* • *Stabilize water by adding polyphosphates or adjust pH.* • *Eliminate cross-connection. Survey for potential cross-connections and install appropriate backflow prevention devices.*
Mouthfeel/Nosefeel	• Galvanized or copper pipes, corrosion, rust, and other corrosion by-products • New plumbing and water coolers • Cross-connections or backflow from carbonation	• *Add stabilizing agents such as polyphosphates, or adjust pH.* • *Fill, hold, and flush new plumbing.* • *Eliminate cross-connection. Survey for potential cross-connections and install appropriate backflow prevention devices.*
Earthy/Musty/Moldy	• Open reservoirs, biological activity • Dead ends or low-flow sections of system • Loss of chlorine and un-masking of background odors • Biomethylation of halogenated phenols • Cross-connections	• *Cover open treated-water reservoirs. Chlorinate open treated-water reservoirs. Install aeration systems.* • *Perform routine flushing of distribution system dead ends. Eliminate dead ends by looping mains.* • *Maintain detectable residual at all times.* • *Treat with chlorine dioxide.* • *Eliminate cross-connections. Survey for potential cross-connections and install appropriate backflow prevention devices.*

(continued)

Table 8.1 (continued)

Category	Possible Source	Possible Corrective Actions
Chlorinous	• New main disinfection	• *For super-chlorinated mains, dechlorinate and discharge.*
	• Residual boosting	• *Optimize weight ratio of chlorine to ammonia-nitrogen for chloraminated water. (Make sure the water is where it should be with respect to the breakpoint curve.)*
	• Blending	• *Eliminate breakpoint-chlorination occurrence in blended water. Reduce blend percentage or residuals of chlorinated water compared to chloraminated water.*
	• Continuing reactions	• *Chloraminate distribution system.*
	• DBP formation	• *Optimize treatment.*
Grassy/Hay/Straw/ Woody	• Open reservoirs, biological activity	• *Cover reservoirs. Treat reservoirs with chlorine. Utilize selective withdrawal from reservoirs. Blend with other sources.*
Marshy/Swampy/ Septic/Sulfurous	• Poorly circulated reservoir	• *Aerate (or aerate hypolimnon of) source-water reservoir. Change withdrawal levels from reservoirs.*
	• Poorly maintained POU devices	• *Replace POU device filters more often.*
	• Poorly maintained hot water systems	• *Flush hot water systems annually.*
	• Poorly maintained sedimentation basins and/or dirty or stagnant filters in treatment plants	• *Clean sedimentation basins frequently. Backwash filters before returning to service.*
	• Dead-end mains	• *Loop or periodically flush dead-end mains.*
	• Oxidation of polysulfides	• *Breakpoint-chlorinate, followed by chloramination.*
	• Cross-connections or backflow	• *Eliminate cross connection. Survey for potential cross-connections and install appropriate backflow prevention devices.*
Fragrant/Vegetable/ Fruity/Flowery	• Open reservoirs, biological activity	• *Aerate source-water reservoirs. Breakpoint-chlorinate, followed by chloramination of distribution system.*
	• Continuing reactions	• *Adjust chloramine-to-ammonia ratios.*
Fishy	• Open reservoirs, biological activity	• *Aerate source-water reservoirs. Pre-oxidate source waters prior to treatment.*

(continued)

Table 8.1 (continued)

Category	Possible Source	Possible Corrective Actions
Medicinal/Phenolic	• Post-chlorination of biofilms	• *Convert to chloramines. Maintain high chloramine residuals.*
	• Slow kinetics and DBP formation	• *Optimize treatment.*
	• Reservoir and tank linings, chlorination of phenolic linings	• *Use coatings approved by a recognized authority. Allow proper curing times on new coatings. Fill and hold, followed by flushing newly coated tanks and pipelines.*
	• Cross-connections or backflow	• *Eliminate cross-connections. Survey for potential cross-connections and install appropriate backflow prevention devices.*
Chemical/Hydrocarbon/ Miscellaneous	• Chlorine dioxide/carpet, volatilization in indoor air	• *Add ferrous salts or reduced sulfur compounds to remove chlorite at the treatment plant or substitute chloramines for chlorine in the distribution system.*
	• New plumbing and pipe materials	• *Use plumbing and fixtures approved by a recognized authority. Fill, hold, and flush plumbing.*
	• Plastic pipes, lubricants, new mains, asphaltic coatings, paint, epoxy/solvent-based linings	• *Use lubricants approved by a recognized authority. Allow proper cure times for new linings. Fill, hold, and flush newly lined mains.*
	• Permeation of buried plastic pipe	• *Replace plastic pipelines or remove contaminated soils surrounding pipelines.*
	• Old coal-tar linings on cast-iron mains	• *Gently flush to refresh water quality in mains. (Aggressive flushing can damage the coal tar lining the mains.)*

odor wheel is intended for use as a guideline by water practitioners to recognize the types of quality changes that may be or are being experienced.

RECORD-KEEPING AND DATABASES

It is strongly recommended that water utilities keep records of taste-and-odor complaints, even when they are easily addressed. A carefully documented and properly compiled database could help water utilities understand and sometimes predict taste-and-odor changes before they occur. Information on location, frequency of occurrence, and type of complaint, along with other relevant water quality parameters, could provide insights into the underlying causes of taste-and-odor changes and help to document trends over time or follow changes in treatment or operations.

Although this report includes representative case histories and examples of potential taste-and-odor changes, new situations will inevitably arise that have not previously been experienced or reported. It is hoped that this report, including the taste-and-odor wheel, sampling protocol, questionnaire for customer complaints, and guidelines for avoiding problems, will assist water utility personnel in providing the best-tasting water possible on a cost-effective basis and at a globally competitive price.

CHAPTER 9
CONCLUSIONS AND RECOMMENDATIONS

The main objectives of this study were to identify the key parameters in the production of tastes and odors in distribution systems, determine the chemical and microbiological causes, assess the potential solutions to specific water quality changes, and, finally, establish guidelines for water utilities to solve these problems or ultimately prevent them from occurring.

CONCLUSIONS

Tastes and Odors Associated with Biological Activity

The following conclusions are based on results obtained from the pipe-loop experiments:

- In the absence of disinfectants, an increase in earthy-musty flavors with residence time was noticed in the pipe-loop facility. No linear correlation between microbial counts (fungi and actinomycetes) in the bulk water and flavor threshold number (FTN) could be found.
- The use of disinfectants (chlorine/chloramines) can induce a delay in the appearance of the earthy-musty flavor and a reduction of its FTN.
- The control of earthy-musty flavors by disinfectants is attributable not only to a chemical masking action, but also to a biostatic effect that induces a lower microbial activity.

This pipe-loop study indicates that chlorine may induce unpleasant flavors, even at very low concentrations. No significant increase in flavor is noticed with monochloramine. Corresponding distribution systems can generate earthy-musty flavors that can be related to biological activity and are mainly dependent on residence time. These flavors can be partially controlled by the use of disinfectants.

The findings from the annular reactor experiments show that nitriles and aldehydes are detected in the effluent, but not the influent water. Generally, the pipe-loop and annular reactor studies show that they generate results similar to what is observed in actual distribution systems.

During a case-study investigation, 2,4,6-tribromoanisole (TBA) was identified as the cause of earthy-musty odors. The odor threshold concentration (OTC) of TBA is approximately 0.03 ng/L at 25°C. Although the biological origin of TBA was not confirmed, it can be assumed that the mechanism leading to its formation is similar to that of 2,4,6-trichloroanisole (TCA). TCA has been reported to be formed by biomethylation of chlorophenols in the presence of fungi.

Tastes and Odors Caused by Disinfectants and Associated Chemicals

The pipe-loop experiments conducted in France led to additional conclusions related to the taste-and-odor behavior of disinfectants:

- Jar tests with a flavor-free water (S.A. Evian Co., Evian, France) have allowed the determination of a flavor threshold concentration (FTC) of 0.025 mg /L for chlorine as Cl_2 and an FTC of 0.28 mg /L for monochloramine as Cl_2. (The FTC is given by the amount of chemical in the greatest dilution volume in which a flavor can be perceived by a panelist, and the dilution ratio is the FTN.) These FTCs were determined at 25°C.
- The FTC is influenced by the initial FTN and flavor description of the water. No significant effect is observed for monochloramine.
- Except for chlorine decay, the distribution system has no specific effect on the chlorinous flavor induced by disinfectants.
- Chlorine flavors became unnoticeable for concentrations below 0.07 mg /L for chlorine as Cl_2 and 0.55 mg /L for chloramines as Cl_2 and were replaced by earthy-musty flavors.
- Field studies showed that chlorinous flavors were perceived down to 0.05 mg /L for free chlorine as Cl_2, and earthy-musty flavors appeared for concentrations below 0.1 mg /L for free chlorine as Cl_2.

Bench-scale experiments performed under various treatment conditions confirmed the sensory masking effect of disinfectants on various waters with different origins and different water-quality characteristics. It was observed that longer detention times and subsequent chlorine decay support the appearance of other taste-and-odor characteristics. When untreated water was compared with treated water at various post-treatment detention times, it was seen that a masking effect of odors (originally present in the untreated water or produced during ozonation) occurred in the presence of a chlorine or chloramine residual. Sometimes the masking effect diminished after an extended detention time and the original odor reappeared, but under other conditions, the original odor was permanently eliminated. In most cases, the initial quality of the water (discounting the chlorine residual) was the determinant factor in the sensory characteristics of the water received at the consumer's tap, especially in the more remote areas of the distribution system.

Bench-scale experiments on three groundwaters, using disinfection only, showed that in general, decay of the disinfectant residual after longer detention times appeared to improve the organoleptic properties of the water. An exception was the appearance of a medicinal odor in one water.

Thus, the shelf life of water with respect to tastes and odors is largely related to the persistence of a total chlorine residual.

Specific conclusions are as follows:

- Very little difference was noted in the characteristics of conventionally treated, chloraminated water at a level of 1.7 mg/L compared to 2.5 mg/L for the water tested.
- Very little difference was noted for ozonated water between the two chloramine residuals for the water tested, and ozonated water appeared to have fewer descriptors and lower intensities than conventionally treated water.
- The occurrence of a single new odor that seemed to be produced by the mixing of two apparently unrelated odors was observed. The decaying vegetation odor at 1 h was replaced by a caramel odor following partial dissipation of the chloramine residual after 48 h.
- Masking effects of the disinfectant residuals were the single most common finding in the studies with simulated full treatment.
- Longer detention times appeared to improve organoleptic properties of the waters tested with simulated full treatment or disinfection only.

- An OTC at 25°C for ammonia was estimated at 0.9 mg/L in this study, with odor descriptors of ammonia, rubbery, plastic, cardboard, and reedy and taste descriptors of bitter and tingling sensation.

Tastes and Odors Caused by Continuing Reactions and Disinfection By-Products (DBPs)

The formation of DBPs is most likely to continue in the distribution system in the presence of disinfectant residuals. The taste-and-odor characteristics of a large number of DBPs have been determined through the course of the study. Descriptors, OTCs at 45°C, and dose-response plots were determined (see Appendix A).

Iodinated trihalomethanes (ITHMs) exhibit medicinal characteristics that become more distinct with the number of iodine atoms in the structure. Tests performed in this study indicate that ITHMs form slowly in the distribution system. It was observed that the chlorine-to-nitrogen ratio during chloramination was an important factor in the formation of ITHMs. Higher concentrations of the compounds are generally formed when the 5:1 ratio is used at the low residual (e.g., 0.5 mg/L monochloramine as Cl_2). Higher concentrations are generally formed when the 3:1 ratio is used at the high residuals (e.g., 4 mg/L monochloramine as Cl_2). An increase in dichloroiodomethane and bromochloroiodomethane concentrations was seen with longer detention times.

Specific conclusions are as follows:

- The Cl_2:N ratio is an important factor in the formation of ITHMs.
- During bench-scale experiments to investigate medicinal and chloramine/swimming pool–type complaints, chlorine dioxide was shown to be the best alternative to avoid taste-and-odor production for the water tested.
- Aldehyde OTCs ranged from 0.03 to 47 µg/L. Aldehydes were detected in the annular reactor experiments and in a case study in Spain.
- OTCs for chloroketones (80–500 µg/L), trihalomethanes (40–90 µg/L), and halonitriles (50–400 µg/L) should not represent a taste-and-odor problem.
- OTCs for nitriles ranged from 0.13 to 64 µg/L and most have fruity-sweet characteristics. Numerous nitriles were identified in chloraminated waters in the annular reactor experiments.

- For ITHMs, the medicinal characteristics appeared to increase with the number of iodine atoms in the compound, with the OTCs ranging from 0.032 to 30 μg/L.
- A longer detention time resulted in less intense chlorine taste and odors, but more intense DBP tastes and odors
- Unidentified chlorinous odors perceived in some networks in Europe (with no or very low chlorine residuals) are not caused by the halogenated DBPs tested in this study.

Tastes and Odors Caused by Leaching of Distribution-System Materials

The bench-scale and case studies indicated that leaching of distribution-system materials may cause the appearance of tastes and odors, either directly from the compounds leached or through the reactions of the disinfectant residuals with these leached chemicals. Every material that is in contact with water has some potential to affect the flavor of the water, given the right conditions. Improper curing of interior asphaltic coatings of pipe are suspected of causing oil-base paint–type tastes and odors in case studies presented. A series of substituted benzenes and naphthalenes were found in samples.

For some of the case studies, the metallic taste seems to be related to the level of iron measured in the water, the astringent mouthfeel to aluminum sulfate and zinc chloride, and the sour taste to zinc chloride. The flowery odor appears to be related to nitriles and the plastic descriptors to alkyl benzenes. The case studies presented real-world examples of what was observed in bench-scale experiments.

Specific conclusions are as follows:

- The rubber gasket without disinfectant and in chlorinated water had a rubbery odor, with a chlorine background odor in the chlorinated water.
- The most common odors in the TYTON™ joint lubricant (TJL) water samples were the same waxy-oily odors as in the original product, which develops a rancid oil note as well as a soapy note. Chlorine disinfection decreased the odor to a sweet, citrus, and soapy odor after 10 days.
- The odor of the Aquaflow VA-SP 10 lubricant (AVA) changed from astringency to oily to pine-phenol. The chloramine disinfection produced only the astringent odor.

- The Blue Lube™ lubricant (BLL) had a sweet odor without disinfectant and with chloramine. Chlorine appeared to oxidize something in the BLL to a kerosene and alcohol odor. BLL odors had the lowest intensity of the lubricants tested.
- A series of straight-chained aldehydes (C-6 to C-10) and a mixture of other aldehydes, ketones, alcohol, hydrocarbons, and bicyclo compounds are indicated as causing the waxy-oil (rancid) odor of the TJL in chlorinated and chloraminated water. The sweet odors included the tentatively identified aldehydes (octanal, E-2-octenal, and nonanal). Other compounds found that may contribute to the sweet odors were 9-methyl-2-heptanone or 4-methyl-2-heptanone, 7-octen-4-ol, E-2-octene-1-ol, and cineole.
- BLL is the least offensive odor former of the three lubricants, followed by AVA and TJL. In chlorinated water in the distribution system near the point of use of the TJL, the odor intensity of waxy-oily (rancid) would decrease to background after about a week. Apparently, chlorine can react with the lubricant and low-molecular-weight compounds, with paint and waxy-oily (rancid) odors forming after 3 days. The waxy-oily odor can then react with chlorine and be removed with time, and a citrus-sweet odor forms after 8 days. A soapy odor was found consistently in most samples.
- Gas chromatography/mass spectrometry (GC/MS) and sensory GC analysis tentatively identified a series of straight-chain C-7 to C-10 aldehydes with odors of TJL off-flavors; however, the true odor of the sample was not matched by the known odors of the chemicals. Also, many other aldehydes, ketones, alcohols, and hydrocarbons in the C-6 to C-20 range in chlorinated and chloraminated water were found after chlorinating water with TJL dissolved in it. The study verifies the conclusions of Fadel et al. (1993) that aldehydes are present and part of the odor mixture, although the FPA panel describes the samples differently. The mechanism of odor production is proposed to be the same oxidative rancidity reaction that is the major cause of food spoilage.
- Polyvinyl chloride (PVC) gray or white pipes alone do not cause odors in the distribution system; however, the primer and the cement, both alone and when applied to gray or white pipes, can cause odor problems of the glue-varnish type.

- The leachate of the ductile iron pipe with cement-lined mortar adds to the odors of the lubricants and gaskets used to hold the pipes together. The data indicate that a complicated mixture of chemicals from the pipes and lubricants is generating odors.
- Iron appears to be a major component of the off-tastes in the leached samples from the gray cast-iron pipe.
- Although linings were not studied as part of this project, there is evidence from the case studies that inadequately cured linings contribute odors.

Tastes and Odors Associated with Blending Factors

It was determined that when blending two chlorinated waters of different origins and different water quality, (1) differences in odors appeared before the differences in flavors (2) differences in odors in the chlorinated waters appeared before differences in the dechlorinated waters, and (3) differences in odors were greater when tested at 45°C than at 25°C. There were no significant differences between blends prepared in the laboratory and tested at zero time and blends collected from the actual distribution system at unknown detention times.

The trends observed for blending chlorinated groundwater with chloraminated surface water generally followed the trends observed by Barrett et al. (1985). The changes in the organoleptic properties of the blended water were very subtle. The organoleptic characteristics of the water changed according to the contact times, with the "swimming pool" and "sweet" characteristics becoming more prevalent at 24 h and 120 h and the "straw" odor decreasing at longer detention times.

Specific conclusions are as follows:

- When comparing the flavor-profile analysis (FPA) of blended waters, it was observed that one source water had a higher influence on the blend than would be expected from its proportion of the mixture.
- Differences in odors in the chlorinated waters appear before differences in the dechlorinated waters.
- Differences between odors appear before differences between flavors.
- For blending of Llobregat and Ter waters, chlorinous and earthy descriptors were found in both sources of water and remained in the blended water.

- There were no characteristics in the blended samples collected from the actual distribution system that were different from those found in the water from the Llobregat and Ter rivers.
- The Llobregat characteristics had a strong influence on the organoleptic profile of the Llobregat/Ter blends, despite a low percentage of Llobregat water compared to Ter water.
- There were no significant differences between blends prepared in the laboratory and tested at zero time and blends collected from the actual distribution system.
- Spider diagrams are a useful way to depict FPA results.

RECOMMENDATIONS

1. A "taste-and-odor formation potential test" similar to the one used for trihalomethanes, but based on sensory evaluation, should be established. Parameters such as detention time, pH, disinfectant residual, and temperature should be taken into consideration. This test should be performed with the specific water matrix and the specific treatment in use by the utility doing the evaluation (site-specific water), and the test should be regularly used in evaluating water quality changes.
2. Certification procedures for materials to be used in the distribution system should require sensory evaluation prior to approval of materials. In addition, all new facilities, including pipes, should be tested for off-odors prior to their release for use.
3. Utilities should use only approved materials in their distribution systems, and certification should be by the recognized authorities (e.g., NSF International or AWWA). The recognized authorities should include taste-and-odor testing in their approval procedures.
4. Methods of taste-and-odor evaluation should be standardized. It is only by means of standardization that the results of a taste-and-odor investigation can be used for comparison with other occurrences.
5. Water utilities should record and compile their taste-and-odor complaints and subsequent investigations in a database that can be accessed for future reference. Such a database could

provide insights into the origins and causes of recurrent quality changes. Ideally, these databases should be made available to other researchers and water practitioners for optimal understanding and solution of problems.

6. Additional work is necessary to identify the causes of chlorinous odors in distribution systems in which the disinfectant residual is very low.
7. Careful flushing of the pipes and minimum use of lubricants will help to minimize the effects of pipes and lubricants on tastes and odors.
8. New techniques are needed to measure lower concentrations of compounds that can cause tastes and odors in the distribution system.
9. More work should be performed to identify the sulfur odors, their mechanisms and kinetics of formation, and possibly their stability in treated waters.
10. More work is necessary to confirm actual relationships of biological activity and tastes and odors generated in the distribution system.

APPENDIX A

Additional organoleptic information for compounds of various classes of potential taste-and-odor importance.

CONTENTS

BUTYRONITRILE

Conc. (µg/L)	32	80	200	350	500	1200	3200	5000
Log conc.	1.5	1.9	2.3	2.5	2.7	3.1	3.5	3.7
Sònia B.	2	1	2	3.5	4	5.5	6.5	8
Núria	1	2	2	4	5	6.5	7	8
Sònia C.	1	1.5	2	2.5	2.5	4	6.5	8.5
Rosa	0	1	2	3.5	4	5	6	7.5
Roser	0	1	3	4	6	6.5	8	10
Diana	0	2	3	4	5.5	6	7	8
Average Intensity	0.7	1.4	2.3	3.6	4.5	5.6	6.8	8.3

exp. OTC: 32 µg/L

Terminal exp.OTCs: 32 µg/L (3) · 80 µg/L (3)

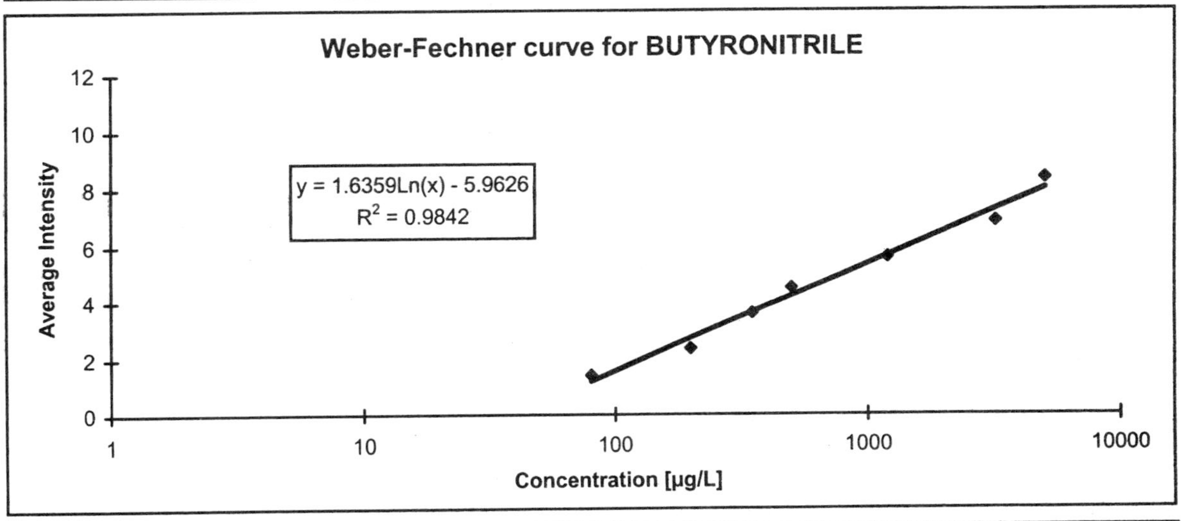

If Intensity = 1 in Weber-Fechner curve, then OTC = 70 µg/L

Descriptors: Below 1200 µg/L: fruity, sweet, citrus, fragrant
Above 1200 µg/L: solvent, sweet organic solvent

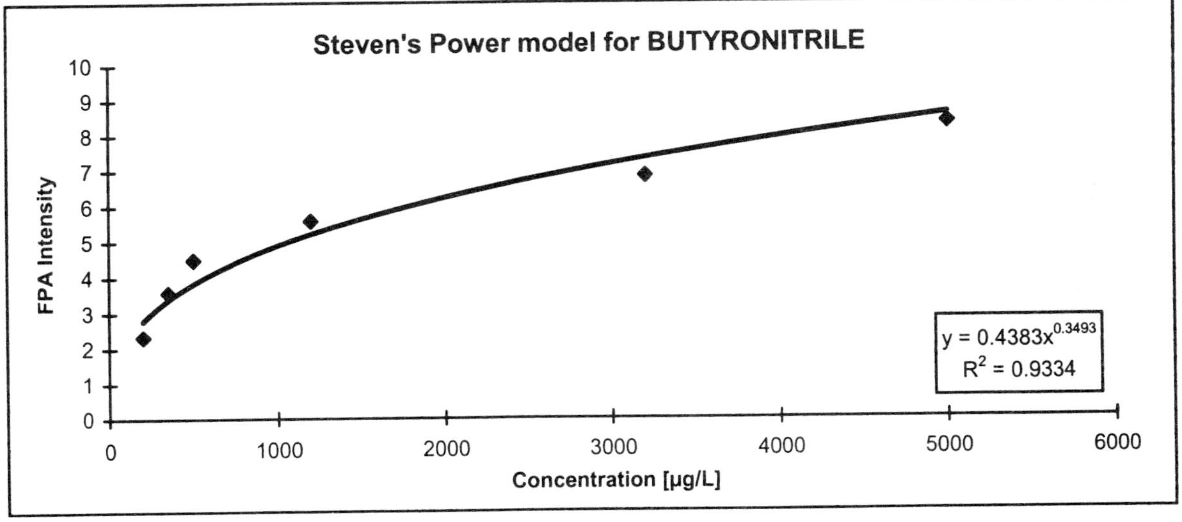

ISOBUTYRONITRILE

Conc. (µg/L)	64	160	400	1000	2600	6400	####
Log conc.	1.8	2.2	2.6	3.0	3.4	3.8	4.0
Mònica	1	1	2	5	6	7	8
Mª José	2.5	1	1.5	4.5	6	7	8
Oscar	0	0	1.5	3	5	6	7.5
Roser	0	0	2.5	4.5	6	7	10
Diana	0	0	2.5	5	6	7.5	9
Sònia B.	0	0	2	4	6	6.5	8.5
Average Intensity	0.6	0.3	2.0	4.3	5.8	6.8	8.5

exp. OTC: 64 µg/L

Terminal exp.OTCs: 64 µg/L (2) - 400 µg/L (4)

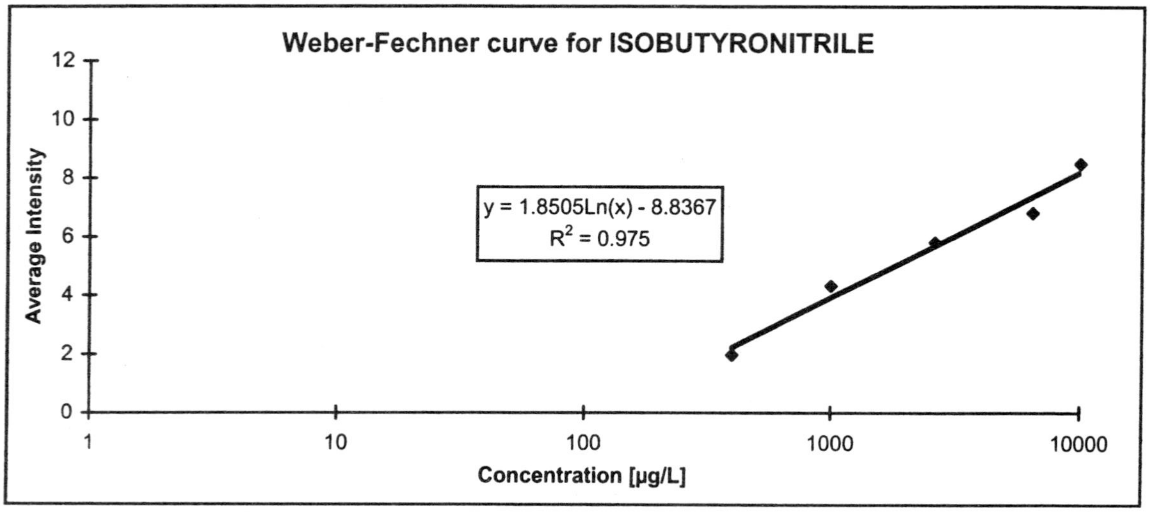

If Intensity = 1 in Weber-Fechner curve, then OTC = 200 µg/L

Descriptors: citrus, sweet
Above 1000 µg/L: rubbery, paint, turpentine, acetone.

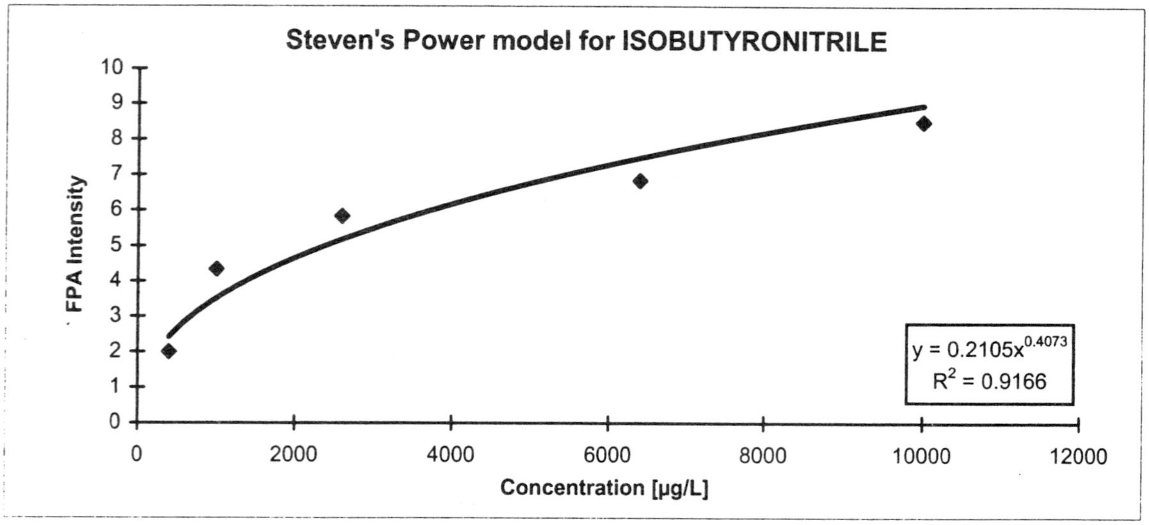

HEXANENITRILE

Conc. (μg/L)	3	8	20	32	50	80	130	320
Log conc.	0.5	0.9	1.3	1.5	1.7	1.9	2.1	2.5
Raquel	1	3.5	6	5.5	8	7	8.5	10
Esther	0	2	4	3.5	5	6.5	6.5	8
Núria	0	3	5	5	7	6	7	9
Mª José	0	1	3	3	5	4.5	7	8.5
Luis	0	0	3	5	4	6	6	7.5
Average Intensity	0.2	1.9	4.2	4.4	5.8	6.0	7.0	8.6

exp. OTC: 3 μg/L

Terminal exp.OTCs: 3 μg/L (1) - 20 μg/L (1)

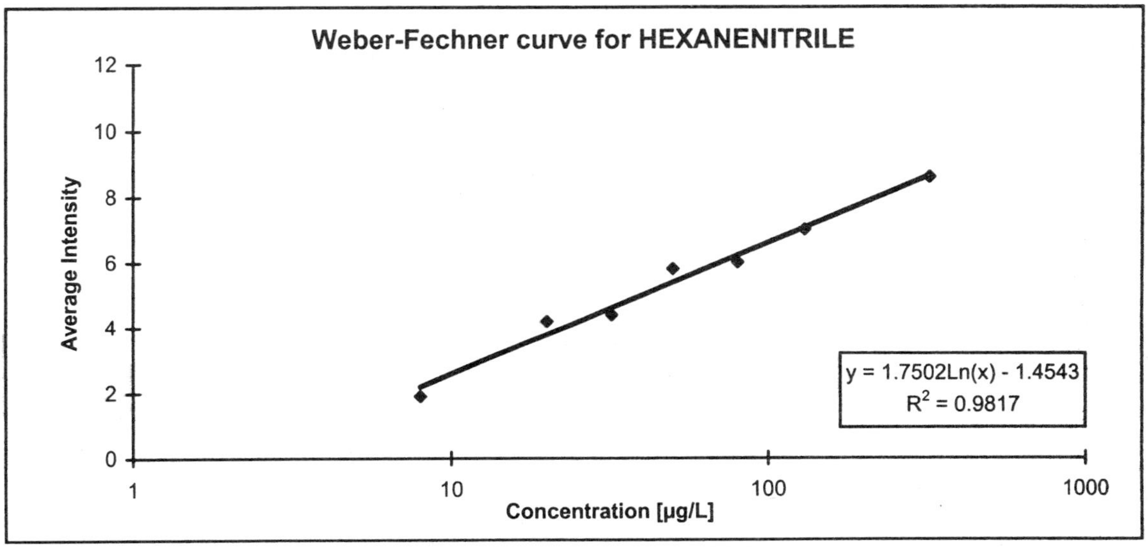

If Intensity = 1 in Weber-Fechner curve, then OTC = 4 μg/L

Descriptors: geranium, plastic, rancid, bitter, gum.
At high concentrations (above 180 μg/L): more green, more fragrant

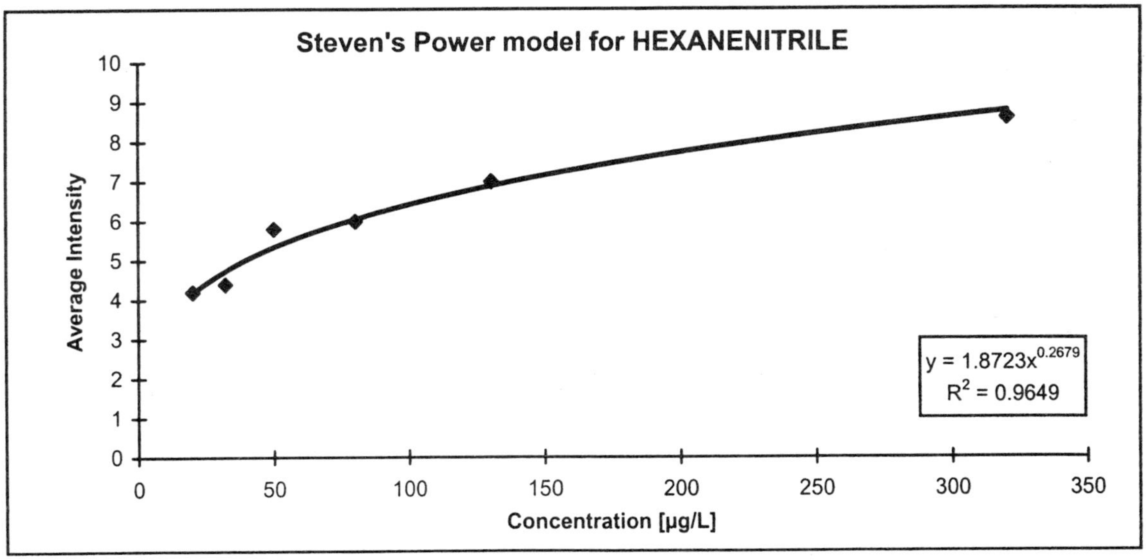

OCTANONENITRILE

Conc. (µg/L)	0.1	0.3	0.8	2	5	13	31
Log conc.	-0.9	-0.5	-0.1	0.3	0.7	1.1	1.5
Iñaki	2	3	4	5.5	7	8.5	10
Montse	0	0	2	3	4	5	6.5
Rosa	0	0	1	5	5.5	6.5	7
Carlota	0	0	1	3	5	7	9
Esther	0	0	2.5	3	5	6	7
Mª José	0	0	0	4	5	6.5	8
Luis	0	0	0	1	3.5	6	7
Average Intensity	0.3	0.4	1.5	3.5	5.0	6.5	7.8

exp. OTC: 0.1 µg/L

Terminal exp.OTCs: 0.1 µg/L (1) - 2 µg/L (2)

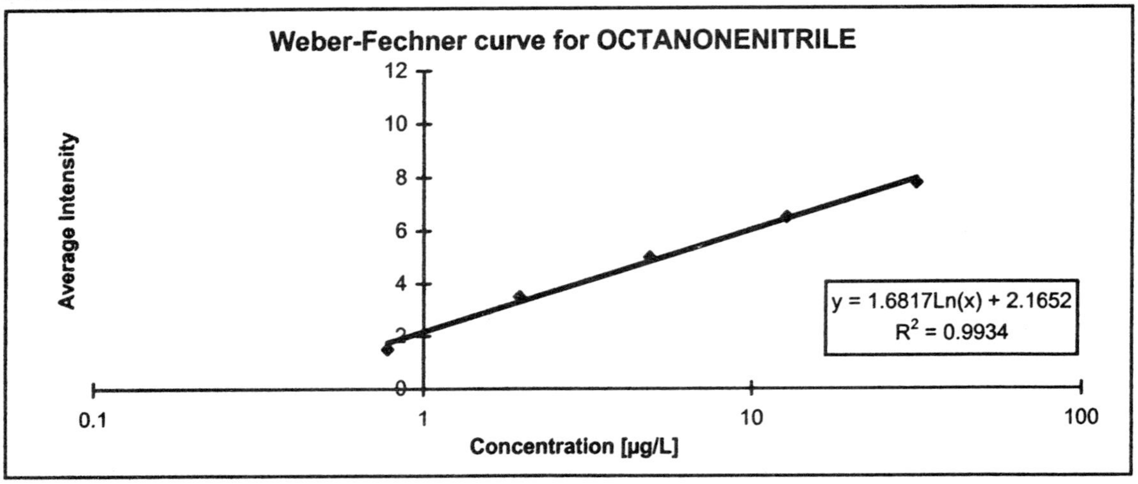

If Intensity = 1 in Weber-Fechner curve, then OTC = 0.5 µg/L

Descriptors:	geranium, plastic, pungent, rancid, sweet, penetrant
	Above 30 µg/L: geranium, plastic, coconut cakes, earthy.

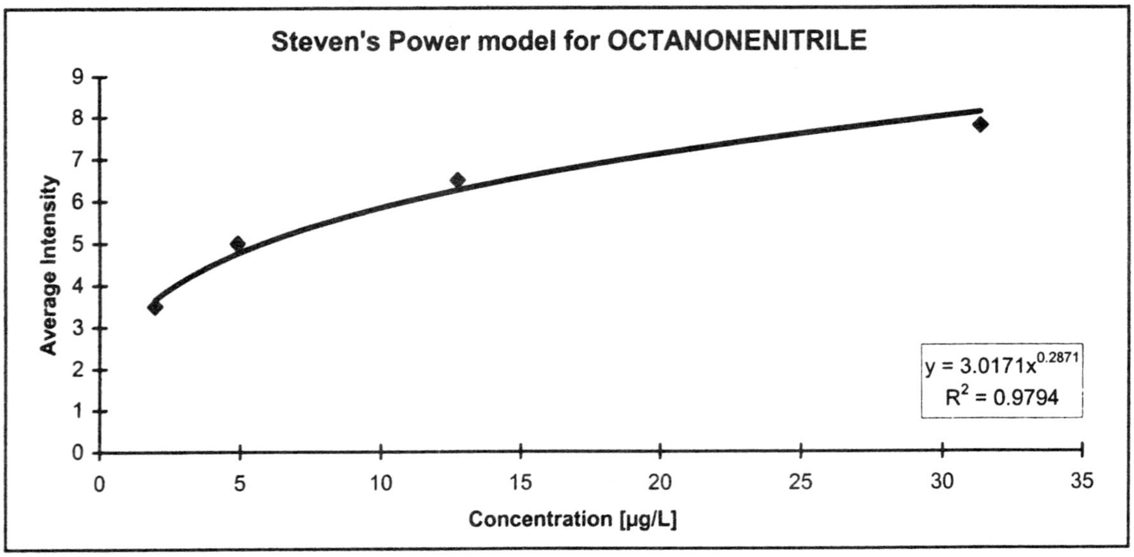

DODECANENITRILE

Conc. (µg/L)	0.3	0.8	2	5	13
Log conc.	-0.5	-0.1	0.3	0.7	1.1
Raquel	3	5	7	9.5	10
Mª José	4	5	6.5	8	9
Montse	0	3	3	4	5
Carlota	0	3.5	5	6.5	8
Esther	0	3	5	6	8
Rosa	0	4.5	5.5	6.5	7
Int. Average	1.2	4.0	5.3	6.8	7.8

exp. OTC: 0.3 µg/L

Terminal exp.OTCs: 0.3 µg/L (2) - 0.8 µg/L (4)

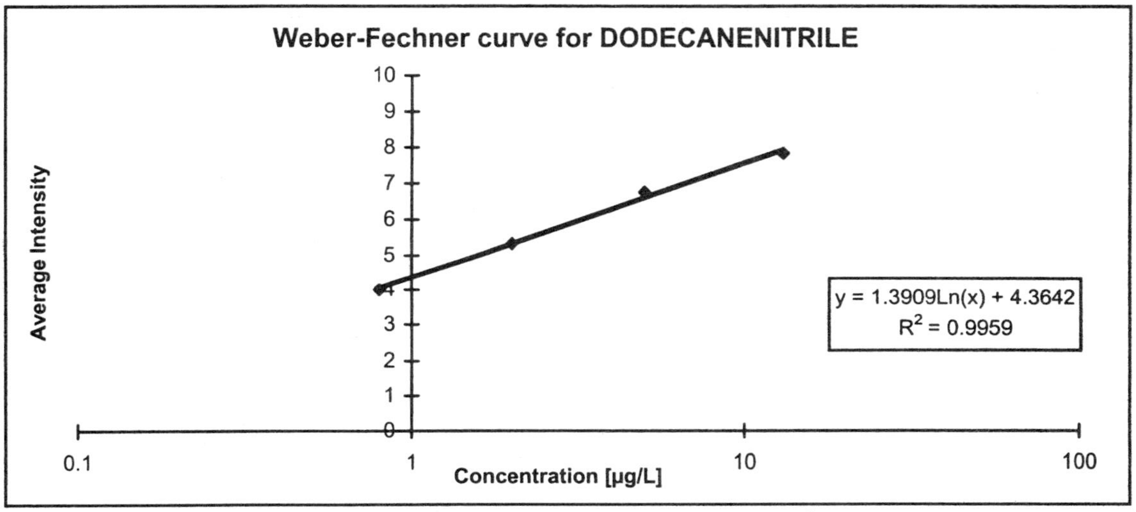

If Intensity = 1 in Weber-Fechner curve, then OTC = 0.1 µg/L

Descriptors: citrus, tangerine, sweet

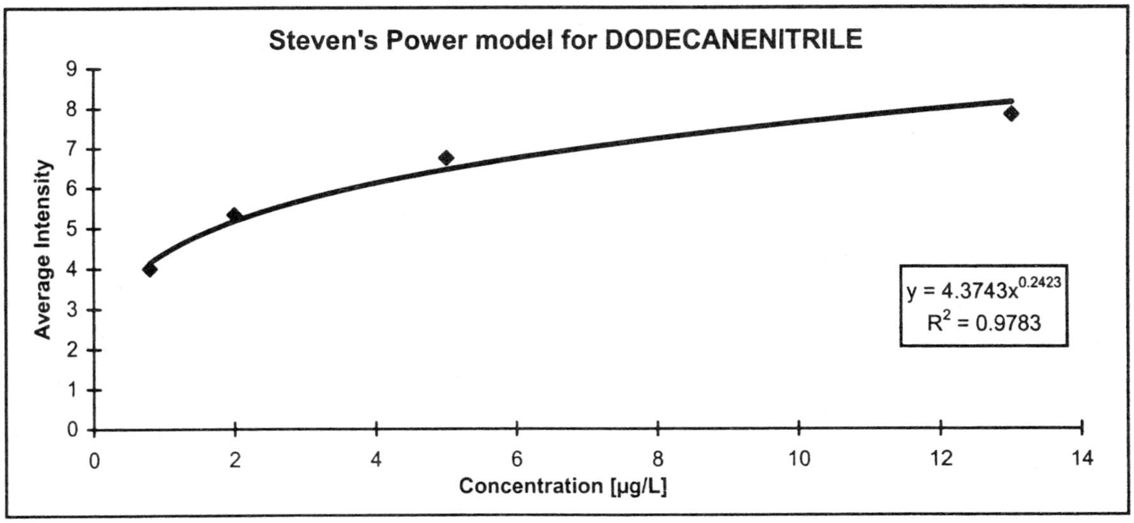

CHLOROACETONITRILE

Conc. (μg/L)	400	1000	2500	6000	16000	25000
Log conc.	2.6	3.0	3.4	3.8	4.2	4.4
Marta	1	2.5	4	6	7	8
Mònica	1	3	4	5	7	8
Núria	1.5	2	5	6	7.5	10
Raquel	1	2	3	5	7	8
Mª José	1	3	4	5	7.5	10
Marisa	0	0	3	4	6.5	8
Average Intensity	0.9	2.1	3.8	5.2	7.1	8.7

exp. OTC: 400 μg/L

Terminal OTC: 400 μg/L (5) - 2500 μg/L (1)

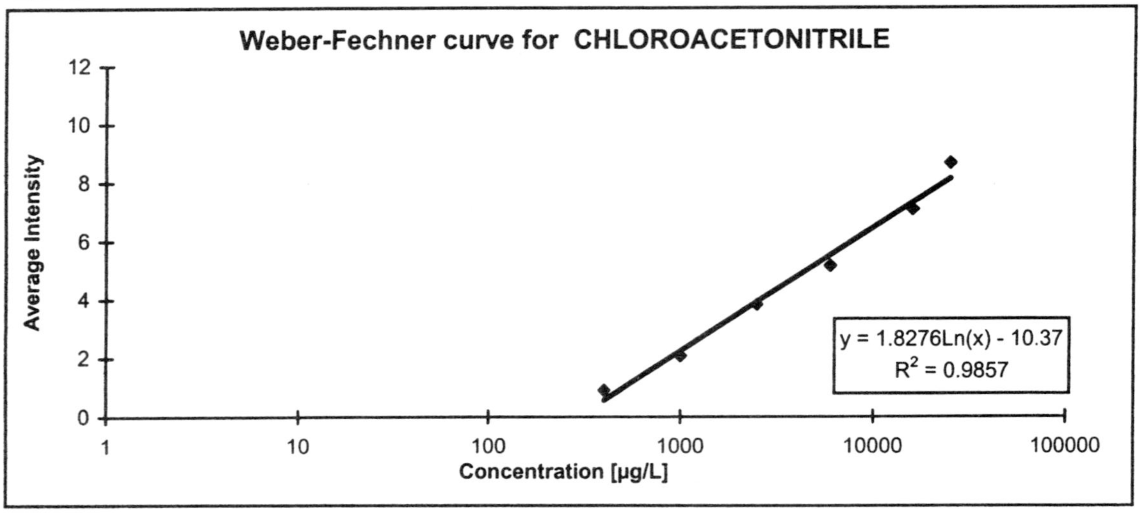

If Intensity = 1 in Weber-Fechner curve, then OTC = 500 μg/L

Descriptors: sweet, rancid, solvent, plastic, rubber, glue, roasted, ginseng, valerian. At high concentrations: very penetrant.

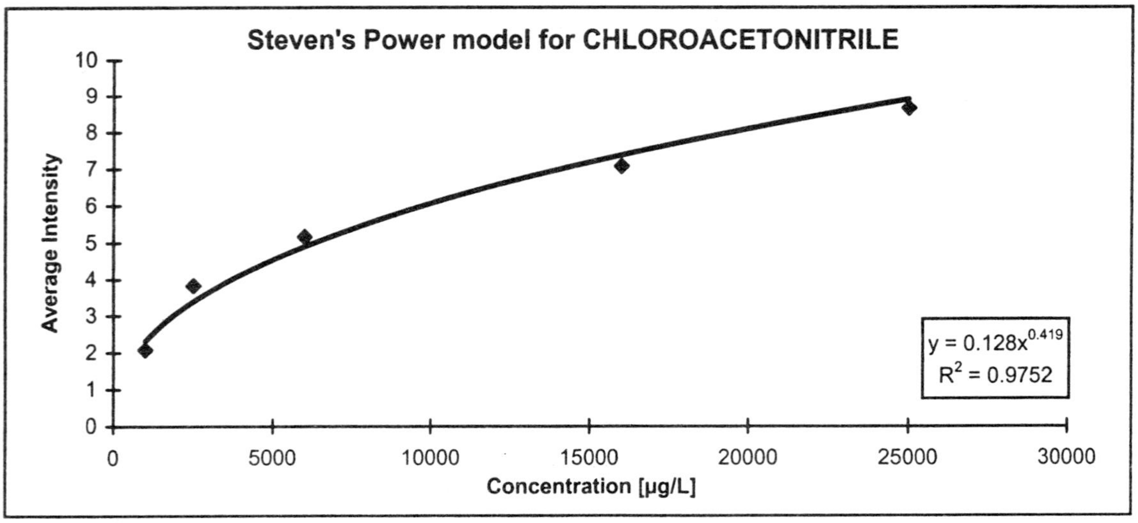

VALERONITRILE

Conc. (µg/L)	13	32	80	200	500	1300	3200
Log conc.	1.1	1.5	1.9	2.3	2.7	3.1	3.5
Rosa	1	2.5	3	4.5	5.5	8	10.5
Esther	0	2	2	3.5	5	6	7
Mª José	0	0	1.5	3	5	6	8
Iñaki	0	0	2	4	5.5	8	9
Sònia B.	0	0	0	4.5	5.5	6.5	11
Sònia R.	0	0	0	3.5	5	8	9
Average Intensity	0.2	0.8	1.4	3.8	5.3	7.1	9.1

exp. OTC: 13 µg/L

Terminal exp. OTCs: 13 µg/L (1) · 200 µg/L (2)

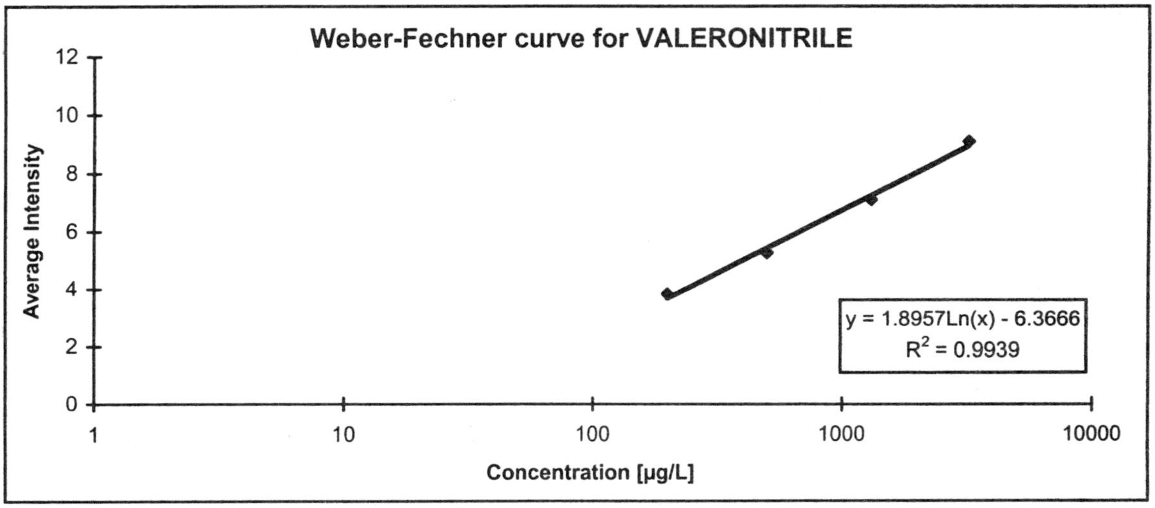

If Intensity = 1 in Weber-Fechner curve, then OTC = 50 µg/L

Descriptors: At low concentrations: putrid, unpleasant, decaying food, rotten
At higher concentrations: sweet, fruity, perfumed
Above 1500 µg/L: sweet almonds, green, sweet, candy

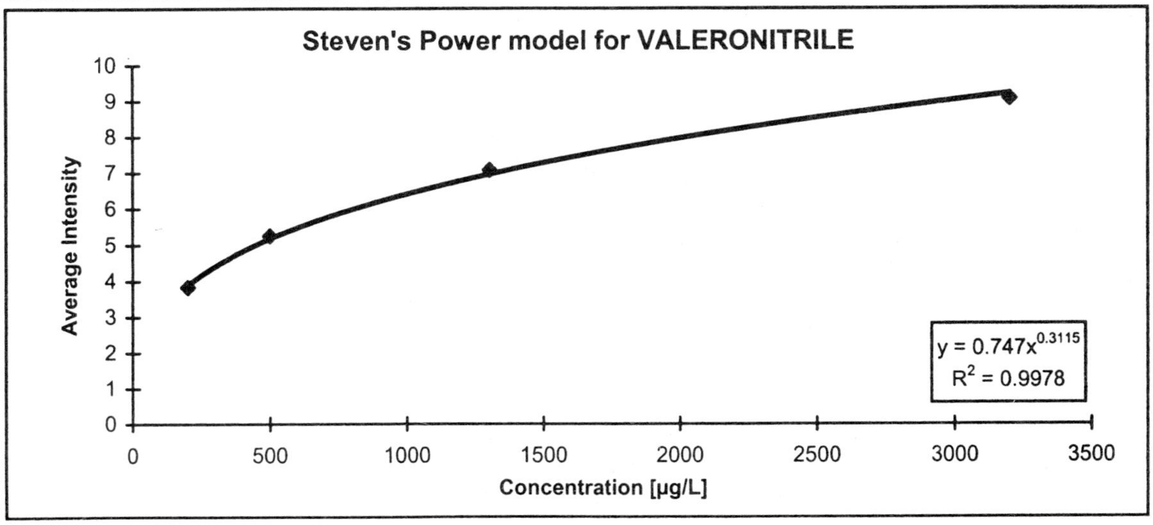

ISOVALERONITRILE

Conc. (µg/L)	3	8	20	50	130	200	320	500	1300
Log conc.	0.5	0.9	1.3	1.7	2.1	2.3	2.5	2.7	3.1
Carlota	3.5	4	6	7	9	9.5	10	11	12
Montse	2	1.5	4	6	7	7	8	9	12
Diana	2	1.5	4.5	6	8	10	9	9	11
Iñaki	2.5	3	5.5	7	9	10	11	11	12
Sònia R.	2	2.5	4	5	7.5	10	9	11	12
Mª José	1.5	2.5	3	5	6.5	9	8	11	12
Average Intensity	2.3	2.5	4.5	6.0	7.8	9.3	9.2	10.3	11.8

exp. OTC: 3 µg/L

Terminal exp.OTCs: 3 µg/L (6)

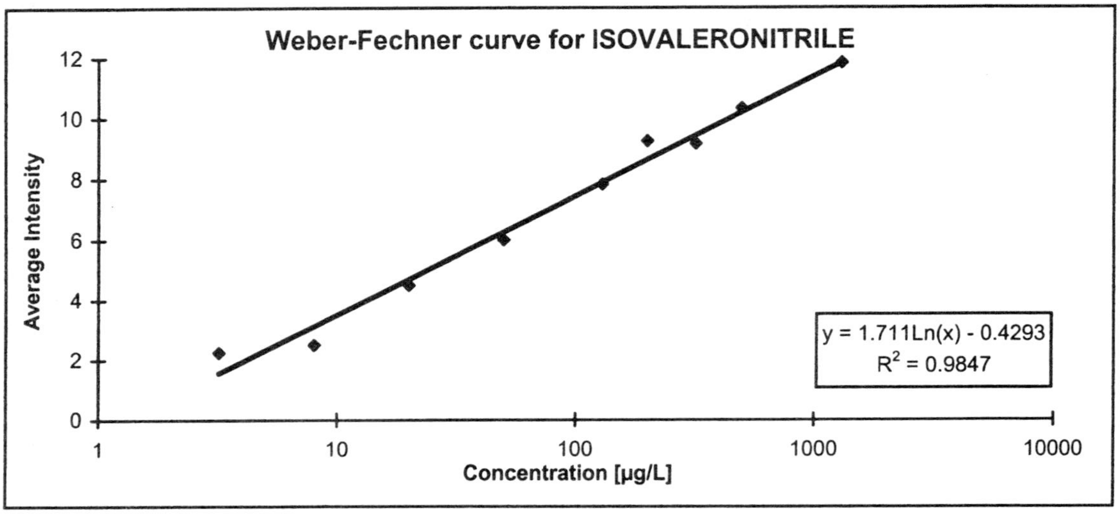

If Intensity = 1 in Weber-Fechner curve, then OTC = 2 µg/L

Descriptors: Onion, solvent (dyer's, gas)
Above 200 µg/L: Solvent, gum, painting, glue, fruity

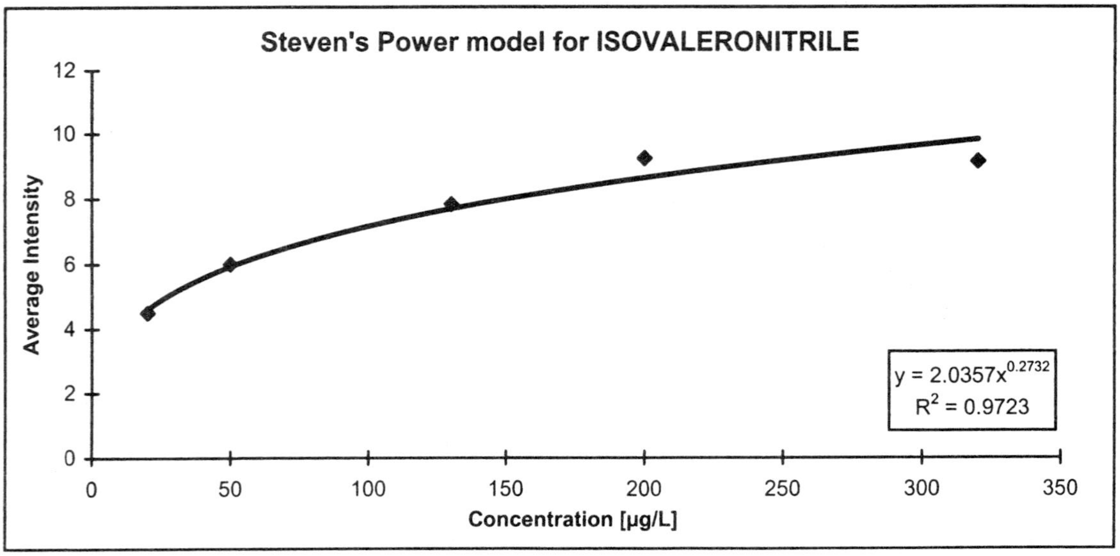

PROPIONITRILE

Conc. (µg/L)	12	32	80	200	500	1200	3200	5000
Log conc.	1.1	1.5	1.9	2.3	2.7	3.1	3.5	3.7
Raquel	1	2	1.5	2.5	3.5	5.5	7	8
Rosa	0	1	1.5	2.5	3.5	5.5	7	8
Sònia B.	0	0	2.5	4	5.5	8	9	10.5
Esther	0	0	3	3.5	5	7	8	9.5
Luis	0	0	0	0.5	2	4	5.5	7
Average Intensity	0.2	0.6	1.7	2.6	3.9	6.0	7.3	8.6

exp. OTC: 12 µg/L

Terminal exp.OTCs: 12 µg/L (1) - 80 µg/L (3)

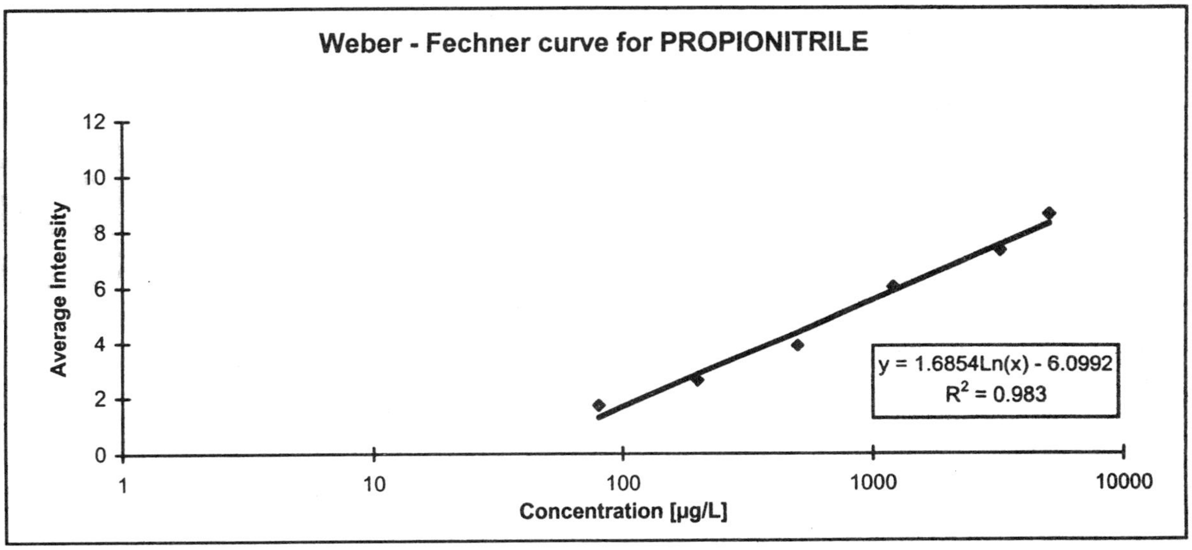

If Intensity = 1 in Weber-Fechner curve, then OTC = 70 µg/L

Descriptors: decaying vegetation, unpleasant
Above 3200 µg/L: paint thinner, decaying vegetation

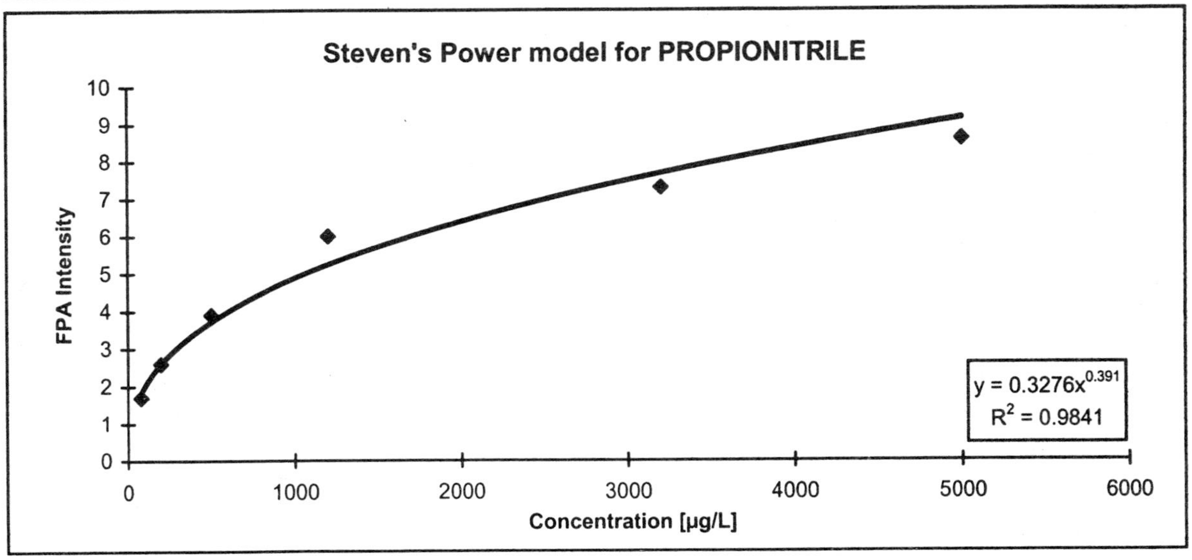

BROMOCHLOROIODOMETHANE

Conc. (µg/L)	5	8	13	21	51	128	320	832	2048
Log conc.	0.7	0.9	1.1	1.3	1.7	2.1	2.5	2.9	3.3
Marta	2.5	3	3	4	5	6	7	8	9
Núria	0	1	1	3	3.5	4	6.5	8	10.5
Raquel	0	1	1	1	2	3.5	6	7.5	10
Sònia R.	0	0	2	4	5.5	6.5	8	9	11.5
Carlota	0	0	1	3	3.5	6	7	8.5	10
Mª José	0	0	0	4	4.5	5.5	6	8	10
Average Intensity	0.4	0.8	1.3	3.2	4.0	5.3	6.8	8.2	10.2

exp. OTC: 5 µg/L

Terminal exp. OTCs: 5 µg/L (1) · 21 µg/L (1)

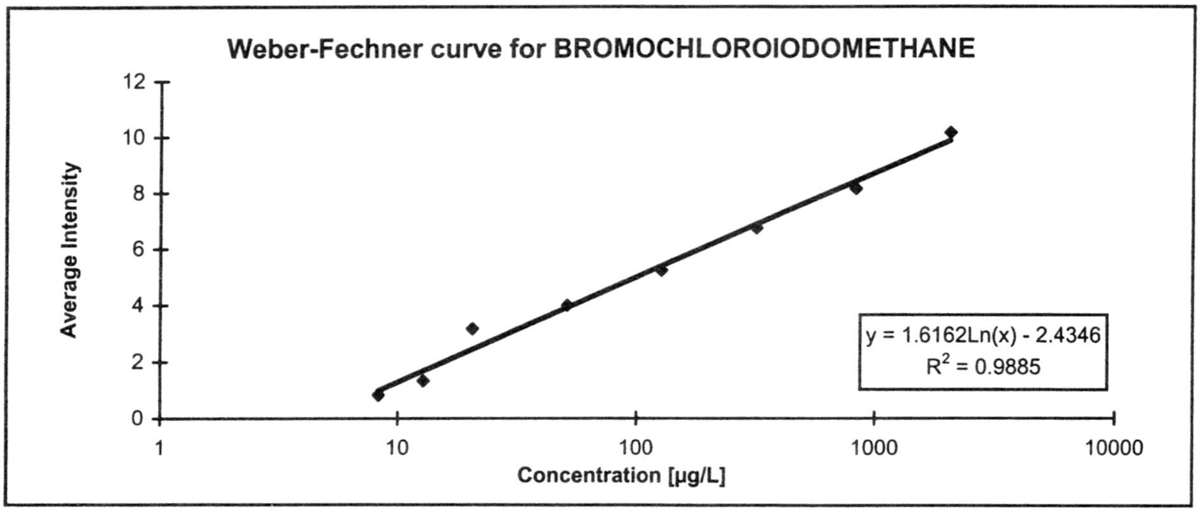

If Intensity = 1 in Weber-Fechner curve, then OTC = 8 µg/L

Descriptors: Below 220 µg/L: Fresh grass, perfumed, sweet, alcoholic
Above 220 µg/L: Solvent, acetone, alcohol.

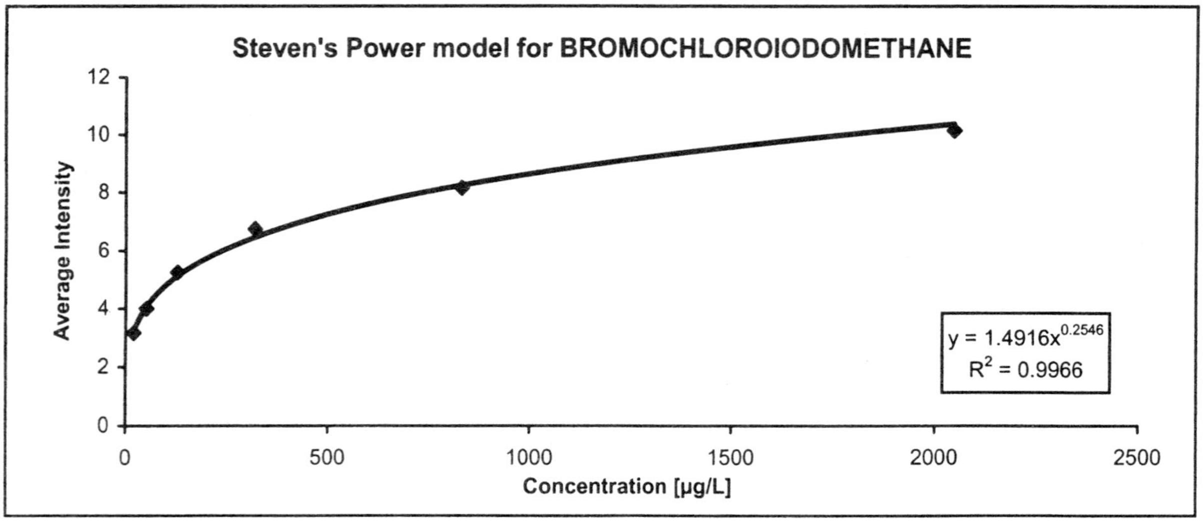

DIBROMOIODOMETHANE

Conc. (µg/L)	3	7	18	45	117	288	451	1171	2883
Log conc.	0.5	0.9	1.3	1.7	2.1	2.5	2.7	3.1	3.5
Mònica	1	2	3	4	4	5	7	9	10.5
Núria	0	1	4	5	5.5	6	7	9	10
Mª José	0	0	2	3.5	5	6	7	8.5	11
Montse	0	0	3	4	5	6	7	8	10
Raquel	0	0	3.5	5.5	7	8	8	11	12
Esther	0	0	2	2.5	3	4	6	7	9
Average Intensity	0.2	0.5	2.9	4.1	4.9	5.8	7.0	8.8	10.4

exp. OTC: 3 µg/L

Terminal exp. OTCs: 3 µg/L (1) - 18 µg/L (4)

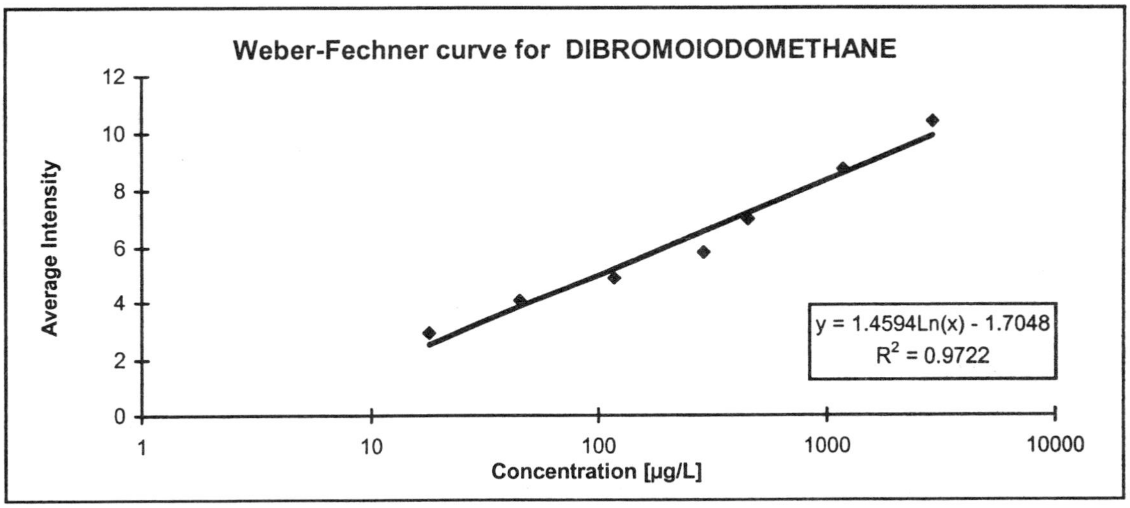

If Intensity = 1 in Weber-Fechner curve, then OTC = 6 µg/L

Descriptors: Bitumen, solvent, parfumed, sweet, fruity, dyeing, rancid. Above 1000 µg/L: not parfumed, not fruity, not sweet.

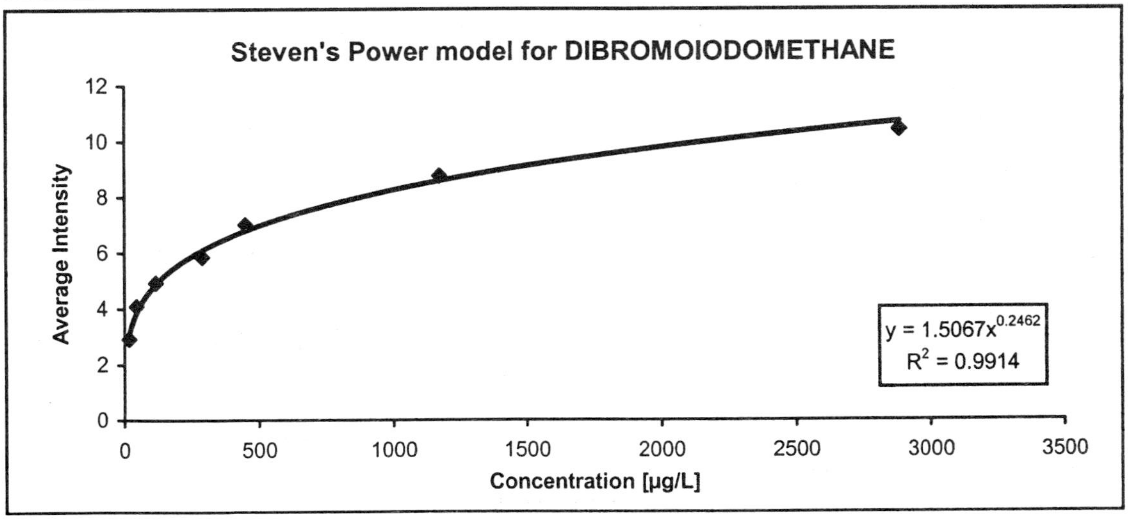

IODOFORM

Conc. (μg/L)	0.03	0.1	0.1	0.2	0.3	0.8	2	5	13	32
Log conc.	-1.5	-1.1	-0.9	-0.7	-0.5	-0.1	0.3	0.7	1.1	1.5
Iñaki	1[(a)]	2	2	2[(b)]	3	4	6	6	8	10
Núria	0	0	0	2	3	4	5	6	7	8
Roser	0	0	0	3	4	5	6	7	8	9
Diana	0	0	0	1	2	2	2	3	4	6
Marisa	0	0	0	0	4	5	6	7	8	9
Sònia R.	0	0	0	0	3.5	4	4.5	5	8	10
Average Intensity	0.0	0.3	0.3	1.2	3.3	4.0	4.9	5.7	7.2	8.7

(a): Detection threshold
(b): Recognition threshold

exp. OTC: 0.03 μg/L

Terminal exp. OTCs: 0.03 μg/L (1) - 0.3 μg/L (2)

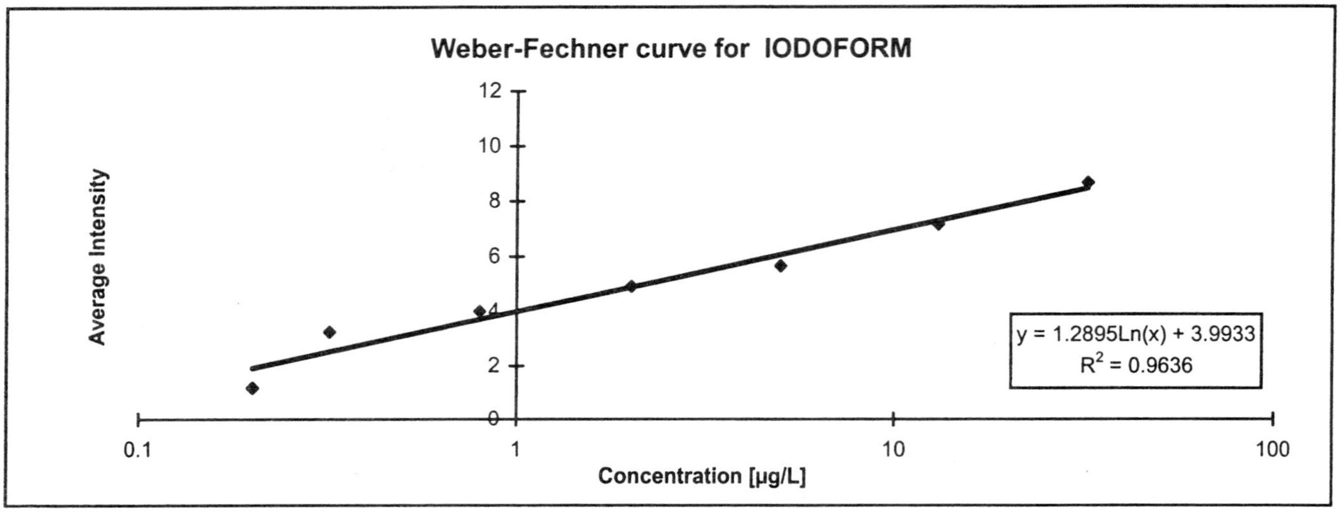

If Intensity = 1 in Weber-Fechner curve, then OTC = 0.1 μg/L

Descriptors: Medicinal (dentist's, aspirin, medicament), gum, sweet, parfumed

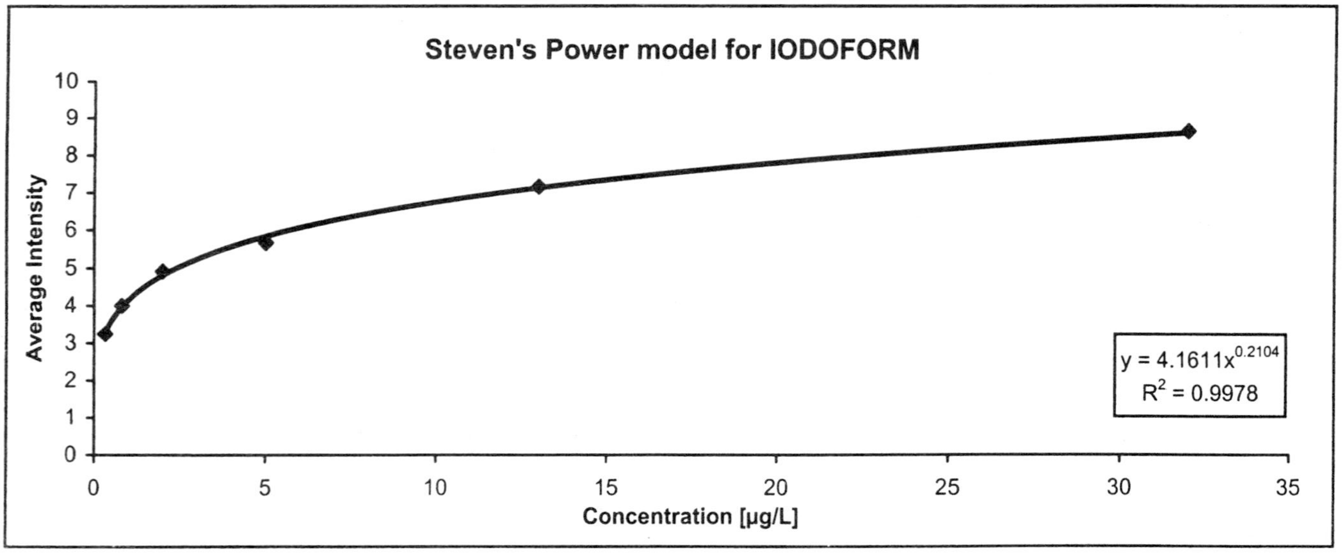

BROMODIIODOMETHANE

Conc. (μg/L)	0.1	0.2	0.5	0.9	2.2	5.5	13.2
Log conc.	-1.1	-0.7	-0.3	-0.1	0.3	0.7	1.1
Carlota	1	1	3	4	5	7	9
Pepa	1.5	2	3	3.5	3.5	4	5
Raquel	0	1	3	4.5	4.5	6	7
Montse	0	1.5	1.5	2	2.5	3	3.5
Núria	0	1	1.5	3	5.5	7.5	10
Roser	0	0	2	3	3	5	6.5
Average Intensity	0.4	1.1	2.3	3.3	4.0	5.4	6.8

exp. OTC: 0.1 μg/L

Terminal exp. OTCs: 0.2 μg/L (2) · 0.5 μg/L (3)

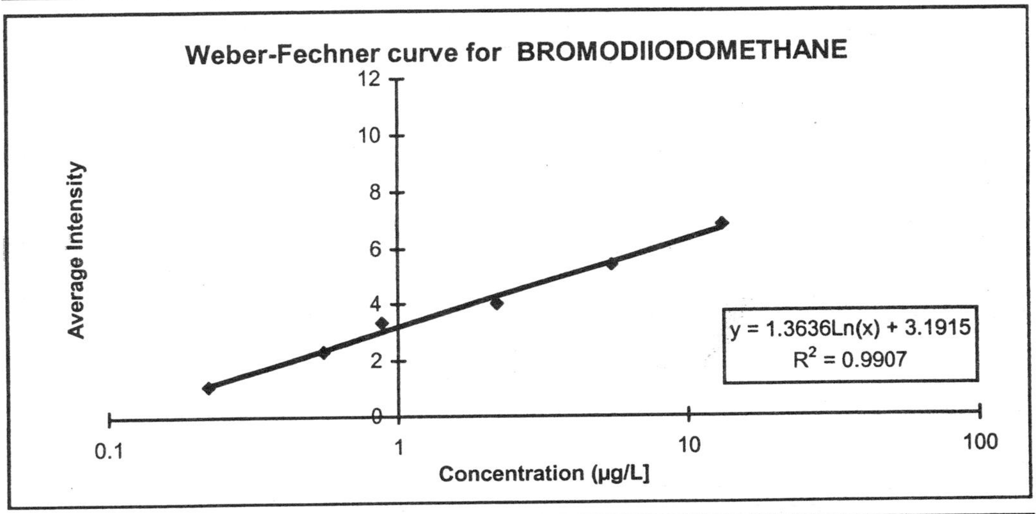

If Intensity = 1 in Weber-Fechner curve, then OTC = 0.2 μg/L

Descriptors: Medicinal product, sweet, solvent, parfumed

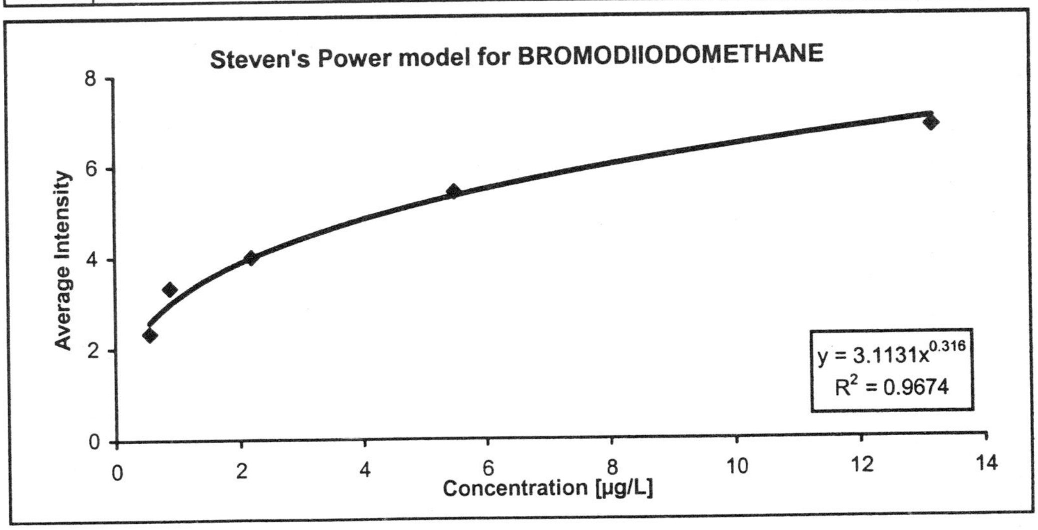

CHLORODIIODOMETHANE

Conc. (µg/L)	0.2	0.6	1	2	6	15
Log conc.	-0.6	-0.2	0.0	0.4	0.8	1.2
Albert	1.5	2	2	2.5	3.5	3
Esther	2	3	3	3	4.5	6.5
Rosa	3	3.5	4	4	6	6.5
Raquel	0	2.5	2.5	2.5	4.5	6.5
Montse	0	2	2	2	3	3
Carlos	0	1	1.5	3	3.5	4.5
Int. Average	1.1	2.3	2.5	2.8	4.2	5.0

exp. OTC: 0.2 µg/L

Terminal exp. OTCs: 0.2 µg/L (3) - 0.6 µg/L (3)

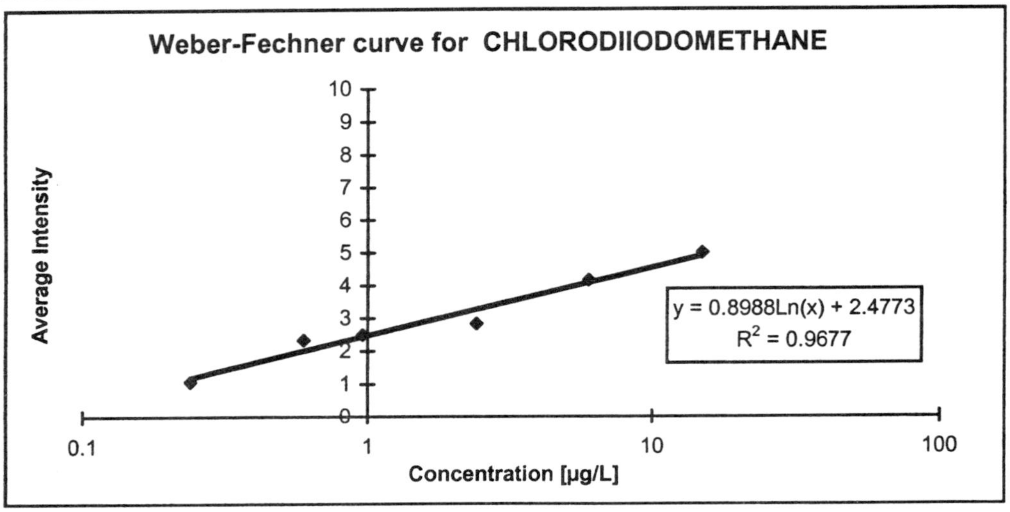

If Intensity = 1 in Weber-Fechner curve, then OTC = 0.2 µg/L

Descriptors: Sweet, candy, rancid, medicinal product, solvent

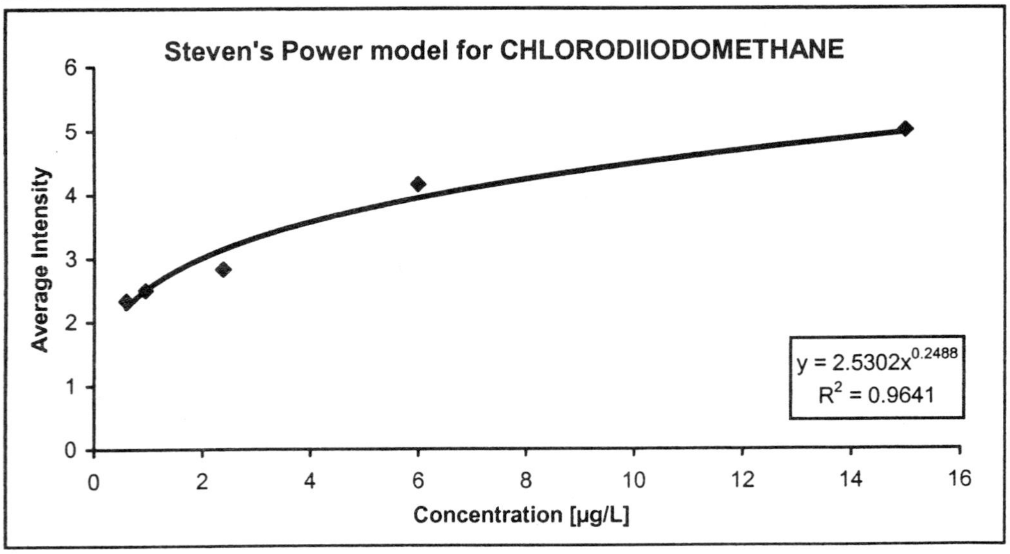

DICHLOROIODOMETHANE

Conc. (μg/L)	6	9	23	58	144	360	936	2304
Log conc.	0.8	1.0	1.4	1.8	2.2	2.6	3.0	3.4
Diana	2	3	3.5	4	5	6	7.5	9
Marta	0	2.5	3	3	4	6	7	8
Rosa	0	0	2	4	5	5.5	6	7
Iñaki	0	0	2	2.5	4	6.5	7	8
Montse	0	0	1.5	2	3	3	5.5	7
Esther	0	0	3	3	4	4	5.5	7
Average Intensity	0.3	0.9	2.5	3.1	4.2	5.2	6.4	7.7

exp. OTC:6 μg/L

Terminal exp. OTCs: 6 μg/L (1) - 23 μg/L (4)

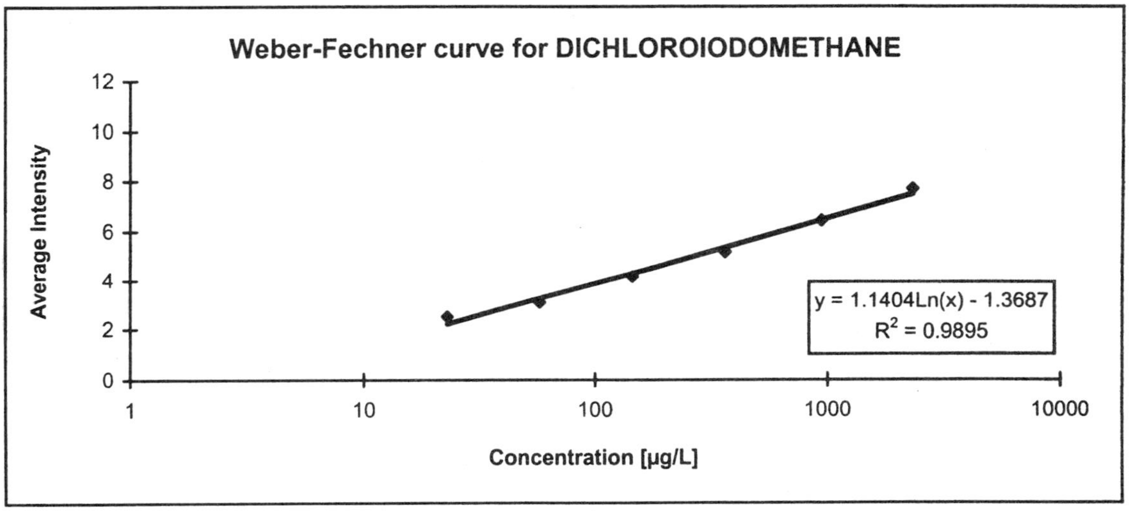

If Intensity = 1 in Weber-Fechner curve, then OTC = 8 μg/L

Descriptors: Below 790 μg/L: Sweet, syrup
Above 790 μg/L: Solvent, glue, paint, varnish .

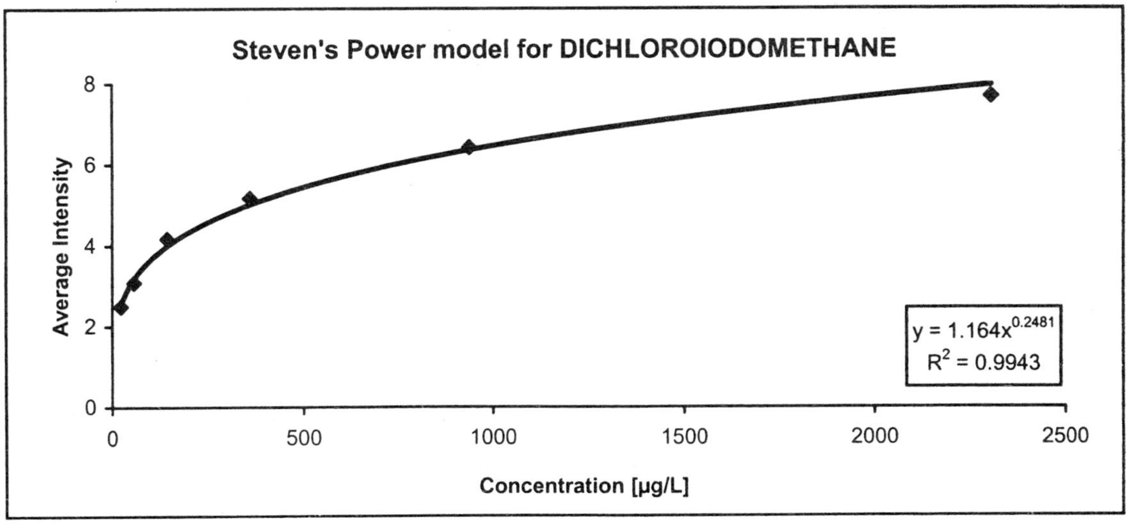

BENZALDEHYDE

Conc. (μg/L)	47	75	86	186	466	1120	2980	4650
Log conc.	1.7	1.9	1.9	2.3	2.7	3.0	3.5	3.7
Luis	3	2	4	5	8	10	11	12
Roser	0	-	2	-	-	5	7	9
Marta	0	0	-	4	7	8	10	12
Marisa	0	0	0	5	7	6	8	10
Diana	0	0	0	5	6	6	8	10
Sònia B.	0	0	-	3	6	9	10	11
Average Intensity	0.5	0.4	1.5	4.4	6.8	7.3	9.0	10.7

exp.OTC: 47 μg/L

Terminal exp. OTCs: 47 μg/L (1) - 186 μg/L (4)

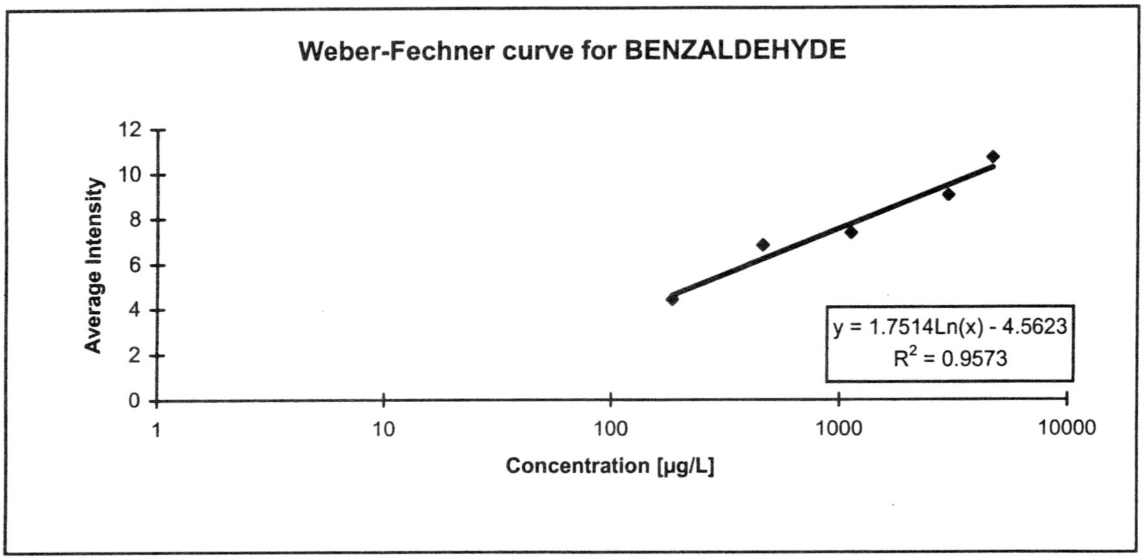

If Intensity = 1 in Weber-Fechner curve, then OTC = 24 μg/L

Descriptors: sweet almonds, rancid, strawberry candy.

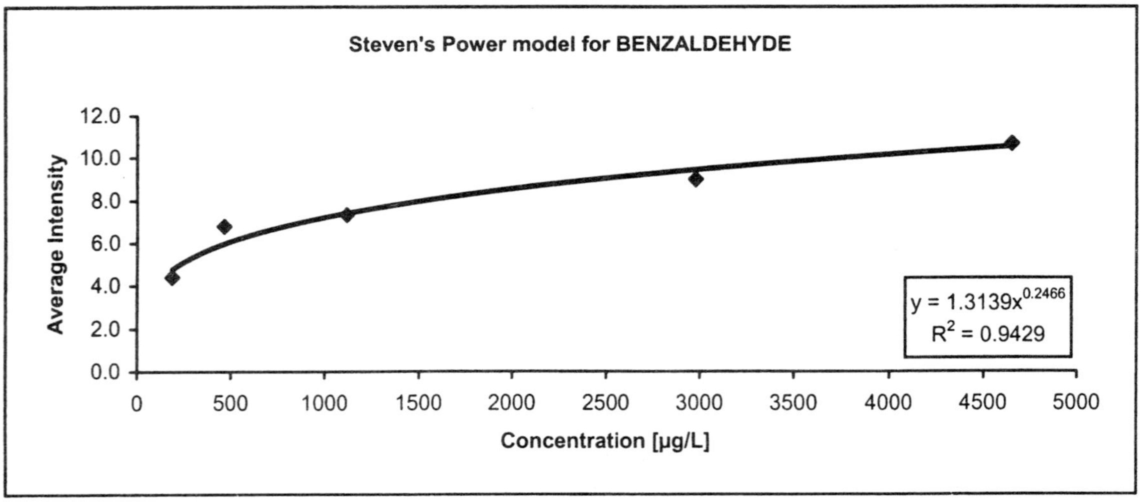

DODECANAL

Conc. (µg/L)	0.1	0.3	0.8	2	5	13	32	80	200
Log conc.	-0.9	-0.5	-0.1	0.3	0.7	1.1	1.5	1.9	2.3
Montse	1	2	2	4	6	7	8	8	10
Carlota	0	2	2.5	4	6	7.5	8.5	9	12
Núria	0	1	2.5	4.5	6	7.5	8	10	11
Roser	0	0	1.5	3.5	5.5	7	8	10	11
Rosa	0	0	3	6	8	8.5	11	11.5	12
Average Intensity	0.2	1.0	2.3	4.4	6.3	7.5	8.7	9.7	11.2

exp. OTC: 0.1 µg/L

Terminal exp. OTCs: 0.1 µg/L (1) - 0.8 µg/L (2)

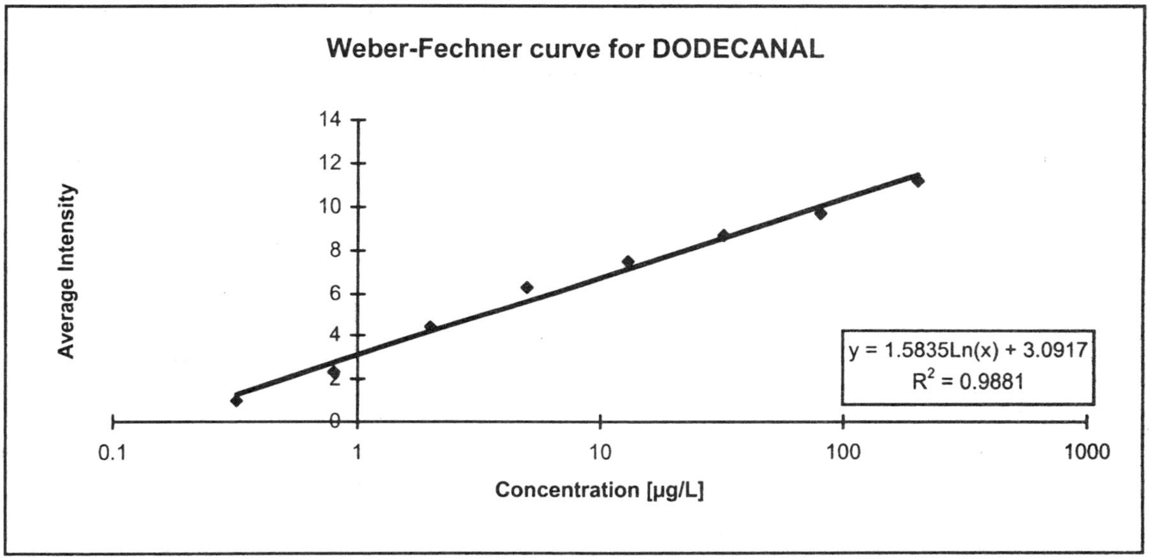

If Intensity = 1 in Weber-Fechner curve, then OTC = 0.3 µg/L

Descriptors: lemon scented deodorizer, insecticide

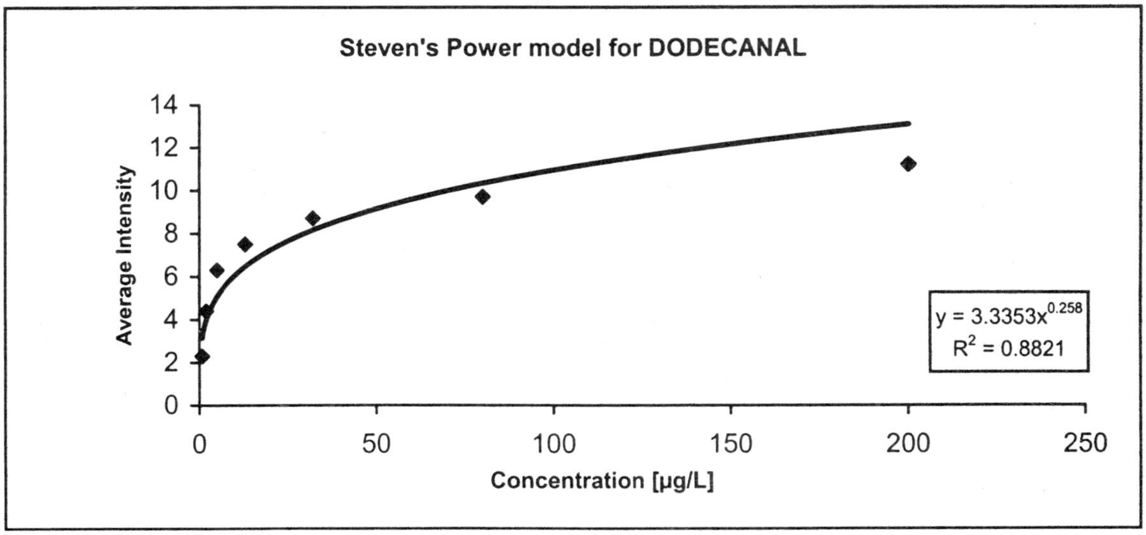

HEPTANAL

Conc. (μg/L)	0.2	0.6	0.9	2	3	4	6	9	16	25	39	62
Log conc.	-0.6	-0.2	0.0	0.2	0.4	0.6	0.8	1.0	1.2	1.4	1.6	1.8
Sònia C.	1	2	-	2.5	-	4	-	8	-	10	-	11
Rosa	0	0	3	2	5	3	6	7	8	7.5	8.5	8.5
Oscar	-	-	4	-	5	-	3	6	6	7	7	10
Raquel	-	-	1	-	2	-	5	5	7	10	8	12
Luis	-	-	3	-	6	-	7	10	8	11	9	12
Average Intensity	0.5	1.0	2.8	2.3	4.5	3.5	5.3	7.2	7.3	9.1	8.1	10.7

exp. OTC: 0.2 μg/L

Terminal exp. OTCs: 0.2 μg/L (1) - 0.9 μg/L (4)

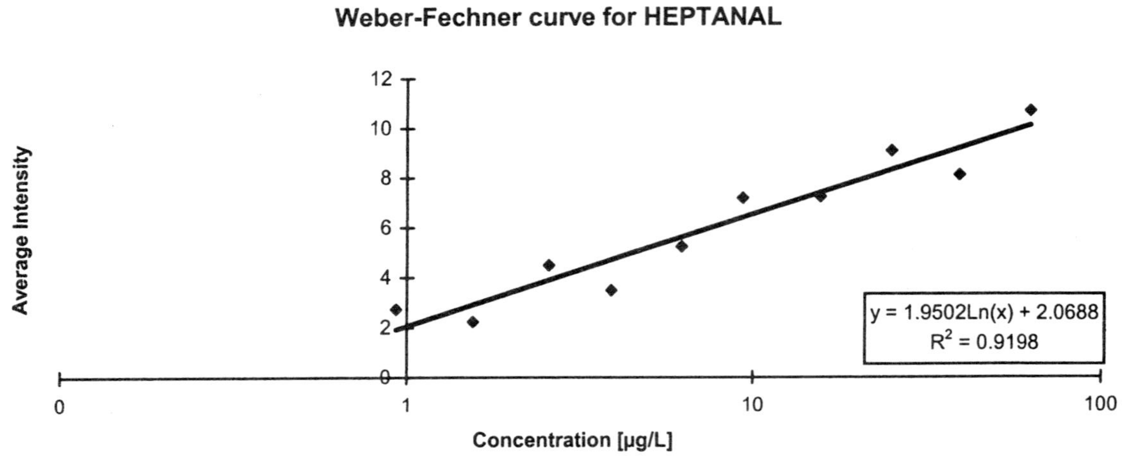

If Intensity = 1 in Weber-Fechner curve, then OTC = 0.6 μg/L

Descriptors: rancid oil + fish, fruity (tangerine, orange)

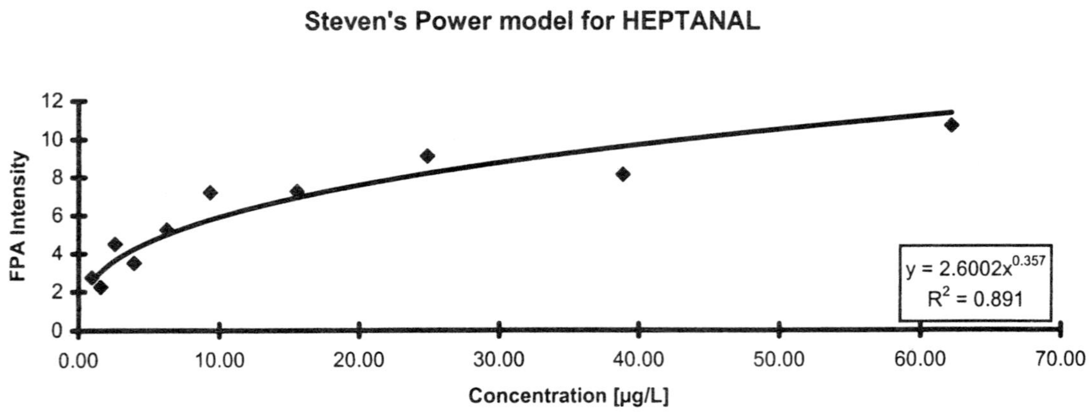

OCTANAL

Conc. (μg/L)	0.3	0.8	2	5	12	32
Log conc.	-0.5	-0.1	0.3	0.7	1.1	1.5
Roser	1	2	4	6	8	9
Rosa	1	2.5	3.5	5	6.5	7.5
Mª José	1	1.5	3	4.5	6	6.5
Sònia B.	1	1.5	2	3	4	5
Núria	0	1	3.5	4	5	7
Average Intensity	0.8	1.7	3.2	4.5	5.9	7.0

exp.OTC: 0.3 μg/L

Terminal exp. OTCs: 0.3 μg/L (4) - 0.8 μg/L (1)

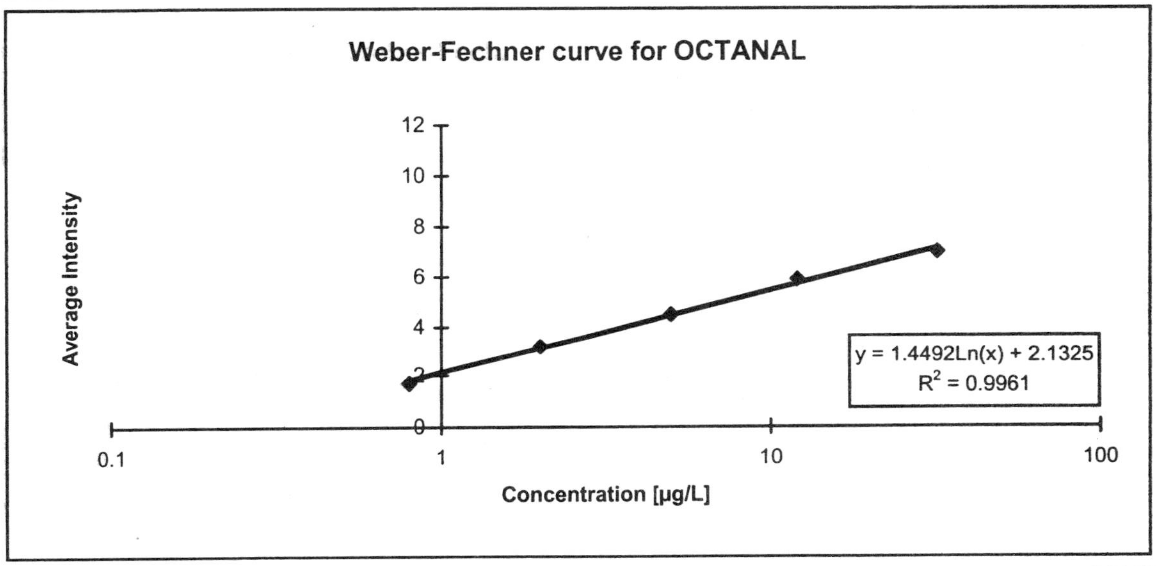

If Intensity = 1 in Weber-Fechner curve, then OTC = 0.5 μg/L

Descriptors: rancid, fruity, citrus, rancid oil, sweet.

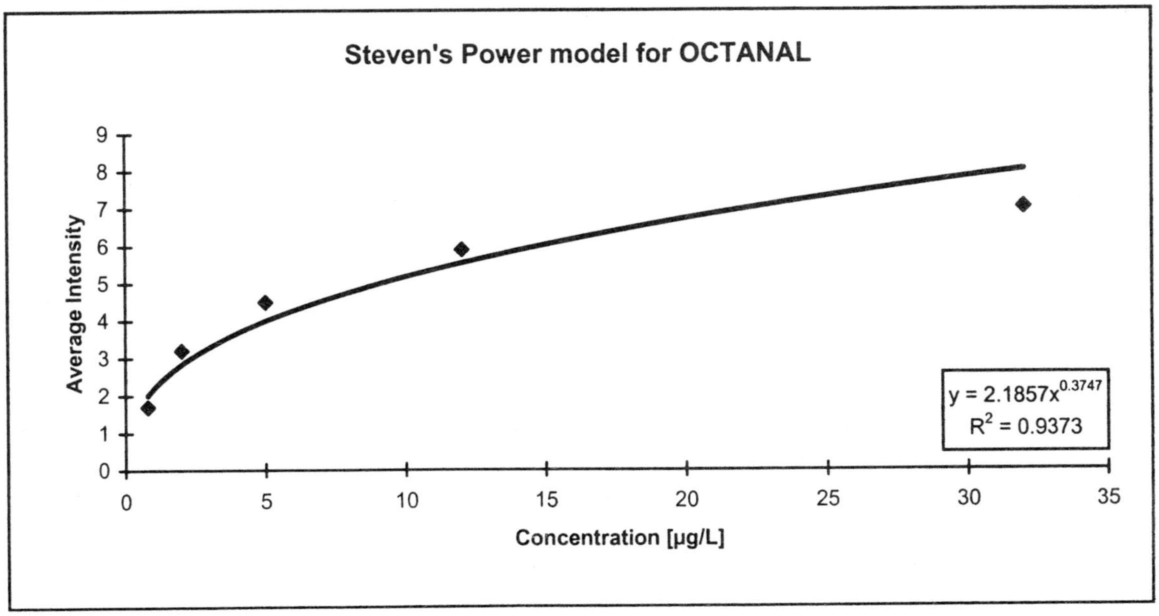

NONANAL

Conc. (μg/L)	0.3	0.8	2	5	12	32	80	200
Log conc.	-0.5	-0.1	0.3	0.7	1.1	1.5	1.9	2.3
Òscar	2	2	3.5	5	5	7.5	9	11
Núria	1	1.5	2.8	5.5	6	7.5	9	10.5
Raquel	1	2	3	4	5.5	7	8	10
Roser	0	2	3	5.5	7	7.5	9	11
Marta	0	1.3	4	5.5	7	8	8.5	11
Mª José	0	1	2.5	4	5	7.5	9.5	11
Luis	0	0.5	2.5	3.5	4	6	7.5	9
Average Intensity	0.6	1.5	3.0	4.7	5.6	7.3	8.6	10.5

exp.OTC: 0.3 μg/L

Terminal exp. OTCs: 0.3 μg/L (3) - 0.8 μg/L (4)

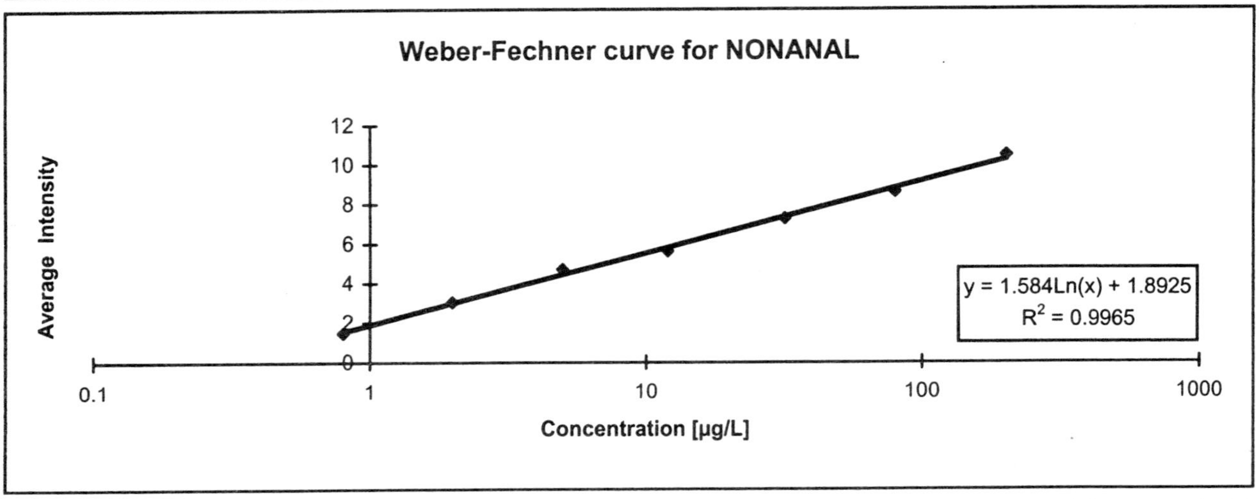

If Intensity = 1in Weber-Fechner curve, then OTC = 0.6 μg/L

Descriptors: rancid, fruity

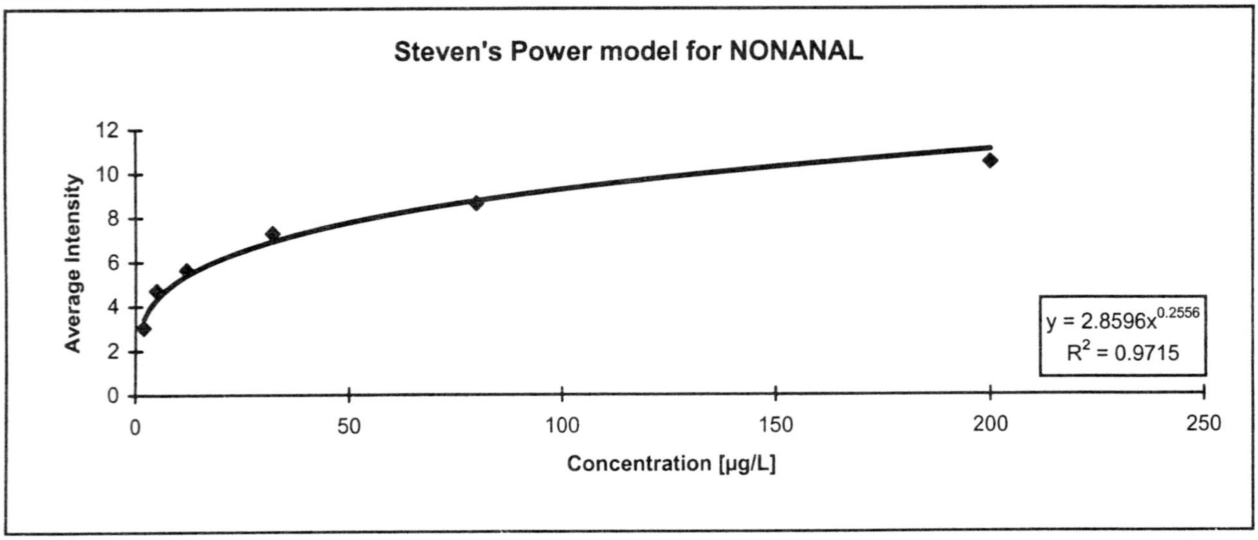

UNDECANAL

Conc. (µg/L)	0.03	0.08	0.1	0.2	0.3	0.8	2	5	12	32
Log conc.	-1.5	-1.1	-0.9	-0.7	-0.5	-0.1	0.3	0.7	1.1	1.5
Montse	1	2	2	3	3	4	5	6	7	8
Núria	1	2	1	3	2.5	4	5.5	7	9	11
Diana	0	1	3	3	4	5	6.5	8	10	11
Mª José	0	2	1	2.5	2.5	3.5	5	5.5	6.5	7.5
Óscar	0	0	0	1	1	1	4	6.5	9	11
Average Intensity	0.4	1.4	1.4	2.5	2.6	3.5	5.2	6.6	8.3	9.7

exp. OTC: 0.03 µg/L

Terminal exp. OTCs: 0.03 µg/L (2) - 0.1 µg/L (1)

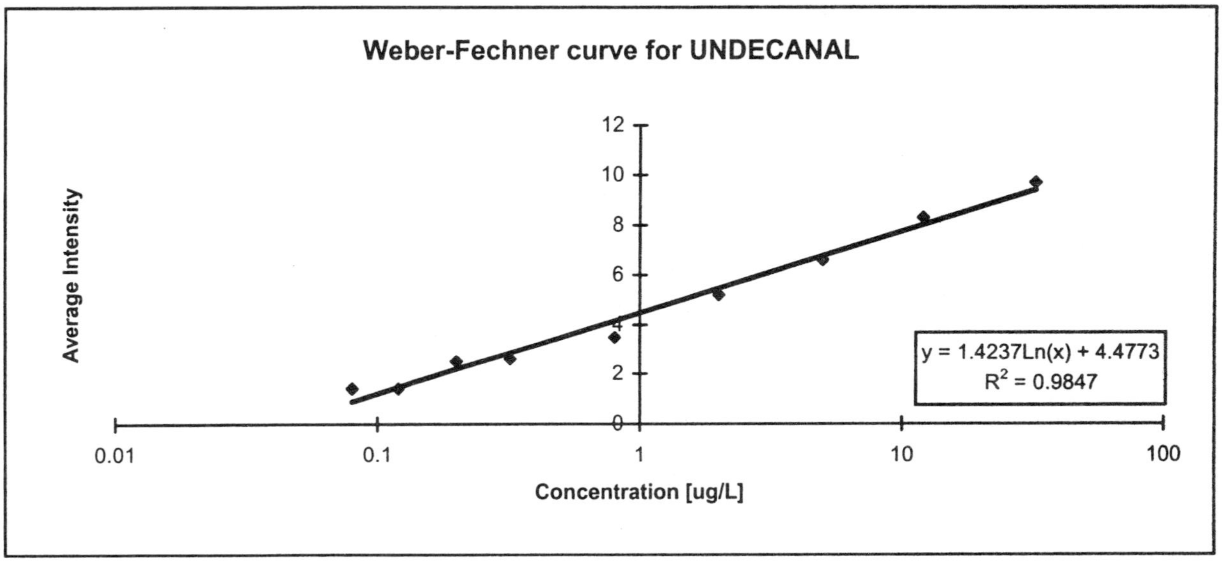

If Intensity = 1 in Weber-Fechner curve, then OTC = 0.09 µg/L

Descriptors: insecticide, citrus (lemon), perfume. irritating, penetrant.

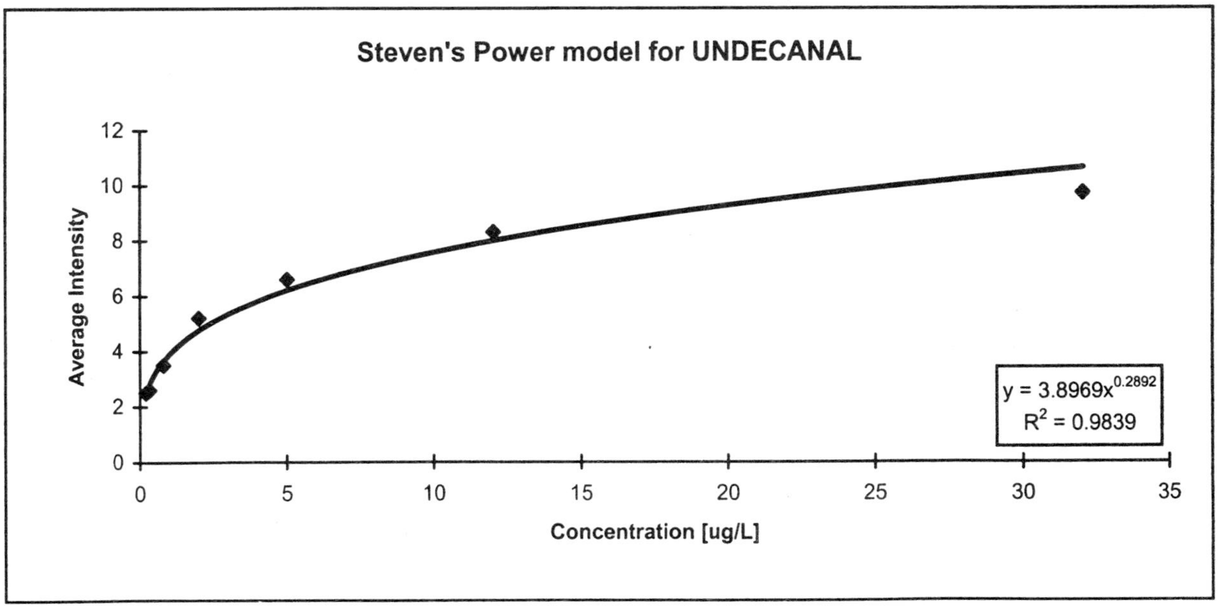

DECANAL

Conc. (μg/L)	0.08	0.1	0.3	0.8	2	5	12	32	80	200	500
Log conc.	-1.1	-0.9	-0.5	-0.1	0.3	0.7	1.1	1.5	1.9	2.3	2.7
Òscar	1	3	4	4	5	5	6.5	8.5	8.5	10	12
Montse	0	1	2	2	3	4	5	7	7	8	9
Roser	0	0	2	2	4	5	7	8	9	10	12
Diana	0	0	2	2	2.5	6	8	9	10	11	12
Luis	0	0	1	2.5	4	6	7	9	10	11	12
Average Intensity	0.2	0.8	2.2	2.5	3.7	5.2	6.7	8.3	8.9	10.0	11.4

exp. OTC: 0.08 μg/L

Terminal exp. OTCs: 0.08 μg/L (1) - 0.3 μg/L (3)

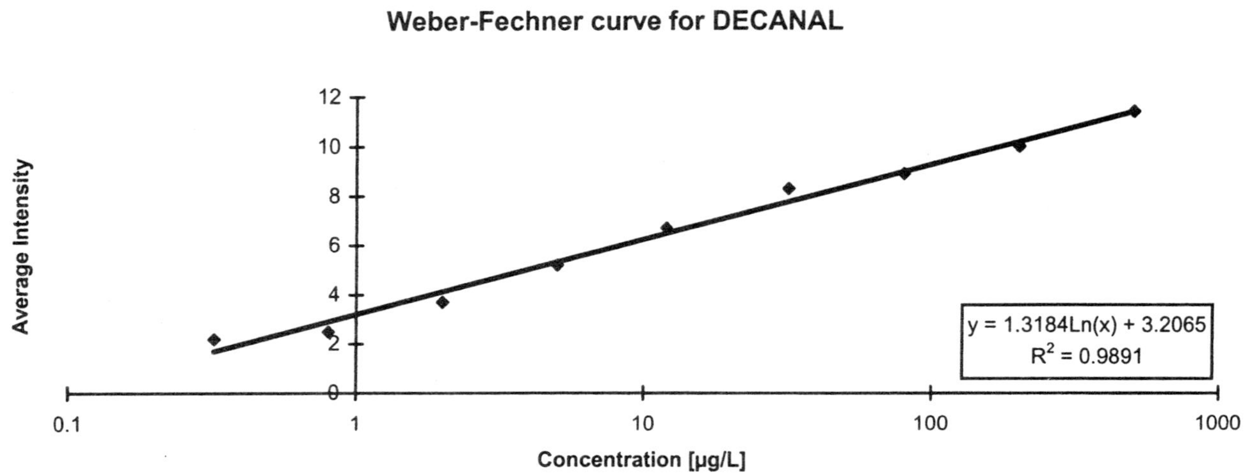

If Intensity = 1 in Weber-Fechner curve, then OTC = 0.2 μg/L

Descriptors: citrus (lemon), rancid

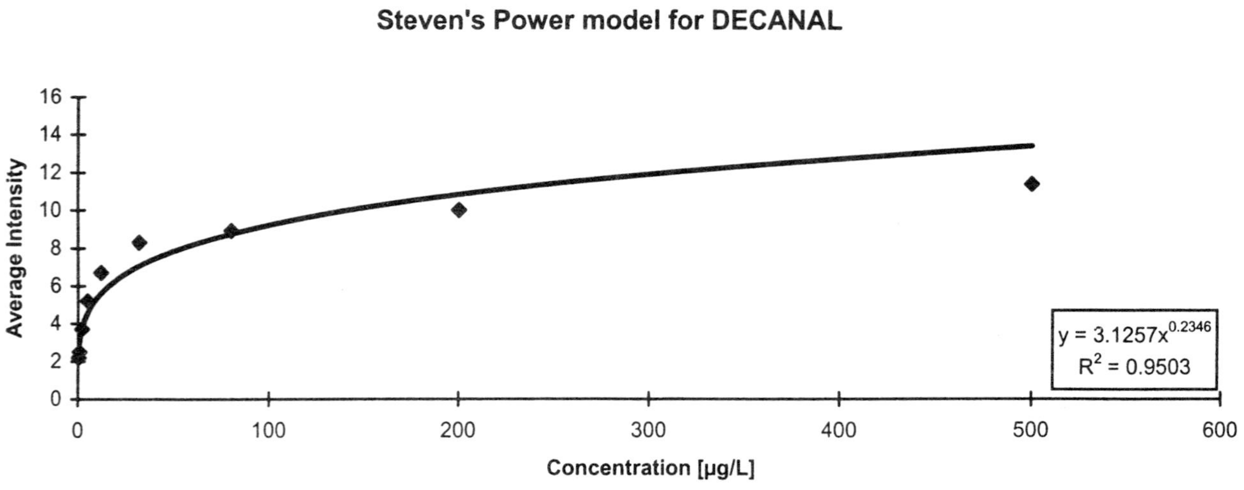

HEXANAL

Conc. (μg/L)	0.3	0.8	2	5	80	500
Log conc.	-0.5	-0.1	0.3	0.7	1.9	2.7
Carlota	1.5	4	6	7.5	8	12
Iñaki	0	1	4	6	8	12
Oscar	0	1	2	3	7	10
Raquel	0	0	5	7	9	12
Núria	0	0	3	5.5	8.5	12
Average Intensity	0.3	1.2	4.0	5.8	8.1	11.6

exp. OTC: 0.3 μg/L

Terminal exp. OTCs: 0.3 μg/L (1) - 2 μg/L (2)

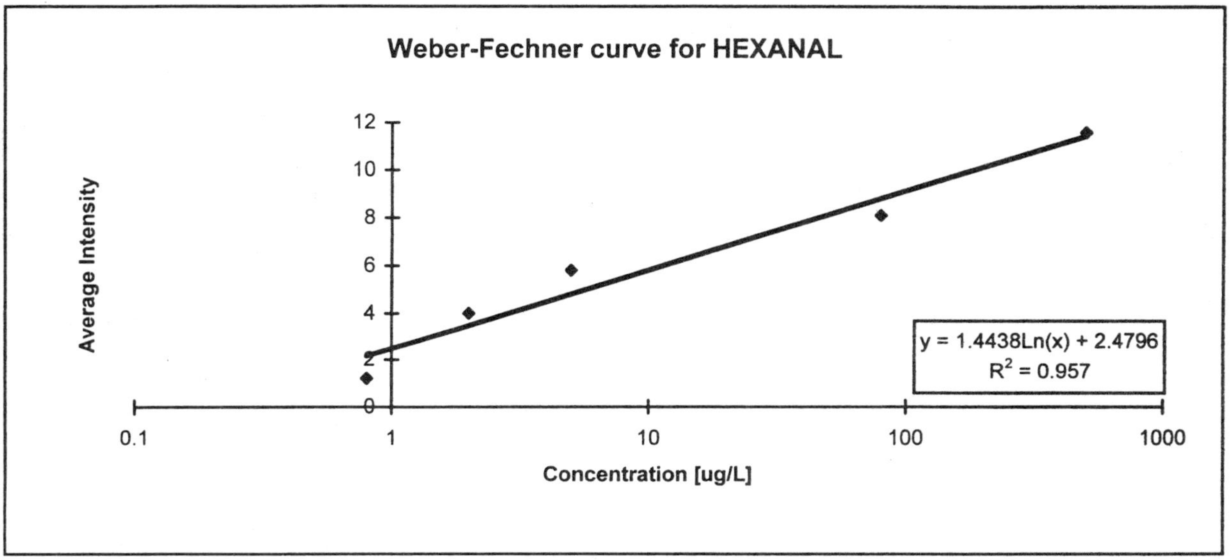

If Intensity=1 in Weber-Fechner curve, then OTC = 0.4 μg/L

Descriptors: Below 80 μg/L: green, grass, earthy
Above 80 μg/L: green apple, fruity

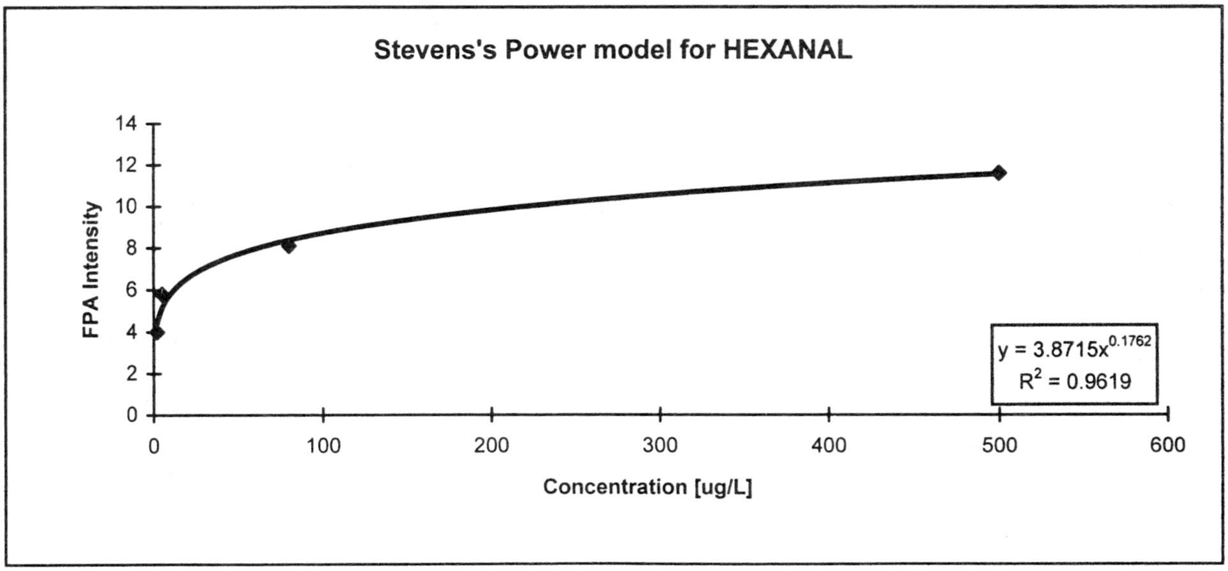

PENTANAL

Conc. (µg/L)	5	12	32	80	200	500	1200
Log conc.	0.7	1.1	1.5	1.9	2.3	2.7	3.1
Mª José	1.5	2	4.5	6	7.5	8.5	10
Montse	0	2.5	2	3	4	6	7
Sònia C.	1	2.5	5	6.5	8	9.5	9
Mònica	1.5	3	7	8	8	8.5	9
Raquel	0	3	5	7	8	9	11
Average Intensity	0.8	2.6	4.7	6.1	7.1	8.3	9.2

exp. OTC: 5 µg/L

Terminal exp.OTCs: 5 µg/L (3) - 12 µg/L (2)

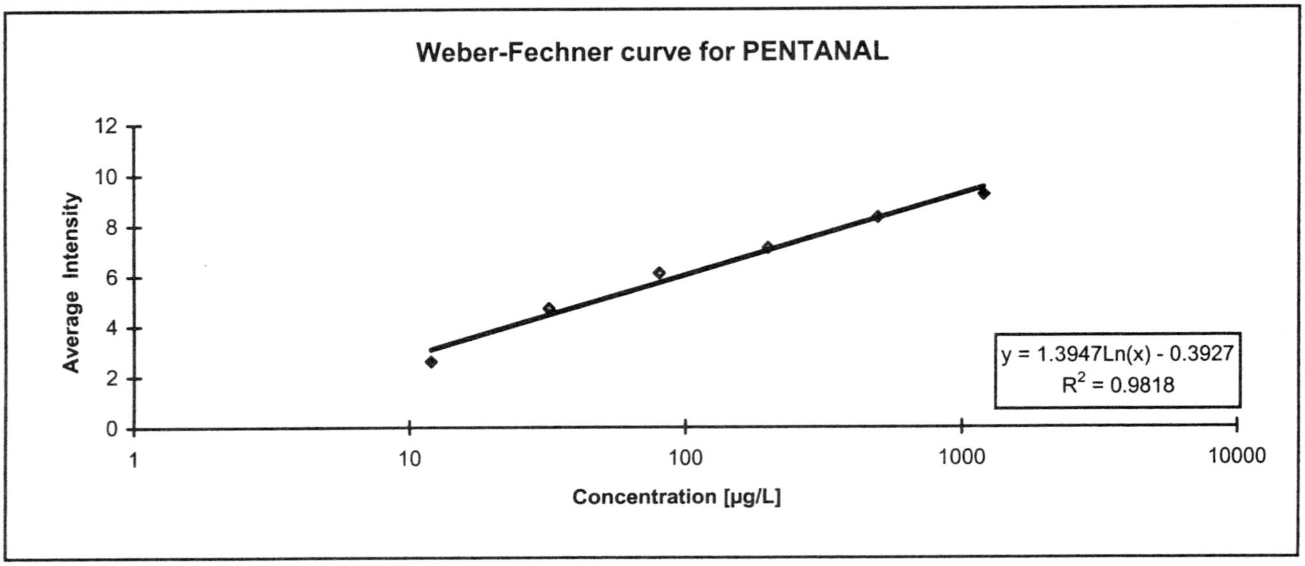

If Intensity = 1 in Weber-Fechner curve, then OTC = 3 µg/L

Descriptors: Below 32 µg/L: sweet, candy, green
Above 32 µg/L: fruity, sour, rancid, candy

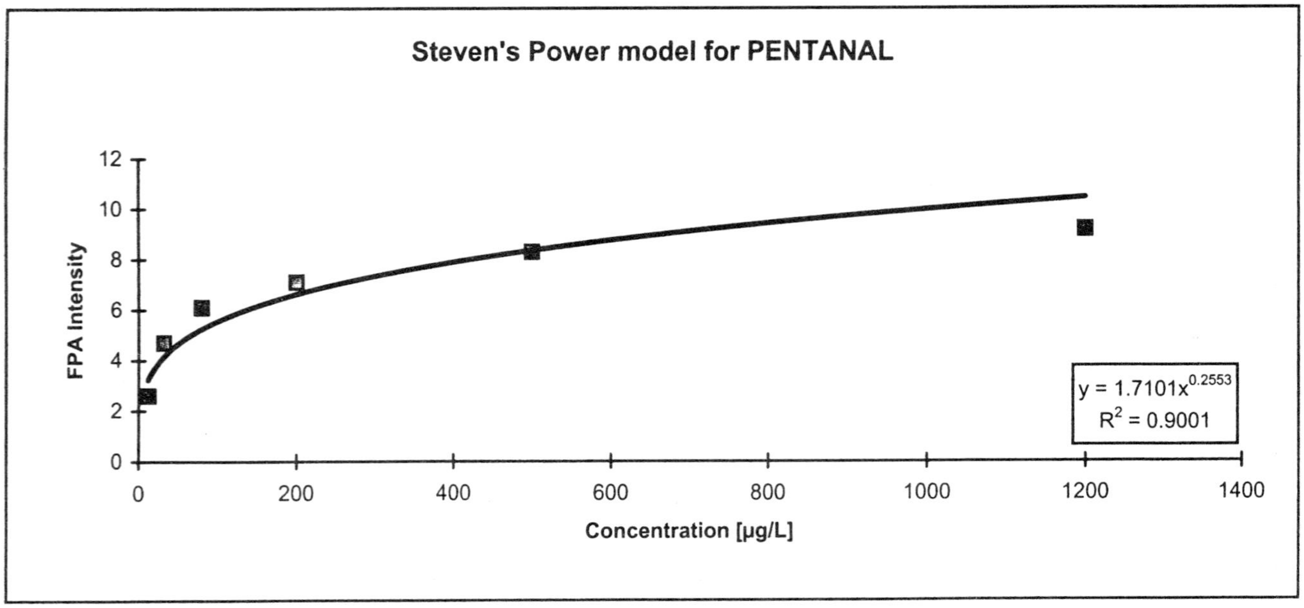

PROPANAL

Conc. (µg/L)	12	32	80	200	500	1200
Log conc.	1.1	1.5	1.9	2.3	2.7	3.1
Sònia C.	2	3	5	7	8.5	10
Mª José	2	3	4.5	6.5	8	9
Raquel	1	2	4.5	6	8	11.5
Luis	0	3	5	7	9	10
Rosa	0	4	5	6	7	8
Montse	0	2	4	5.5	6	8
Average Intensity	0.8	2.8	4.7	6.3	7.8	9.4

exp. OTC: 12 µg/L

Terminal exp.OTCs: 12 µg/L (3) - 32 µg/L (3)

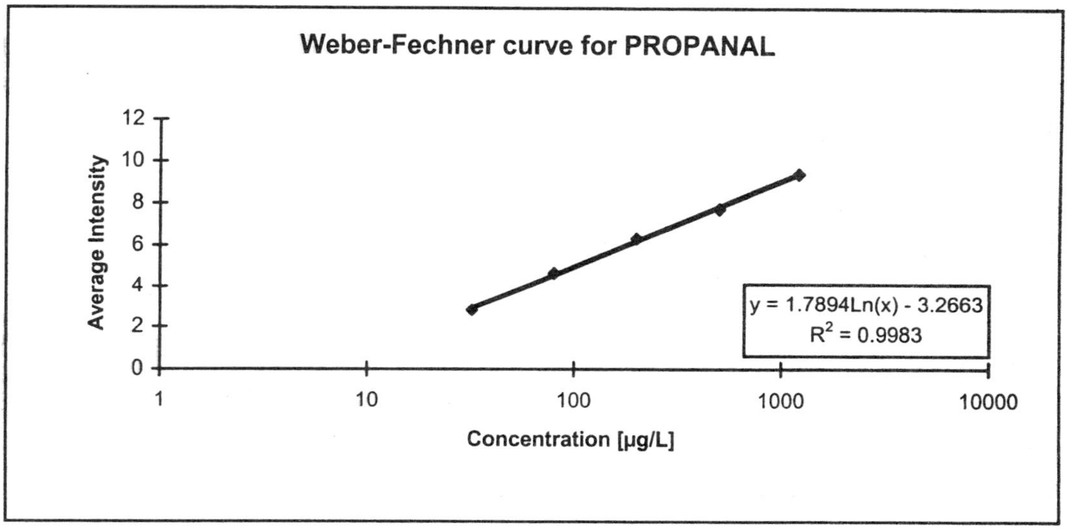

If Intensity = 1 in Weber-Fechner curve, then OTC = 11 µg/L

Descriptors: Below 200 µg/L: lemon, citrus, rancid, milky
Above 200 µg/L: solvent, gum, paint, glue, fruity

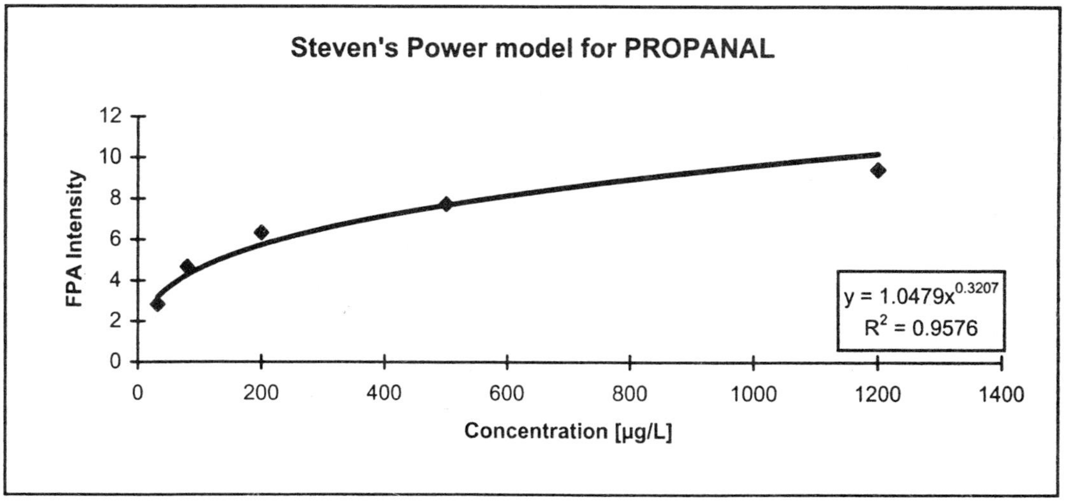

BUTANAL

Conc. (μg/L)	2	5	12	32	80	200	500	1200
Log conc.	0.3	0.7	1.1	1.5	1.9	2.3	2.7	3.1
Raquel	1	3	4.5	7	8	8.5	10	12
Roser	0	1	2.5	3	4	5.5	7	9
Marisa	0	1.5	3.5	5	6	6.5	7	8
Mª José	0	2.5	3	5	6	7.5	9	10.5
Average Intensity	0.3	2.0	3.4	5.0	6.0	7.0	8.3	9.9

exp. OTC = 2 μg/L

Terminal exp. OTCs: 2 μg/L (1) - 5 μg/L (3)

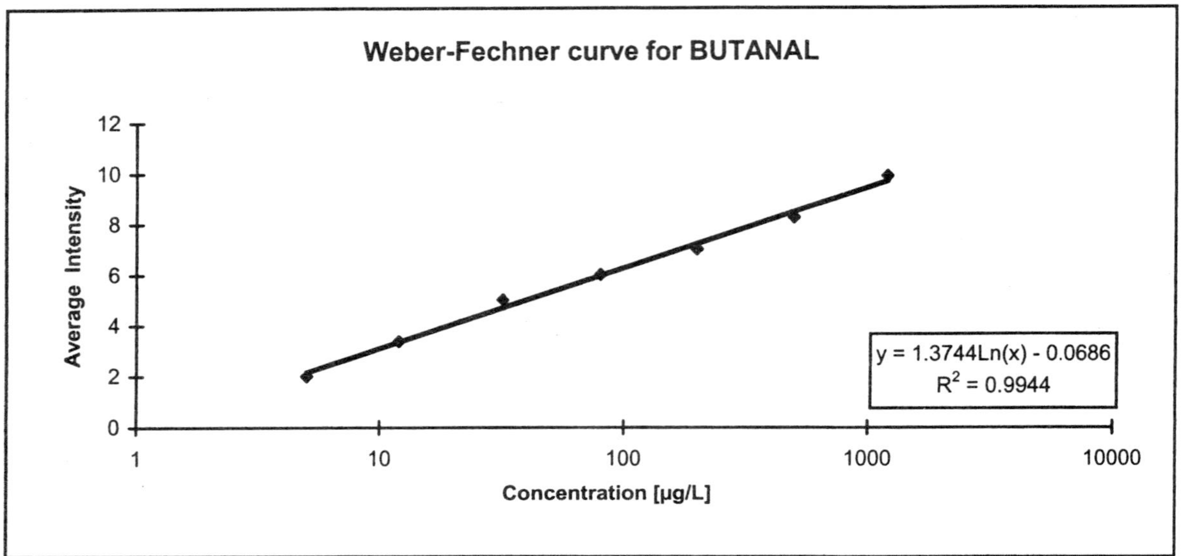

If Intensity = 1 in Weber-Fechner curve, then OTC= 2 μg/L

Descriptors: sweet, marking pen, solvent, fruity, dehydrated milk.

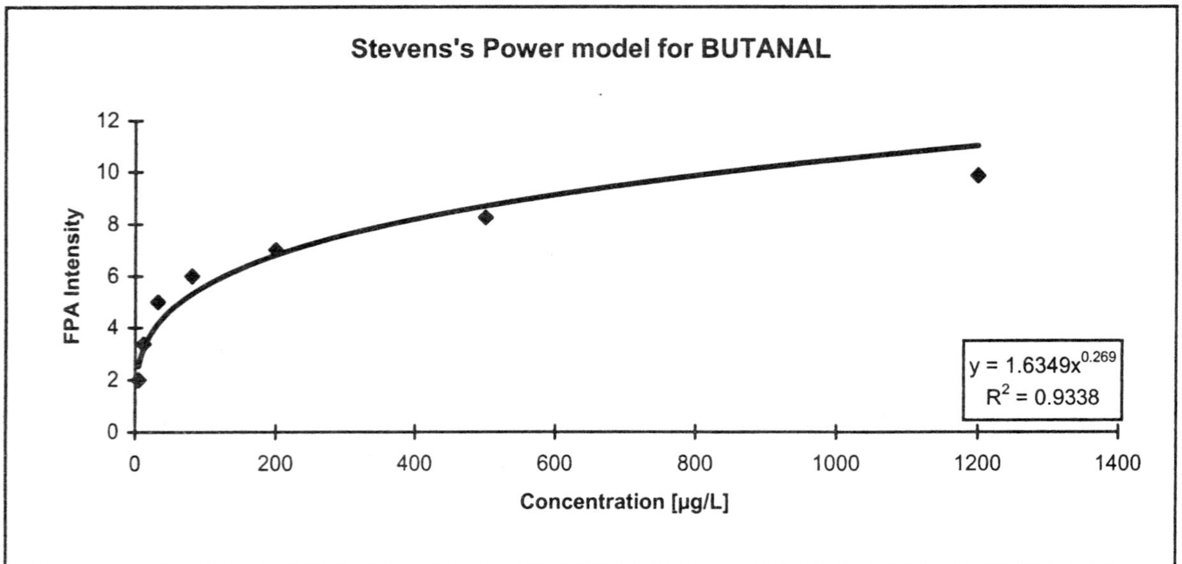

APPENDIX B
DISTRIBUTION SYSTEM TASTE AND ODOR COMPLAINT INVESTIGATION (PROTOCOL)

The purpose of the investigation is to identify the tastes and odors, generated within the distribution system, which are causing consumer's complaints or which potentially may cause complaints . It is important to distinguish problems that were already present in the source water or formed during treatment from the problems that were produced in the distribution system or the house plumbing.

The goal of such an investigation is to determine the nature of the problem (by chemical or microbiological analyses of appropriate samples) and to understand the conditions that favored the production of these chemicals and/or microorganisms.A good knowledge of the distribution system, its operation, and practical experience are the three major requirements in the investigation of a taste and odor problem in the distribution system. The first step is to collect all the relevant information. The second step is to collect representative samples when the problem is occurring. If the problem is identified in the consumer's system, the investigator should work back from the customer's tap, to the service connection, to the local main, to a distribution-system section, and finally back to the treatment plant or well.

POSSIBLE CAUSES FOR COMPLAINTS IN THE DISTRIBUTION SYSTEM

To help determine the possible causes of complaints, you should first locate the site of the complaint on the distribution system map. Keeping track of the complaints on this map may help estimate the magnitude and complexity of the problem. The following list highlights some of the conditions that may be linked to the taste and odor problem:

- Disinfection of new mains, tanks, or reservoirs
- Recent maintenance work
- Dead-end mains
- Cross-connections

- Main in shallow trenches
- Other miscellaneous causes

OTHER POSSIBLE CAUSES FOR COMPLAINTS IN DOMESTIC PLUMBING SYSTEM

In addition to problems that may be occurring in the utilities' distribution system, the cause of the problem may be in the consumer's system. You should consider the following possible sources of problems:

- Hot water tanks
- Cross-connections
- Old piping (including service line)
- Exposed water lines
- Compounds added by consumer
- Kitchen/bathroom sink odors
- Point of use devices
- Long house line runs from meter
- Other miscellaneous causes
- New carpeting when chlorine dioxide is used

Taste and Odor Complaint Investigation Form

Date first complaint was received: __________

INFORMATION COLLECTION

The following questionnaire can be filled in by the investigator through conversation with the complainant, either by phone or during the site visit. Additional information may be supplied by the water utility.

1. Location and description of complaint site (address, water utility, residential or commercial...)	
2. When was the taste or odor first detected? Does the problem last? How long ?	
3. Was it a unique occurrence or was it a recurrent problem? If recurrent, describe the conditions?	
4. Describe recent water activity [how long has it been since the tap was last used (overnight, weekend..)?]	
5. Is the problem still occurring?	
6. How does the customer describe the problem? When is it the most noticeable? Is it mainly a taste or odor (drinking water, coffee, ice cubes...) Does everyone at the household describe the problem in the same manner?)	

Taste and Odor Complaint Investigation Form (continued)

7. Is the taste or odor in both the hot and cold water?	
8. Are neighbors experiencing the same problem? (*It is always a good idea to obtain a sample from the nearest neighbor, even if no complaint has been reported. No complaint does not always mean no problem.*) If the answer is no and the neighbors have no problems, the problem appears to originate within the consumer's premises. If the answer is yes and other locations are experiencing a similar problem, the origin may be in the distribution system.	
9. Does the consumer have a point of use device or a point of entry device? How long has it been operating? Has maintenance been done on the device?	
10. Does the consumer have a pressurized hose (gun-type nozzle)?	
The following questions need to be answered by the water utility or wholesaler	
11. What is source of water to the consumer's premises (has a taste and odor problem been reported in the source water supply?) Contact the wholesaler (if appropriate) to check on the status of the source water.	
12. Is the service line new or old? Copper or galvanized?	
13. Is the flow through the service line adequate to prevent loss of pressure on the upper floors of the house?	
14. Where in the distribution network is (are) the complaint(s) located (proximity of a dead end, tanks, etc.). Locate the complaint (s) on the distribution system map.	

Taste and Odor Complaint Investigation Form (continued)

SAMPLE COLLECTION AND ON-SITE SAMPLE ANALYSIS

Date of field visit: __________

1. Check the water at consumer's premises (do an on site taste and odor analysis).	______________________ ______________________
2. Does the field analysis agree with the consumer's description? Have customer taste and smell the water.	______________________ ______________________ ______________________
3. Compare the taste and odor of the sample being investigated to odor-free (or plant effluent water). Have the consumer do the same and note the results.	______________________ ______________________ ______________________ ______________________
4. Measure • Temperature • pH • Chlorine residual (free and total) • Dissolved oxygen (when possible)	 ______________________ ______________________ ______________________ ______________________
5. Collect samples at all pertinent locations and take them to laboratory for analysis. (Locations sampled will include customer tap, if the problem is still occurring, others sites in the distribution system experiencing a similar taste and odor problem and plant effluent and other sources which may be supplying the site).	

Taste and Odor Complaint Investigation (continued)

EQUIPMENT AND GLASSWARE NEEDED

Field analysis
Thermometer
Portable pH meter
Portable dissolved oxygen probe (optional)
Residual chlorine kit

Sampling for laboratory analysis
Ice chests with Blue Ice
Glassware:

1. Flavor profile analysis (FPA): 1 liter Wheaton bottle withTeflon lined caps (2 bottles per site)

2. Closed loop stripping analysis (CLSA): 1 liter Wheaton bottle with Teflon lined caps (2 bottles per site)

3. Simultaneous distillation extraction (SDE): 1 liter Wheaton bottle with Teflon lined caps (4 bottles per site)

4. Inorganic analysis:
   General parameters (hardness, alkalinity, pH, electroconductivity, turbidity: 1/2 gallon plastic bottle (1 bottle per site)
   Metals: 250 ml plastic bottles (1 bottle per site)

5. Microbiological analysis:

- Heterotrophic plate count (HPC): 250 mL autoclaved plastic bottle with thiosulfate as preservative (1 bottle per site)
- Fungi and actinomycetes: 1 L autoclaved plastic bottle (1 bottle per site)

When possible, an aditional 8-10 liters of sample should be collected in amber glass bottles to provide the opportunity for extra analyses when needed. The bottles of sample should be stored in a refrigerator.

REFERENCES

Acree, T.E., J. Barnard, and D.G. Cunningham. 1984. The Analysis of Odor-Active Volatiles in Gas-Chromatographic Effluent. In *Analysis of Volatiles*. Edited by P. Schreier. Berlin: De Gruyter.

Acree, T.E. 1976. Sniffer to Determine the Odor of Gas Chromatographic Effluents. *Analyt. Chem.*, 48:1821.

Ainsworth, R.G. 1981. *A Guide to Solving Water Quality Problems in Distribution Systems*. Tech. Rept. No. TR 167. Medmenham, UK: Water Research Centre.

Alfrey, A.C. 1993. Aluminum and Renal Disease—Moving Points in Nephrology. *Contrib. Nephrol.*, 102:110.

Anselme, C., A. Bruchet, J. Mallevialle, and F. Fiessinger. 1986. Influence of Polyethylene Pipes on Tastes and Odors of Supplied Water. In *Proc. 1986 AWWA Annual Conference*. Denver, Colo.: AWWA.

Anselme, C., J. Mallevialle, and I.H. Suffet. 1985a. Removal of Tastes and Odors by the Ozone–Granular Activated Carbon Water Treatment Processes. In *Proc. Seventh Ozone World Congress*. Tokyo: International Ozone Association.

Anselme, C., K. N'Guyen, A. Bruchet, and J. Mallevialle. 1985b. Can Polyethylene Pipes Impart Odors in Drinking Water? *Environ. Technol. Lett.*, 6:477.

Anselme, C., I.H. Suffet, and J. Mallevialle. 1988. Effects of Ozonation on Tastes and Odors, *Jour. AWWA*, 80(10):45-51.

APHA, AWWA, and WEF (American Public Health Association, American Water Works Association, and Water Environment Federation). 1995. *Standard Methods for the Examination of Water and Wastewater*. 19th ed. Washington, D.C.: APHA.

APHA, AWWA, and WEF (American Public Health Association, American Water Works Association, and Water Environment Federation). 1998. *Standard Methods for the Examination of Water and Wastewater*. 20th ed. Washington, D.C.: APHA.

ASTM (American Society for Testing and Materials). 1996. *Sensory Testing Methods*. 2nd ed. Edited by E. Chambers IV and M.B. Wolf. Philadelphia, Pa.: ASTM.

AWWA (American Water Works Association). 1996. *Water Transmission and Distribution: Principles and Practices of Water Supply Operations.* 2nd ed. Denver, Colo.: AWWA.

AWWA (American Water Works Association). 1999. *Standard for Disinfecting Water Mains.* C651-99. Denver, Colo.: AWWA.

Barrett, S.E., M.K. Davis, and M.J. McGuire. 1985. Blending Chloraminated and Chlorinated Water. *Jour. AWWA*, 77(1):50-61.

Bartels, J.H.M., G.A. Burlingame, and I.H. Suffet. 1986. The Flavor Profile Analysis Method: Taste and Odor Control of the Future. *Jour. AWWA*, 78(3):50.

Berend, K., and T. Trouwborst. 1999. Cement–Mortar Pipes as a Source of Aluminum. *Jour. AWWA*, 91(7):91-100.

Berens, A.R. 1985. Prediction of Organic Chemical Permeation Through PVC Pipe. *Jour. AWWA*, 77(11):57-64.

Blyth, A.E., and E.S. Johnston. 1997. What's the Stink in What We Drink? *Envir. Prot.*, 8(6):12-14.

Bruchet, A. 1999. Solved and Unsolved Cases of Taste and Odor Episodes in the Inspector Clouseau. *Water Sci. Technol.*, 40(6):15-21.

Bruchet, A., K. N'Guyen, J. Mallevialle, and C. Anselme. 1989. Identification and Behavior of Iodinated Haloform Medicinal Odor. Presented at the AWWA Water Quality Technology Conference (Sunday Seminar), Philadelphia, Pa.

Bruchet, A., D. Khiari, and I.H. Suffet. 1995. *Monitoring and Analysis: Advances in Taste-and-Odor Treatment and Control.* Denver, Colo: AwwaRF and Lyonnaise des Eaux.

Burlingame, G.A. 1993. *Laboratory Comparison of Lubricants: Independent Utility Study.* Philadelphia: Philadelphia Water Department.

Burlingame, G.A., and C. Anselme. 1995. Distribution System Tastes and Odors. In *Advances in Taste-and-Odor Treatment and Control.* Cooperative Research Report. Denver, Colo.: AwwaRF and Lyonnaise des Eaux.

Burnam, N.P. 1970. Taste and Odor Due to Stagnation and Local Warming in Long Lengths of Piping: Symposium on Consumer Complaints. *Proc. Soc. Water Treat. Exam.*, 14:125-131.

Burttshell, R.H., A.A. Rosen, M.F. Middleton, and B.M. Ettinger. 1959. Chlorine Derivatives of Phenol Causing Taste and Odor. *Jour. AWWA*, 51:205-221.

Caul, J.F. 1957. The Profile Method of Flavor Analysis. *Adv. Food Res.*, 7:50.

Chedal, H. 1987. L'Incidence des Divers Traitement Sur la Saveru de l'Eau Mise en Distribution. *T. S. M. L'Eau*, 82(10):419.

Desmond, A. 1985. Leaching of Tetrachloroethylene from Vinyl-Lined Pipe. *J. Environ. Eng.*, 111(1):1-9.

Dietrich, A.M., J.P. Dodd, S. Mirlohi, W.F. DaCosta, R. Sauer, M. Holman, and J. Schultz. 1999. Flavor Profile Analysis and GC/MS Detection of Phenolic Iodinated Disinfection By-Products in Drinking Water for the USA Space Program. *Water Sci. Technol.*, 40(6):45-51.

Ellenberger, C.S., R.C. Hoehn, D.L. Gallegher, W.R. Knocke, E.V. Wiseman, R.W. Benninger, and A. Rosenblatt. 1998. Water Quality Impacts of Pure Chlorine Dioxide Pretreatment at the Roanoke County (Virginia) Water Treatment Plant. In *Proc. 1998 AWWA Annual Conference* (Part 4). Denver, Colo.: AWWA.

Fadel, M., G. A. Burlingame, J. Choi,, and L. Gammie. 1993. Odor Problem Caused by Oleate-Based Pipe Joint Lubricants. In *Proc. 1993 AWWA Water Quality Technology Conference*. Denver, Colo.: AWWA.

Gallagher, D.L., R.C. Hoehn, and A.M. Dietrich. 1994. *Sources, Occurrence, and Control of Chlorine Dioxide By-Product Residuals in Drinking Water*. Denver, Colo.: AwwaRF and AWWA.

Gittelman, T.S., and T.L. Yohe. 1989. Treatment of Iodinated Haloform Medicinal Odors in Drinking Water. In *Proc. 1989 AWWA Annual Conference*. Denver, Colo.: AWWA.

Gould, J.P., M. Giabbai, and J. Kim. 1985. Formation of Iodinated Trihalomethanes. In *Water Chlorination: Chemistry, Environmental Impact and Health Effects*, Vol. 5. Edited by R.L. Jolley et al. Chelsea, Mich.: Lewis Publishers, Inc.

Guo, Q. 1997. Increases of Lead and Chromium in Drinking Water from Using Cement–Mortar-Lined Pipes: Initial Modeling and Assessment. *J. Haz. Mat.*, 56:188-213.

Guo, Q., P.J. Toomuluri, and J.O. Eckert Jr. 1998. Leachability of Regulated Metals from Cement–Mortar Linings. *Jour. AWWA*, 90(3):62-73.

Gurol, M.D., A. Wowk, S. Myers, and I.H. Suffet. 1983. Kinetics and Mechanism of Haloform Formation: Chloroform Formation from Trichloroacetone. In *Water Chlorination: Environmental Impact and Health Effects*, Vol. 4. Edited by R.L. Jolley et al. Ann Arbor, Mich.: Ann Arbor Science Publishers.

Hanson, C.R., M.J. Henderson, P. Jack, and R.D. Taylor. 1987. Iodoform Taste Complaints in Chloramination. *Water Res.*, 21(10):1265-1271.

Hoehn, R.C., A.M. Dietrich, W.S. Farmer, R.G. Lee, E.M. Aieta, D.W. Wood, and G. Gordon. 1990. Household Odors Associated with the Use of Chlorine Dioxide During Drinking Water Treatment, *Jour. AWWA*, 82(4):166-172.

Hrudey, S.E., A. Gac, and S.A. Daignault. 1988. Potent Odor Causing Chemicals Arising from Drinking Water Disinfection. *Water Sci. Technol.*, 20(8/9):55-61.

Ibrahim, E.A., I.H. Suffet, and A.B. Sakla. 1987. Evaporative Concentration System (EVACS) for Trace Organic Analysis. *Analyt. Chem.*, 59:2091.

Izaguirre, G., and J. Devall. 1995. Resource Control for Management of Taste-and-Odor Problems. In *Advances in Taste-and-Odor Treatment and Control.* Denver, Colo.: AwwaRF and Lyonnaise des Eaux.

Jolley, R.L., G. Jones, W.W. Pitt, and J.E. Thompson. 1978. Chlorination of Organics in Cooling Waters and Process Effluents. In *Water Chlorination: Environmental Impact and Health Effects*, Vol. 1. Edited by R.L. Jolley et al. Ann Arbor, Mich.: Ann Arbor Science Publishers.

Jolley R.L., W. Lyman, J. Condie, D. Johnson, S. Katz, R. Minear, J.S. Mattice, V. A. Jocobs. 1983. *Water Chlorination: Environmental Impact and Health Effects*, Vol. 4 (Book 1, *Chemistry and Water Treatment*). Ann Arbor, Mich.: Ann Arbor Science Publishers.

Khiari, D., S.E. Barrett, and I.H. Suffet. 1997. Sensory GC Analysis of Decaying Vegetation and Septic Odors. *Jour. AWWA*, 89(4):150-161.

Khiari, D., L. Brenner, G.A. Burlingame, and I.H. Suffet. 1992. Sensory Gas Chromatography for Evaluation of Taste and Odor Events in Drinking Water. *Water Sci. Technol.*, 25(2):97.

Khiari, D., F. Ventura, and S.E. Barrett. 1999. The Occurrence of Iodinated Trihalomethanes in Drinking Water. Presented at Symposium on Natural Organic Matter and Disinfection By-Products: Characterization and Control in Drinking Water, American Chemical Society (ACS) Division of Environmental Chemistry, Anaheim, Calif.

Koch, B., S.W. Krasner, M.J. Sclimenti, and W.K. Schimpff. 1991. Predicting the Formation of DBPs by the Simulated Distribution System. *Jour. AWWA*, 83(10):62.

Kochhar, S.P. 1996. Oxidative Pathways to the Formation of Off-Flavors. In *Food Taints and Off-Flavours*. 2nd ed. Edited by M.J. Saxby. London: Blackie Academic and Professionals, An Imprint of Chapman and Hall.

Krasner, S.W, S.E. Barrett, M.S. Dale, and C.J. Hwang. 1989. Free Chlorine Versus Monochloramine for Controlling Off-Tastes and Off-Odors. *Jour. AWWA*, 81(2):86-93.

Krasner, S.W., M.J. McGuire, and V.B. Ferguson. 1985. Tastes and Odors: The Flavor Profile Method. *Jour. AWWA*, 77(3):34.

Krasner, S.W., and S.E. Barrett. 1984. Aroma and Flavor Characteristics of Free Chlorine and Chloramines. In *Proc. 1984 AWWA Water Quality Technology Conference*. Denver, Colo.: AWWA.

LeCloirec, C., and G. Martin. 1985. Evolution of Amino Acids in Water Treatment Plants and the Effect of Chlorination on Amino Acids. In *Water Chlorination: Chemistry, Environmental Impact and Health Effects*, Vol. 5. Edited by R.L. Jolley et al. Chelsea, Mich.: Lewis Publishers, Inc.

Likens, S.T., and G.B. Nickerson. 1964. Detection of Certain Hop Oil Constituents in Brewing Products. In *Proc. Am. Soc. Brew. Chem. (ASBC)*. St. Paul, Minn.: ASBC.

MacKenzie, E.F.W (1938). Research in the Bacteriological, Chemical and Biological Examination of London Waters. 33:26.

Mallevialle, J., and I.H. Suffet, eds. 1987. *Identification and Treatment of Tastes and Odors in Drinking Water*. Cooperative Research Report. Denver, Colo.: AwwaRF and Lyonnaise des Eaux.

Matia, L. 1995. Treatment of Tastes in Drinking Water: Causes and Control. In *Advances in Taste and Odor Treatment and Control.* Edited by I.H. Suffet, J. Mallevialle, and E. Kawczynski. Denver, Colo.: AwwaRF.

McGuire, M. J., S.W. Krasner, C.J. Hwang, and G. Izaguirre. 1981. Closed-Loop Stripping Analysis as a Tool for Solving Taste-and-Odor Problems. *Jour. AWWA*, 73(10):530-537.

Meng, A. 1990. Correlation Analysis of Chromatographic and Sensory Profiles of Drinking Waters. Ph.D. diss., Drexel University, Philadelphia, Pa.

Meng A., L. Brenner and I.H. Suffet. 1992. Correlation of Chemical and Sensory Data by Principal Component Factor Analysis. *Water Sci. Technol.*, 25(2):97

Montiel, A., S. Rigal, and B. Welté. 1999. Study of the Origin of Musty Taste in the Drinking Water Supply. *Water Sci. Technol.*, 40(6):171-177.

Montiel, A., S. Rigal, and J. Ouvrard. 1991. Study Investigating the Origin and Mechanism of Formation of Sapid Compounds Producing a Musty Taste in Drinking Water Supplies. In *Proc. 1991 AWWA Water Quality Technology Conference*. Denver, Colo.: AWWA.

Morris, J.C. 1975. Formation of Halogenated Organics by Chlorination of Water Supplies (A Review). EPA 600/1-75-002. Washington, D.C.: USEPA Office of Research and Development.

Nayyar, M.L., ed. 1992. *Piping Handbook.* 7th ed. New York: McGraw-Hill Book Group.

Nickerson, G.B., and S.T. Likens. 1966. Gas Chromatography Evidence for the Occurrence of Hop Oil Components in Beer. *Jour. Chromatogr.*, 21:1.

Nyström, A., A. Grimvall, C. Krantz-Rülker, R. Sävenhed, and K. Åkerstrand. 1992. Drinking Water Off-Flavor Caused by 2,4,6-Trichloroanisoles. *Water Sci. Technol.*, 25:214-249.

Palin, A.T. 1975. Water Disinfection—Chemical Aspects and Analytical Control. In *Disinfection: Water and Wastewater.* Edited by J.D. Johnson, T.F. LaPointe, and G. Inman. Ann Arbor, Mich.: Ann Arbor Science Publishers.

Parliment, T.H., and R. Scarpellino. 1977. Organoleptic Techniques in Chromatographic Food Flavor Analysis. *J. Agric. Food Chem.*, 25(1):97-101.

Piriou, P., and Y. Levi. 1994. A New Tool for the Study of the Evoluation of Water Quality in Distribution Systems: Design of a Network Pilot. In *Proc. 1994 AWWA Annual Conference*. Denver, Colo.: AWWA.

Reasoner, D.J., and E.E. Geldreich. 1985. A New Medium for the Enumeration and Subculture of Bacteria from Potable Water. *Appl. Environ. Microbiol.*, 49(1):1-7.

Richard, J.J., and G.A. Junk. 1984. Steam Distillation, Solvent Extraction, and Ion Exchange for Determining Polar Organics in Shale Process Waters. *Analyt. Chem.*, 56(9):1625-1628.

Rigal, S., and J. Danjou. 1999. Tastes and Odors in Drinking Water Distribution Systems Related to the Use of Synthetic Materials. *Water Sci. Technol.*, 40(6):103-208.

Sävenhed, R., H. Bören, A. Grimvall, and A. Nyström. 1991. Identification of Odorous Organic Compounds in Swedish Surface Waters. *Water Supply*, 4:4-8.

Snoeyink, V.L., and D. Jenkins. 1980. *Water Chemistry*. New York: John Wiley & Sons.

Snyder, J.M., E.N. Frankel, and E. Selke. 1985. Capillary Gas Chromatographic Analyses of Headspace Volatiles from Vegetable Oils. *J. Am. Oil Chemists Soc.*, 62(12):1675-1679.

Soderhjelm, L., and S. Eskelinen. 1985. Characterization of Packaging Material with Respect to Taints and Odours. *Appita*, 38(3):205-209.

Suffet, I.H., L. Brenner, and B. Silver. 1976. The Identification of 1,1,1-Trichloroacetone (1,1,1-Trichloropropanone) in Drinking Water: A Known Precursor in the Haloform Reaction. *Environ. Sci. Technol.*, 10:1273.

Suffet, I.H., A. Corado, D. Chou, S. Butterworth, and M.J. McGuire. 1993. AWWA Taste and Odor Survey. In *Proc. 1993 AWWA Water Quality Technology Conference* (Part II). Denver, Colo.: AWWA.

Suffet, I.H., J. Ho, D. Chou, D. Khiari, and J. Mallevialle. 1995. Taste-and-Odor Problems Observed During Drinking Water Treatment. In *Advances in Taste-and-Odor Treatment and Control.* Edited by I.H. Suffet, J. Mallevialle, and E. Kawczynski. Cooperative Research Report. Denver, Colo.: AwwaRF and Lyonnaise des Eaux.

UKWIR (United Kingdom Water Industry Research, Ltd.). 1996. *Formation and Occurrence of Bromophenols, Iodophenols, Bromoanisoles and Iodoanisoles in Drinking Water: An Investigation of Taste and Odour Potential*. Report No. 97/DW/05/13. London: UKWIR.

Wajon, J.E., B.V. Kavanagh, R.I. Kagi, R.S. Rosich, R. Alexander, and B.J. Fleay. 1986. Swampy Odor in the Drinking Water of Perth, Western Australia. *Water*, 13(3):28-32.

Wajon, J.E., R. Alexander, and R.I. Kagi. 1985. Dimethyl Trisulfide and Objectionable Odors in Potable Water. *Chemosphere*, 14(1):85.

Weete, J.D., Blevins, W.T., Wilt, G.R. and D. Durham. 1977. Chemical, Biological and Environmental Factors Responsible for the Earthy Odor in the Auburn City Water Supply. *Auburn Univ. Agric. Exp. Sta. Bull.*, 490:1.

Wei, I.W., and J.C. Morris. 1975. Dynamics of Breakpoint Chlorination. In *Chemistry of Water Supply, Treatment and Distribution*. Edited by A.J. Rubin. Ann Arbor, Mich.: Ann Arbor Science Publishers.

Welté, B., and A. Montiel. 1999. Study of the Possible Origins of Chlorinous Taste and Odor Episodes in a Distribution Network. *Water Sci. Technol.*, 40(6):257-263.

White, G.C. 1986. *The Handbook of Chlorination.* 2nd ed. New York: Van Nostrand Reinhold.

White, G.C. 1992. *The Handbook of Chlorination and Alternative Disinfectants.* 3rd ed. New York: Van Nostrand Reinhold.

White, G.C. 1999. *Handbook of Chlorination and Alternative Disinfectants.* 4th ed. New York: Wiley–Interscience.

Whitfield, F.B. 1988. Chemistry of Off-Flavors in Marine Organisms. *Water Sci. Technol.*, 20(8/9):63-74.

Whittle, G.P., and A. Lapteff Jr. 1975. New Analytical Techniques for the Study of Water Disinfection. In *Chemistry of Water Supply, Treatment and Distribution*. Edited by A.J. Rubin. Ann Arbor, Mich.: Ann Arbor Science Publishers.

Williams, D.B. 1949. A New Method of Odor Control. *Jour. AWWA*, 41(5):441.

Williams, D.B. 1966. Elimination of Nitrogen Trichloride in Dechlorination Practice. *Jour. AWWA*, 58(2):248.

Winnen, H., M.T. Suidan, P.V. Scarpino, B. Wrenn, N. Cicek, V. Urbain, and J. Manem. 1996. Effectiveness of the Membrane Bioreactor in the Biodegradation of High MW Compounds. *Water Sci. Technol.*, 34(9):197.

Yates, R.S., and M.K. Stenstrom. 2000. Gravimetric Sampling Procedure for Aqueous Ozone Concentrations. *Water Res.*, 34:1714-1717.

Yoshida, F., and Y. Miura. 1963. Gas Absorption in Agitated Gas-Liquid Contactors. *Ind. Eng. Chem. Proc. Des. Dev.*, 2(4):263-268.

Young, C.C., and I.H. Suffet. 1999. Development of a Standard Method: Analysis of Compounds Causing Tastes and Odors in Drinking Water. *Water Sci. Technol.*, 40(6):279-285.

Zurer, P.S. 1987. GC Technique Quantitates Odor-Active Substances in Food. *Chem. Eng. News*, 9/28/87:21-22.

ABBREVIATIONS

μS	microSiemens
ANOVA	analysis of variance
APHA	American Public Health Association
AR	annular reactor
AVA, AVA 10	Aquaflow VA-SP 10 lubricant
BDOC	biodegradable organic carbon
BET	best-estimate threshold
BHT	2,6-*tert*-butyl-4-methylphenol (butylated hydroxytoluene)
BLL	Blue Lube™ lubricant
CFU	colony-forming unit
CIP	cast-iron pipe
CLSA	closed-loop stripping analysis
CRW	Colorado River water
CSTR	continuously stirred tank reactor
DBP	disinfection by-product
DCM	dichloromethane
DI	dilution index
DIP	ductile iron pipe
DOC	dissolved organic carbon
DPD	*N,N*-diethyl-*p*-phenylenediamine
DPD-FAS	*N,N*-diethyl-*p*-phenylenediamine–ferrous ammonium sulfate
dt	detention time
EVACS	evaporative concentration system
FAS	ferrous ammonium sulfate
FPA	flavor-profile analysis
FTC	flavor threshold concentration
FTN	flavor threshold number
G	velocity gradient
GAC	granular activated carbon
GC	gas chromatography
GC/MS	gas chromatography/mass spectrometry
gpm	gallons per minute
HDPE	high-density polyethylene
HOCl	hypochlorous acid
HPC	heterotrophic plate count

ICP	inductively coupled plasma
ITHM	iodinated trihalomethane
LLE	liquid/liquid extraction
Llo	Llobregat
LVI	large-volume injection
MCL	maximum contaminant level
MIB	2-methylisoborneol
MS	mass spectrometry
MWDSC	Metropolitan Water District of Southern California
NCl_3	trichloramine
ND	non-detected
NH_2Cl	monochloramine
$NHCl_2$	dichloramine
NOM	natural organic matter
NSF	NSF International, Inc.
NTU	Nephelometric turbidity units
OCl^-	hypochlorite ion
OF	odor-free
OTC	odor threshold concentration
OTN	odor threshold number
PFM	perfluoromeric
pg	picograms
POU	point of use
PVC	polyvinyl chloride
RDL	reported detection limit
Sch. R.	Schuylkill River water
SDE	simultaneous distillation/extraction
SDS	simulated distribution system
SNK	Student-Newman-Keuls
SoCal. Gr.	a southern California groundwater
SPW	(California) state project water
T	Ter
t	time
TBA	2,4,6-tribromoanisole
TCA	2,4,6-trichloroanisole
TDS	total dissolved solids
THM	trihalomethane

TJL	TYTON® joint lubricant
TOC	total organic carbon
TRG	TYTON® rubber gasket
UCM	unresolved chromatographic mixture
UKWIR	UK Water Industry Research, Ltd.
UV	ultraviolet
UVA_{254}	ultraviolet absorbance at 254 nm
VOC	volatile organic compound